Combinatorial Aspects of
LIE SUPERALGEBRAS

Combinatorial Aspects of

LIE SUPERALGEBRAS

Alexander A. Mikhalev
Andrej A. Zolotykh

CRC Press
Boca Raton New York London Tokyo

Library of Congress Cataloging-in-Publication Data

Mikhalev, Alexander A., 1965–
Combinatorial aspects of Lie superalgebras / Alexander A. Mikhalev, Andrej A. Zolotykh.
p. cm.
Includes bibliographical references and index.
ISBN 0-8493-8960-7
1. Lie superalgebras. 2. Combinatorial analysis. I. Zolotykh, Andrej A., 1964– . II. Title.
QA252.3.M54 1995
512′.559--dc20

95-15995
CIP

Direct all inquiries to CRC Press, Inc., 2000 Corporate Blvd., N.W., Boca Raton, Florida 33431.

International Standard Book Number 0-8493-8960-7
Library of Congress Card Number 95-15995
Printed in the United States of America 1 2 3 4 5 6 7 8 9 0
Printed on acid-free paper

Contents

Introduction

The theory of Lie algebras and Lie groups is one of the most developed and advanced branches of modern algebra which has a lot of applications in mathematics and in theoretical physics. At the same time there are many reasons to work together with the usual pairwise commuting coordinates and also with anti-commmuting ones. We mention only Grassmann and Clifford algebras.

Be aware that there are different sources of Lie superalgebras. For example, they appeared in homotopy theory in connection with Whitehead product on homotopy groups, in particle physics for describing the boson-fermion systems where a fundamental role is played by the supersymmetry mixing even and odd coordinates, in the theory of deformations of graded algebraic systems. In fact, supersymmetry is described by Lie supergroups or, infinitesimally, by Lie superalgebras.

Since the 1950s the theory of Lie superalgebras was developed systematically (Killing-Cartan theory, representation theory, classification of simple Lie superalgebras, Lie superalgebras of vector fields, varieties of Lie superalgebras etc.).

At the same time in many situations in physics we have to consider not only $\mathbb{Z}_2$- or $\mathbb{Z}$-gradings of algebras but also G-gradings where G is an Abelian group equipped by a skew symmetric bilinear form given by a 2-cocycle. In the 1970s we got some generalized Lie superalgebras under different names: color Lie superalgebras; ε-Lie algebras; generalized Lie algebras, and so on. In fact, these algebras (generalized Lie algebras) were introduced by R. Ree in 1960.

Algebra has a long tradition of constructive tools going back to Euclid, Kronecker, H. Weil and others. For many reasons computer algebra has attracted an increasing interest in computer science, mathematics and applications. For example, the theory of canonical bases (Gröbner bases) of ideals in polynomial rings is very useful in commutative algebra, in algebraic geometry, in the theory of systems of algebraic equations, etc. Now noncommutative theories of canonical bases are becoming also very popular in producing calculations with computers in noncommutative rings, Lie (super)algebras, Hopf algebras, and noncommutative geometry.

This book is in some sense an introduction to the combinatorial theory of (color) Lie superalgebras (given by generators and defining relations). The main object under consideration is the free color Lie (p-)superalgebra. We consider combinatorics on words, construct linear bases of free Lie superalgebras, consider free Lie superalgebras as S_n-modules, describe the structure of subalgebras, give canonical bases of ideals (composition techniques, noncommutative Gröbner bases), study endomorphisms and automorphisms, introduce free differential calculus, describe primitive systems of elements. We present a number of examples on these topics and describe possible directions of the research. Some interesting problems are pointed out.

The main part of the results of the book is combinatorial. It gives a possibility to obtain a number of algorithms in combinatorial theory of Lie superalgebras and to give their computer realization. For some basic facts of the theory of Lie superalgebras which can be found in the monograph "Infinite Dimensional Lie Superalgebras" by Yu. A. Bahturin, A. A. Mikhalev, M. V. Petrogradsky and M. V. Zaicev, Walter de Gruyter, Berlin, New York, 1992, we give only sketches of the proofs. In the present book we are looking at the whole theory from the point of view of Computer Algebra. In addition a diskette with a program for combinatorial calculations is enclosed. On this base it could be possible to create an effective program for combinatorial calculations. However such project exceeds any limits for a monograph. By the same reasons some interesting topics in Lie superalgebras theory were not included in this book. In any case (understanding that the growth of the number of published articles is overexponential) we present a rather wide bibliography on the subject and on related topics.

The book is organized as follows. In Chapter 1 we start with necessary definitions and examples, deal with actions of the symmetric group, introduce some universal constructions (free associative algebras and free color Lie superalgebras, universal enveloping algebras and free products with the amalgamated subalgebra).

Chapter 2 is devoted to the linear structure of free Lie superalgebras. We go in some depth into certain special aspects of combinatorics on Lyndon-Shirshov words, construct some bases of free Lie color Lie (p-)superalgebras, present various generalizations of Witt formula, consider images of inner derivations, prove Dynkin-Specht-Wever criterion, describe some algorithms and examples connected with canonical forms of elements of free color Lie superalgebras and of free color Lie p-superalgebras. In particular, we get an algorithm to reach the canonical form of an element; an algorithm to decide whether an associative element is a super Lie element, or not; an algorithm for extracting a square root from an even element.

In Chapter 3 we concentrate on subalgebras of free color Lie (p-)superalgebras. The main theorem is focused on the Nielsen property of such algebras (i.e., each reduced subset is independent). We prove that these algebras possess by Schreier property (a subalgebra of the free algebra is also free). Using this theorem we obtain a number of results on subalgebras of free Lie superalgebras. In particular, we prove in the case of prime characteristic that the intersection of two subalgebras of finite rank is a subalgebra of finite rank. We concern also fixed points and present algorithms and examples, culminating in algorithms to solve the entry problem in a finitely generated subalgebra and to decide whether an endomorphism of a free color Lie (p-)superalgebra is an automorphism, or not. Some open problems are formulated.

Chapter 4 concentrates on canonical bases in Lie superalgebras. Using Diamond Lemma (Junction Lemma) we prove Poincaré-Birkhoff-Witt theorems for Lie superalgebras. We show that the free Lie superalgebras are finitely separable algebras. We deal with the structure of graded Hopf algebras on universal enveloping algebras, prove Friedrichs criteria, and introduce super shuffle algebras, and the Baker-Campbell-Hausdorff formula. Then we define compositions for elements of free color Lie (p-)superalgebras and formulate Composition Lemma. It gives us a possibility to define noncommutative Gröbner bases of ideals of free Lie superalgebras and to construct canonical bases of Lie superalgebras. This part ends with illustrations of linear bases of free products with the amalgamated subalgebra. We point out some open problems and present algorithms and examples in this field. In particular, we give an algorithm to construct Gröbner basis for an ideal given by a homogeneous set of defining relations and an algorithm to solve the equality problem using Gröbner basis.

In Chapter 5 we introduce the free differential calculus (the universal derivation) for free Lie superalgebras and prove the Jacobian conjecture for these algebras. We consider the orbits of elements of free Lie superalgebras under the action of the automorphism group and give a criterion for an element to have the given rank. A criterion for a system of elements to be primitive (i.e., to be a part of some free generating set) is included. It gives a possibility to present an algorithm to construct a complement of a primitive system with respect to a free generating set. An example of a nonfree Lie algebra of cohomological dimension one and such that its universal enveloping algebra is the free associative algebra of the rank two is presented. We prove that an endomorphism of the free color Lie p-superalgebra preserving primitivity of elements is an automorphism, present a series of algorithms, examples and open problems, and discuss possible directions of the research. In particular, we get an algorithm to calculate the rank of an element or of a system of elements, an algorithm to decide whether a system of elements is primitive, or not.

In the Appendix one can find the description of the program for combinatorial calculations; the diskette with this program is enclosed.

This work was done when the authors were members of the Department of Mechanics and Mathematics of Moscow State University. The topics which are under consideration in the book were discussed at the algebraic seminars of Moscow State University (in particular, on Scientific Research Seminar of the Chair of Higher algebra, on seminars "Lie algebras and superalgebras", "Rings and modules", "Computer Algebra"), at many conferences and symposia, for example, at International Conferences on Algebra (Novosibirsk, 1989; Barnaul, 1991; Krasnoyarsk, 1993), at the 1st European Congress in Mathematics (Paris, 1992), at the 3rd International Conference on non associative algebra (Oviedo, 1993), and at the International Congress of Mathematicians (Zürich, 1994). The first author used the material of the book in his special lecture course "Lie algebras and superalgebras" for graduate and post graduate students of the Department of Mechanics and Mathematics of Moscow State University.

We thank Professors Yu. A. Bahturin, K. I. Beidar, L. A. Bokut', E. S. Golod, A. I. Kostrikin, V. N. Latyshev, A. V. Mikhalev, Yu. P. Razmyslov, I. P. Shestakov, A. L. Shmelkin, A. M. Slinko, V. Shpilrain, U. U. Umirbaev, M. V. Zaicev, and E. I. Zelmanov for many helpful discussions.

Our greatest debt is to L. I. Mikhaleva, T. L. Mikhaleva, and A. L. Zolotykh: it is they who have sacrificed most!

The first author is thankful to the Ministry of Science, Education and Technical Policy of Russia for the research grant no 2-11-1-2 (1992/93); to the Pro Mathematica Foundation, France, for the research grant (1992/93, a half of a year); to the American Mathematical Society for the research grant in 1993, and to the Soros Foundation for Soros Fellowship in Mathematics (1993/94). He is also thankful to the International Science Foundation for the grant to attend the International Congress of Mathematicians (Zürich, 1994). The second author is thankful to the Government of Russian Federation for the State Science Fellowship (1994–1996). Both authors thank the Russian Foundation of Fundamental Researches for the grant 93-011-1543 (1993–1995), and the International Science Foundation for the grant M22000 (1994).

Finally, our profound thanks are due to S. V. Landau, the acquiring editor at CRC Press for her kind help and cheerful demeanor in all the project and to the staff of CRC Press for the grant for preparation the manuscript, their care and helpfulness.

Moscow State University
October 1994

Alexander A. Mikhalev
Andrej A. Zolotykh

Chapter 1

Preliminaries

In this chapter we start with basic notions of the theory of Lie superalgebras and present some basic facts. In Section 1 we define the notions of Lie algebra and Lie superalgebra and give some examples. In Section 2 we introduce color Lie superalgebras and color Lie p-superalgebras. We consider their skew envelopes and show off a few examples that occur naturally. In Section 3 we present the following universal constructions: free G-graded associative algebras and free color Lie superalgebras, universal enveloping algebras and free products with the amalgamated subalgebra. In Section 4 the free Lie superalgebras as S_n-modules are under consideration.

We note here some, largely standard, notations and conventions to be used throughout in all what follows in the book. Let $\emptyset$ be the empty set, $\mathbb{Z}$ the set of integers, $\mathbb{N} = \{k \in \mathbb{Z} \mid k > 0\}$ the set of natural numbers, $\mathbb{Z}_+ = \{k \in \mathbb{Z} \mid k \geq 0\}$, $\mathbb{Q}$ the field of rational numbers, $\mathbb{R}$ the field of real numbers, $\mathbb{R}_+ = \{\alpha \in \mathbb{R} \mid \alpha \geq 0\}$, $\mathbb{C}$ the field of complex numbers, $n! = 1 \cdot 2 \cdot \ldots \cdot n$ with $n \in \mathbb{Z}_+$, $0! = 1! = 1$, $C_n^k = \frac{n!}{k!(n-k)!}$, δ_{ij} the Kronecker delta. For $n \in \mathbb{N}$ by $\mathbb{Z}_n$ we denote the residue class ring modulo n (n is said to be the module of residue class ring $\mathbb{Z}_n$).

1 Lie superalgebras

In this section we define Lie algebras and Lie superalgebras and give some examples.

Algebras

Let R be a nonempty set with two binary operations: addition and multiplication, such that it is an Abelian group relative the addition and $(a + b)c = ac + bc$, $a(b + c) = ab + ac$ for all $a, b, c \in R$. Then R is said to be *a ring*. If there exists an element $e \in R$ such that $ea = ae = a$ for all $a \in R$, then e is called *the unity element* of R, and R is called *a ring with the unity element.* If $(ab)c = a(bc)$ for all $a, b, c \in R$, then R is said to be associative. If $ab = ba$ for all $a, b \in R$, then R is said to be commutative.

Let K be an associative commutative ring with the unity element. A ring R is *a left K-module* if there is a mapping $K \times R \to R$ such that:

$$e \cdot a = a; \qquad \alpha \cdot (a + b) = \alpha \cdot a + \alpha \cdot b;$$
$$(\alpha + \beta) \cdot a = \alpha \cdot a + \beta \cdot a; \qquad (\alpha\beta) \cdot a = \alpha \cdot (\beta \cdot a)$$

for all $\alpha, \beta \in K$, $a, b \in R$, where e is the unity element of K. A ring R with a structure of a left K-module is said to be *a K-algebra* if

$$(\alpha a)b = \alpha(ab) = a(\alpha b)$$

for all $\alpha \in K$, $a, b \in R$.

A mapping from K-algebra to another K-algebra which is at the same time a ring homomorphism and a homomorphism of K-modules is called *a homomorphism of K-algebras.*

Let X be a nonempty set, $\Gamma(X)$, *the free groupoid of nonassociative monomials* without unity element on the alphabet X, i.e. $X \subseteq \Gamma(X)$; if $u, v \in \Gamma(X)$, then $u \circ v \in \Gamma(X)$, where $u \circ v$ is the formal multiplication on nonassociative words, for example $x \circ y = xy$, $(xy)z \circ xy = ((xy)z)(xy)$. We consider the free left K-module $F(X)$ with the basis consisting of $1_{F(X)}$ and elements of $\Gamma(X)$ with the multiplication given by

$$1 \cdot a = a \cdot 1 = a, \qquad (\alpha \cdot a) \cdot (\beta \cdot b) = (\alpha\beta) \cdot (a \cdot b)$$

for all $\alpha, \beta \in K$, $a, b \in \Gamma(X)$. It is clear that the algebra $F(X)$ is *the free nonassociative K-algebra with the unity element 1_A on the set X*, i.e. if A is a K-algebra with the unity element and $\xi: X \to A$ is some mapping, then there exists a homomorphism of K-algebras $\psi: F(X) \to A$ (ψ is a K-linear mapping, $\psi(1_{F(X)}) = 1_A$, $\psi(a \circ b) = \psi(a)\psi(b)$ for all $a, b \in F(X)$) such that $\psi|_X = \xi$.

It is clear that the algebra $F(X)$ is determined up to isomorphism of K-algebras by this universal property. Note that in the case when we consider algebras without unity elements we define by the same way *the free nonassociative algebra without unity element* $F^+(X)$. In fact, $F(X) = K \cdot 1_{F(X)} \oplus F^+(X)$.

A K-submodule I of $F(X)$ is said to be a (two sided) *ideal* of $F(X)$ if $a \circ b, b \circ a \in I$ for all $a \in F(X)$ and $b \in I$. Any K-algebra A is a homomorphic image of some free nonassociative K-algebra. Indeed, let $Y = \{y_i \mid i \in \Omega\}$ be a set of generators of A (i.e., each element of A is a K-linear combination of nonassociative monomials on Y; one can set $Y = A$). We consider the set X, $|X| = |Y|$, $X = \{x_i \mid i \in \Omega\}$, and the mapping $\xi: X \to A$ given by $\xi(x_i) = y_i$ for all $i \in \Omega$. Then by the universal property there exists a homomorphism $\psi: F(X) \to A$ such that $\psi(x_i) = y_i$. By the Homomorphism Theorem $A \cong F(X)/I$ where $I = \operatorname{Ker}\psi$.

For a subset $Z = \{z_j \mid j \in \Theta\}$ of $F(X)$ by $I(Z)$ we denote the ideal of $F(X)$ generated by $\mathbb{Z}$, i.e. the least ideal of $F(X)$ such that $Z \subseteq I(Z)$. The algebra $F(X)/I(Z)$ is called *the algebra given by the set X of generators and the set Z of defining relations*, sometimes it is convenient to use for it the following notation: $F(X; Z = 0)$. Each K-algebra has such a presentation, but it is not unique.

Let A be a K-algebra, $x_1, \dots, x_n \in X$, $h_1, \dots, h_n \in A$. We can consider the homomorphism $\psi: F(X) \to A$ such that $\psi(x_i) = h_i$ for $i = 1, \dots, n$ and $\psi(x) = 0$ for all $x \in X \backslash \{x_1, \dots, x_n\}$. For any element $a(x_1, \dots, x_n) \in F(X)$ depending only on $x_1, \dots, x_n$ the image of this element with respect to ψ will be denoted by

$$\psi(a) = a(h_1, \dots, h_n) = a\Big|_{\substack{x_1 = h_1 \\ \cdots \\ x_n = h_n}} .$$

This element can be obtained by the formal replacing of $x_1, \dots, x_n$ with $h_1, \dots, h_n$. In the case when $A = F(X)$, $m < n$, and $h_i = x_i$ for $i = m+1, \dots, n$, we write

$$a\Big|_{\substack{x_1 = h_1 \\ \cdots \\ x_m = h_m}}$$

instead of $a(h_1, \dots, h_m, x_{m+1}, \dots, x_n)$.

The equality $a(x_1, \dots, x_n) = 0$ is called *an identity* of the K-algebra A if $\varphi(a) = 0$ for any homomorphism $\varphi: F(X) \to A$. In other words, for $a = a(x_1, \dots, x_n) \in F(X)$ depending only on $x_1, \dots, x_n$, the equality $a(x_1, \dots, x_n) = 0$ is an identity of A if and only if $a(h_1, \dots, h_n) = 0$ for any $h_1, \dots, h_n \in A$.

Similar notations will be used for other types of free algebras.

For basic facts on rings, algebras and modules see for example textbooks [14] by M. Artin, [95] by N. Bourbaki, [128], [131] by P. M. Cohn, [229] by I. N. Herstein, [240], [243], [244] by N. Jacobson, [281] by A. I. Kostrikin, [311] by S. Lang, and [578] by B. L. Van der Waerden.

Lie algebras

Let K be a commutative associative ring with the unity element. A K-algebra L is said to be *a Lie algebra* if for its multiplication $[\,,\,]$ the following conditions are satisfied:

$$[x, x] = 0,$$

$$[x, [y, z]] + [z, [x, y]] + [y, [z, x]] = 0 \qquad \textit{(Jacobi identity)}$$

for all $x, y, z \in L$.

If 2 is an invertible element of K, then the identity $[x, x] = 0$ is equivalent to the identity $[x, y] = -[y, x]$ (*anti-commutativity*).

Example 1.1
Let $K = \mathbb{R}$. The vector space $\mathbb{R}^3$ is a Lie algebra relatively to the multiplication given by the vector product. □

Example 1.2
Let $K = \mathbb{R}$. Consider the set L of infinite differentiable functions $f: \mathbb{R} \to \mathbb{R}$. On this vector space over $\mathbb{R}$ we introduce the multiplication by the following way:

$$[f_1, f_2] = f_1 f_2' - f_1' f_2$$

where f_1', f_2' are derivatives of f_1, f_2, respectively. Then L is a Lie algebra over $\mathbb{R}$. □

Now the theory of Lie algebras is one of the well developed branches of the modern algebra. It has a big number of applications in mathematics and physics. We mention only the monographs [7] by R. Amayo and I. Stewart, [26] by Yu. A. Bahturin, [87] by L. A. Bokut' and G. P. Kukin, [96] by N. Bourbaki, [155] by J. Dixmier, [187] by D. B. Fuchs, [203] by M. Goto and F. D. Grosshans, [235] by J. E. Humphreys, [241] by N. Jacobson, [250] by V. G. Kac, [282] by A. I. Kostrikin, [473] by Yu. P. Razmyslov, [481] by Ch. Reutenauer, and [500] by J.-P. Serre.

G-graded algebras

Let G be an additively written monoid (i.e., a semigroup with unity). We say that a K-algebra A is *G-graded* if G-graded algebra

$$A = \bigoplus_{g \in G} A_g$$

of K-submodules A_g and $A_g A_h \subseteq A_{g+h}$. If A has a unity element 1 we suppose that $1 \in A_0$. An element $a \in A$ is said to be *homogeneous* if $a \in A_g$ for some $g \in G$ (in this case we write $d(a) = g$).

Let A and B be G-graded K-algebras. We say that a homomorphism $\psi: A \to B$ is *a homomorphism (of degree zero) of G-graded K-algebras* if $\psi(A_g) \subseteq B_g$ for all $g \in G$.

If I is a G-graded ideal of a G-graded K-algebra A, i.e.

$$I = \bigoplus_{g \in G} I_g,$$

where $I_g = I \cap A_g$, then *the factor algebra A/I* is naturally G-graded and the canonical mapping $\theta: A \to A/I$, $\theta(a) = a + I$, is a homomorphism of G-graded K-algebras.

Lie superalgebras

A $\mathbb{Z}_2$-graded algebra $A = A_0 \oplus A_1$ is called *a superalgebra*. If $d(a) = 0$, i.e. $a \in A_0$, then a is said to be even; if $d(a) = 1$, i.e. $a \in A_1$, then a is said to be odd.

Example 1.3
Let K be a field, $\operatorname{char} K \neq 2$. *The Grassmann algebra* (or *the exterior algebra*) Λ over K is the associative algebra given by generators $x_1, \ldots, x_n$ and defining relations $x_i x_j + x_j x_i = 0$, $i, j = 1, \ldots, n$ (which imply, of course, $x_i^2 = 0$ for all i). One can easily see that 1 and monomials $x_{i_1} x_{i_2} \cdots x_{i_k}$, where $x_{i_j} \in X$, $i_1 < \cdots < i_k$, $k \leq n$, form a basis of the free K-module Λ. We define

$$d(x_{i_1} x_{i_2} \cdots x_{i_k}) = \sum_{j=1}^{k} d(x_{i_j}) \in \mathbb{Z}_2.$$

Then $\Lambda = \Lambda_0 \oplus \Lambda_1$ is a superalgebra. Moreover, for homogeneous elements a, b of Λ we have $ab = (-1)^{d(a)d(b)} ba$. □

Example 1.4
Let $V = V_0 \oplus V_1$ be a Z_2-graded finite dimensional linear space over a field K, $L(V) = \operatorname{End}(V)$ the algebra of all linear transformations of V,

$$L(V)_i = \{f \in L(V) \mid f(V_j) \subseteq V_{i+j},\ j \in \mathbb{Z}_2\},$$

$i \in \mathbb{Z}_2$. Then $L(V) = L(V)_0 \oplus L(V)_1$ is an associative superalgebra. If $\{e_1, \ldots, e_m\}$ is a basis of V_0 and $\{e_{m+1}, \ldots, e_n\}$ is a basis of V_1, then each operator f of $L(V)$ has the matrix of the form

$$F = \begin{pmatrix} F_{11} & F_{12} \\ F_{21} & F_{22} \end{pmatrix},$$

where F_{11} is an $(m \times m)$-matrix, and F_{22} is an $((n-m) \times (n-m))$-matrix, $d(F) = 0$ if and only if $F_{12} = 0$ and $F_{21} = 0$; $d(F) = 1$ if and only if $F_{11} = 0$ and $F_{22} = 0$. □

In linear algebra over a $\mathbb{Z}_2$-graded space we need to make some changes. For example, instead of the usual trace of a matrix $\operatorname{tr}(F)$ we have to use *the supertrace* function $\operatorname{str}(F) = \operatorname{tr}(F_{11}) - \operatorname{tr}(F_{22})$. For the details see [55]. If $A = A_0 \oplus A_1$ is an associative K-superalgebra and $\Lambda = \Lambda_0 \oplus \Lambda_1$ is a Grassmann algebra over K, then the algebra

$$\Lambda(A) = A_0 \otimes_K \Lambda_0 \ \oplus \ A_1 \otimes_K \Lambda_1$$

with multiplication, defined by $(a \otimes a') \cdot (b \otimes b') = ab \otimes a'b'$, is called *the Grassmann envelope* of A. It is easy to see that A is again an associative algebra. If F is a matrix of an element $f \in \Lambda(L(V))$ in some homogeneous basis, F_{22} is invertible, then the function of *superdeterminant* is given by

$$\operatorname{sdet}(F) = \det(F_{11} - F_{12}F_{22}^{-1}F_{21}) \det F_{22}^{-1},$$

where det is the usual determinant. In particular, $\operatorname{sdet}(FH) = \operatorname{sdet}(F) \cdot \operatorname{sdet}(H)$. In [437] one can find the Cayley-Hamilton theorem for supermatrices.

Definition 1.5
Let K have no 2-torsion, i.e. $2x \neq 0$ for all $x \neq 0$, $x \in K$. A $\mathbb{Z}_2$-graded algebra $L_0 \oplus L_1$ with the multiplication $[\,,\,]$ over K is called *a Lie superalgebra* if for any homogeneous elements $x, y, z \in L$ the following conditions are satisfied:

$$[x, y] = -(-1)^{d(x)d(y)}[y, x] \tag{1.1}$$

(*skew anti-commutativity*) and

$$(-1)^{d(x)d(y)}\big[x, [y, z]\big] + (-1)^{d(z)d(y)}\big[z, [x, y]\big] + (-1)^{d(y)d(x)}\big[y, [z, x]\big] = 0 \tag{1.2}$$

(*super Jacobi identity*). If $\operatorname{char} K = 3$, then additionally

$$\big[x, [x, x]\big] = 0 \qquad \text{for } x \text{ with } d(x) = 1.$$

It is clear that if L is a Lie superalgebra and $x \in L$, $y \in L_1$, then $\big[[y, y], x\big] = 2\big[y, [y, x]\big]$ and $\big[x, [y, y]\big] = 2\big[[x, y], y\big]$. Therefore in the case when $\operatorname{char} K = 2$ under a Lie superalgebra it is appropriate to consider an algebra with the identities (1.1) and (1.2) and an operator $q\colon L_1 \to L_0$ such that for all $\alpha \in K$, $a, b \in L_1$

$$q(\alpha a) = \alpha^2 q(a); \qquad \big[a, [a, b]\big] = [q(a), b];$$
$$[a, b] = q(a + b) - q(a) - q(b).$$

Example 1.6
If $\operatorname{char} K \neq 2$, $A = A_0 \oplus A_1$ is a $\mathbb{Z}_2$-graded associative K-algebra, then we consider the new operation $[\,,\,]$ on A given by

$$[a, b] = ab - (-1)^{d(a)d(b)}ba$$

for homogeneous elements $a, b \in A$. Then A with the multiplication $[\,,\,]$ (denoted by $[A]$) is a Lie superalgebra. If V is a finite dimensional $\mathbb{Z}_2$-graded vector space over a field K, then $[L(V)]$ is called *the general linear Lie superalgebra* and is denoted by $\operatorname{gl}(V)$. If we consider all matrices F of $L(V)$ with the conditions $\operatorname{str}(F) = 0$, then it is easy to see that this set is a subalgebra of $[L(V)]$. This subalgebra is called *the special linear Lie superalgebra* and is denoted by $\operatorname{spl}(V)$. □

Example 1.7
Let V be a linear space over a field K with a basis $X = \{x_1, \ldots, x_n\}$, Λ the Grassmann algebra on X, $A = A_0 \oplus A_1$ a K-superalgebra. A homogeneous element D with the degree $d(D) \in \mathbb{Z}_2$ of $\mathrm{gl}(A)$ is called *a derivation* if $D(xy) = D(x)y + (-1)^{d(D)d(x)} x D(y)$ for any $y \in A$ and homogeneous $x \in A$. It is clear that the set of derivations of A is an associative superalgebra (notation $\mathrm{Der}(A)$). Then $[\mathrm{Der}(A)]$ is a Lie superalgebra. By $W(V)$ we denote the Lie superalgebra $[\mathrm{Der}(\Lambda)]$. □

A subset I of a Lie superalgebra L is *an ideal* of L if $x \in I$ and $y \in L$ implies $[x, y] \in I$. We say that L is *simple* if it contains no nontrivial ideals and if its multiplication is not trivial. The classification of the finite dimensional simple Lie superalgebras over an algebraically closed field of zero characteristic was given by V. Kac (see [247], [249], and also [489]). For example, Lie superalgebras $W(V)$ with $\dim V \geq 3$, and $\mathrm{spl}(V)$ with $V = V_0 \oplus V_1$, $\dim V_0 > \dim V_1 \geq 1$, are simple.

2 Color Lie superalgebras

In this section we define color Lie superalgebras, color Lie p-superalgebras and (restricted) differential color Lie superalgebras. The basic notions of Lie superalgebras theory (subalgebras, ideals, representations etc.) are described first. We consider the free ε-commutative algebra and introduce the skew envelope of color Lie superalgebra. A series of examples are presented to illustrate the theory.

Commutation factor

Definition 2.1
Let K be a commutative associative ring with the unity element, K^* the group of invertible elements of K, G an Abelian semigroup. A mapping $\varepsilon: G \times G \to K^*$ is called *a commutation factor* on G if:

$$\begin{aligned} \varepsilon(g,h)\,\varepsilon(h,g) &= 1; \\ \varepsilon(g,g) &= \pm 1; \\ \varepsilon(g,h+f) &= \varepsilon(g,h)\,\varepsilon(g,f); \\ \varepsilon(g+h,f) &= \varepsilon(g,f)\,\varepsilon(h,f) \end{aligned}$$

for all $f, g, h \in G$.

Let also

$$G_+ = \{g \in G \mid \varepsilon(g,g) = 1\}, \qquad G_- = \{g \in G \mid \varepsilon(g,g) = -1\}.$$

A mapping $\sigma: G \times G \to K^*$ is said to be *a multiplier* on G if

$$\sigma(g,h+f)\,\sigma(h,f) = \sigma(g,h)\,\sigma(g+h,f)$$

for all $f, g, h \in G$, and $\sigma(0,0) = 1$ (the normalization condition).

A G-graded algebra A is said to be *ε-commutative* if $ab = \varepsilon(d(a), d(b))ba$.

It is clear that G_+ is a subsemigroup of G. If G is a group, then G_+ is a subgroup of G.

If ε is a commutation factor on an Abelian group G, then the mapping ε' given by $\varepsilon'(f,g) = \varepsilon(g,f)$ is also a commutation factor. If σ is a multiplier on G, then the mapping $\delta: G \times G \to K^*$ defined by $\delta(g,h) = \sigma(g,h)/\sigma(h,g)$ for all $g, h \in G$ is a commutation factor on G such that $\delta(g,g) = 1$. If G is finitely generated, δ is a commutation factor on G such that $\delta(g,g) = 1$ for all $g \in G$, then there exists a multiplier σ on G such that $\delta(g,h) = \sigma(g,h)/\sigma(h,g)$ for all $g, h \in G$.

Two multipliers σ_1 and σ_2 on the Abelian group G yield the same commutation factor if and only if the mapping $\tau = \sigma_1(\sigma_2)^{-1}$ is symmetric, i.e. $\tau(g,h) = \tau(h,g)$ for all $g, h \in G$. If G is a finitely generated Abelian group and the field K is algebraically closed, then any symmetric multiplier τ on G has the form

$$\tau(g,h) = \frac{\omega(g+h)}{\omega(g)\,\omega(h)}$$

for some mapping $\omega: G \to K^*$.

If ε is a commutation factor on G, σ is a multiplier on G, then the mapping $\delta: G \times G \to K^*$ given by $\delta(g,h) = \varepsilon(g,h)\sigma(g,h)/\sigma(h,g)$ is a commutation factor on G. Let $\pi(g) = (1-\varepsilon(g,g))/2$ for all $g \in G$. For every commutation factor ε on G we consider the mapping $\delta(g,h) = (-1)^{\pi(g)\pi(h)}$. Then δ is a commutation factor on G.

The details can be found, for example, in [488].

Example 2.2

For some Abelian groups G we give examples of commutation factors.

1) For any Abelian group G one can consider the trivial commutation factor ε given by $\varepsilon(g,h) = 1$ for all $g, h \in G$.

2) $G = \mathbb{Z}_2$ or $G = \mathbb{Z}$, $\varepsilon(g,h) = (-1)^{gh}$ for $g, h \in G$.

3) Let $G = \mathbb{Z}_2 \oplus \mathbb{Z}_2$, $K = \mathbb{C}$, $f = (f_1, f_2), g = (g_1, g_2) \in G$. The following forms are commutation factors on G:

$$\begin{aligned}
\varepsilon_1(g,h) &= (-1)^{(f_1+f_2)(g_1+g_2)}, \qquad G_- = \{(0,1),(1,0)\}; \\
\varepsilon_2(g,h) &= \exp\left(2\pi\mathbf{i}((f_1g_2 - f_2g_1)/2 + (f_1g_1 + f_2g_2))\right), \\
&\qquad\qquad G_- = \emptyset, \quad G_+ = G, \quad \varepsilon_2 \text{ is not trivial}; \\
\varepsilon_3(g,h) &= (-1)^{(f_1g_2+f_2g_1)}, \qquad G_- = \emptyset, \quad G_+ = G; \\
\varepsilon_4(g,h) &= (-1)^{(f_2g_1-f_1g_2)}, \qquad G_- = \emptyset, \quad G_+ = G.
\end{aligned}$$

If we consider the $\mathbb{R}$-algebra H of quaternions with the standard generators $\{\mathbf{1}, \mathbf{i}, \mathbf{j}, \mathbf{k}\}$, then

$$H = \bigoplus_{g \in \mathbb{Z}_2 \oplus \mathbb{Z}_2} H_g$$

is an ε_4-commutative G-graded $\mathbb{R}$-algebra,

$$H_{(0,0)} = \mathbb{R}\cdot\mathbf{1};\quad H_{(1,0)} = \mathbb{R}\cdot\mathbf{i};\quad H_{(0,1)} = \mathbb{R}\cdot\mathbf{j};\quad H_{(1,1)} = \mathbb{R}\cdot\mathbf{k}.$$

4) $G = \mathbb{Z}_n \oplus \mathbb{Z}_n \oplus Z_2$, $n \in \mathbb{N}$, $\alpha_1, \alpha_2, \beta_1, \beta_2 \in \mathbb{Z}_m$, $\alpha_3, \beta_3 \in \mathbb{Z}_2$, $\alpha, \beta \in G$, $\alpha = (\alpha_1, \alpha_2, \alpha_3), \beta = (\beta_1, \beta_2, \beta_3) \in G$,

$$\varepsilon(\alpha,\beta) = (-1)^{2(\alpha_1\beta_2-\alpha_2\beta_1)/n+\alpha_3\beta_3}.$$

5) $G = \underbrace{\mathbb{Z}_2 \oplus \cdots \oplus \mathbb{Z}_2}_{k}$, $k \in \mathbb{N}$, $\alpha = (\alpha_1, \ldots, \alpha_k), \beta = (\beta_1, \ldots, \beta_k) \in G$,

$$\begin{aligned} \varepsilon_1(\alpha,\beta) &= (-1)^{\alpha_1\beta_1+\cdots+\alpha_k\beta_k}; \\ \varepsilon_2(\alpha,\beta) &= (-1)^{\alpha_2\beta_1+\alpha_4\beta_3+\cdots+\alpha_{2k}\beta_{2k-1}-\alpha_1\beta_2-\alpha_3\beta_4-\cdots-\alpha_{2k-1}\beta_{2k}}. \end{aligned}$$

6) $G = \underbrace{\mathbb{Z}_n \oplus \cdots \oplus \mathbb{Z}_n}_{2l}$, $n, l \in \mathbb{N}$, $e_i = (0, \ldots, 0, 1, 0, \ldots, 0)$ the i-th unit row,

$$\varepsilon(e_i, e_j) = \begin{cases} \exp(2\pi\mathbf{i}/n), & \text{if } j - i = l \\ 1, & \text{otherwise.} \end{cases}$$

7) We consider *the Clifford algebra* $C(m,n)$. It is the associative algebra with the unity element e_0 given by generators $e_1, \ldots, e_{m+n}$ and defining relations $e_i^2 = 1$ with $i = 1, \ldots, m$, $e_j^2 = -1$ with $j = m+1, \ldots, m+n$, $e_ie_j + e_je_i = 0$ with $i \neq j$.

If $G = \underbrace{\mathbb{Z}_2 \oplus \cdots \oplus \mathbb{Z}_2}_{m+n}$, $g = (g_1, \ldots, g_{m+n})$, $h = (h_1, \ldots, h_{m+n}) \in G$, $a = e_1^{g_1} \cdots e_{m+n}^{g_{m+n}}$, $b = e_1^{f_1} \cdots e_{m+n}^{f_{m+n}}$,

$$\varepsilon(g,h) = (-1)^{g_1f_2+g_3f_4+\cdots+g_{2k-1}f_{2k}-g_2f_1-g_4f_3-\cdots-g_{2k}f_{2k-1}},$$

then $C(m,n)$ is an ε-commutative associative algebra. □

Theorem 2.3 (M. Scheunert [488])
Let $K = \mathbb{C}$,

$$G = \bigoplus_{i=1}^{N} \mathbb{Z}_{p_i^{n_i}}$$

a finite Abelian group (p_i are prime numbers, $n_i \in \mathbb{N}$), e_i generators of the cyclic subgroups $\mathbb{Z}_{p_i^{n_i}}$, $i = 1, \ldots, N$.

Any commutation factor $\varepsilon\colon G \times G \to \mathbb{C}^*$ has the following presentation. If

$$g = \sum_{i=1}^{N} \alpha_i e_i, \qquad h = \sum_{j=1}^{N} \beta_j e_j,$$

then

$$\varepsilon(g,h) = \prod_{i,j=1}^{N} c_{ij}^{\alpha_i \beta_j}$$

where $c_{ij} \in \mathbb{C}$, $c_{ij} = c_{ji}^{-1}$, $c_{ij}^{\gamma} = 1$, γ *is the greatest common divisor of* $p_i^{n_i}$ *and* $p_j^{n_j}$.

Color tensor product

Let Q and R be G-graded associative algebras over a commutative associative ring K with 1, G an Abelian group, $\varepsilon: G \times G \to K^*$ a commutation factor. Consider the tensor product $Q \otimes_K R$ as a K-module and define the multiplication on it setting $(a \otimes b)(c \otimes d) = \varepsilon(b,c)\, ac \otimes bd$ for homogeneous elements $a, c \in Q$, $b, d \in R$ (for other elements use bilinearity). This algebra $Q \mathbin{\widehat{\otimes}} R$ is G-graded with $d(a \otimes b) = d(a) + d(b)$ for homogeneous elements $a \in Q$, $b \in R$ and it is associative because of

$$\begin{aligned}
(a \otimes b)((c \otimes g)(e \otimes f)) &= \varepsilon(b, ce)\, \varepsilon(g,e)\, ace \otimes bgf \\
&= \varepsilon(b,c)\, \varepsilon(b,e)\, \varepsilon(g,e)\, ace \otimes bgf \\
&= \varepsilon(b,c)\, \varepsilon(bg,e)\, ace \otimes bgf \\
&= ((a \otimes b)(c \otimes g))(e \otimes f)
\end{aligned}$$

for homogeneous elements $a, c, e \in Q$, $b, g, f \in R$.

We call $Q \mathbin{\widehat{\otimes}} R$ *the color tensor product of the* G*-graded associative algebras* Q and R.

Free ε-commutative associative algebra

Let X be a G-graded set, i.e.

$$X = \bigcup_{g \in G} X_g, \qquad X_f \cap X_h = \emptyset$$

for all $f \neq h$, ε a commutation factor on G. If $x \in X_g$, then $d(x) = g$. Let also

$$X_+ = \bigcup_{g \in G_+} X_g, \qquad X_- = \bigcup_{g \in G_-} X_g.$$

We consider the associative K-algebra $K[X]$ given by generators X and defining relations $xy - \varepsilon(d(x), d(y))\, yx = 0$ with $x, y \in X$.

It is clear that 1 and the monomials $x_{i_1}^{\alpha_1} \cdots x_{i_k}^{\alpha_k} y_{j_1} \cdots y_{j_l}$ with $x_{i_j} \in X_+$, $y_{j_t} \in X_-$, $x_{i_1} < \cdots < x_{i_k}$; $y_{j_1} < \cdots < y_{j_l}$ form a basis of the free K-module $K[X]$.

For a monomial $x_{i_1} \cdots x_{i_n}$ we set

$$d(x_{i_1} \cdots x_{i_n}) = \sum_{j=1}^{n} d(x_{i_j}).$$

In addition, we assume that $d(1) = 0$. It is clear that $K[X]$ is a G-graded algebra. Now we show that $K[X]$ is ε-commutative. It is sufficient to consider two monomials $x_{i_1} \cdots x_{i_m}$ and $x_{j_1} \cdots x_{j_n}$ with $x_{i_l}, x_{j_t} \in X$. We use the induction on n. The basis of induction is the case $n = 0$. For $n \geq 1$ we get

$$\begin{aligned} x_{i_1} \cdots x_{i_m} \cdot x_{j_1} \cdots x_{j_n} \quad &= \quad (x_{i_1} \cdots x_{i_m} \cdot x_{j_1} \cdots x_{j_{n-1}}) \cdot x_{j_n} \\ = \varepsilon\left(\sum_{l=1}^{m} d(x_{i_l}), \sum_{t=1}^{n-1} d(x_{j_t})\right) & (x_{j_1} \cdots x_{j_{n-1}} \cdot x_{i_1} \cdots x_{i_m}) \cdot x_{j_n} \\ = \varepsilon\left(\sum_{l=1}^{m} d(x_{i_l}), \sum_{t=1}^{n-1} d(x_{j_t})\right) & \varepsilon\left(\sum_{l=1}^{m} d(x_{i_l}, d(x_{j_n})\right) x_{j_1} \cdots x_{j_n} \cdot x_{i_1} \cdots x_{i_m} \\ = \varepsilon(d(x_{i_1} \cdots x_{i_m}), d(x_{j_1} & \cdots x_{j_n})) x_{j_1} \cdots x_{j_n} \cdot x_{i_1} \cdots x_{i_m}. \end{aligned}$$

Here we use the associative law, our induction assumption, and some properties of the commutation factor ε.

It is clear that $K[X]$ is *the free ε-commutative associative K-algebra* on X, i.e. if A is an ε-commutative associative K-algebra and $\varphi: X \to A$ is a mapping of degree zero (for each homogeneous $x \in X$ the element $\varphi(x)$ is also a homogeneous element of A and $d(\varphi(x)) = d(x)$), then there exists the unique homomorphism of zero degree of ε-commutative associative K-algebras $\psi: K[X] \to A$ such that $\psi|_X = \varphi$. If $X = X_+$ then we say that $K[X]$ is the algebra of color polynomials on X; if $X = X_-$ then we say that $K[X]$ is the color Grassmann algebra on X and will denote it by $\Lambda(X)$.

Remark 2.4
Let $X = \{x_1, \ldots, x_n\}$, K be a field, $\operatorname{char} K \neq 2$, $Q = (q_{ij})$ an $(n \times n)$-matrix, $q_{ij} \in K^*$, $q_{ii} = q_{ij}q_{ji}$, $1 \leq i, j \leq n$. We consider the free associative K-algebra $A(X)/I$ where I is the ideal of $A(X)$ generated by the elements

$$x_i x_j - q_{ij} x_j x_i, \qquad 1 \leq i, j \leq n.$$

The algebra $A(X)/I$ is called *the algebra of quantum polynomials*. Note that it is exactly the free skew commutative associative algebra. Indeed, it is clear that the monomials $x_1^{\alpha_1} \cdots x_n^{\alpha_n}$, $\alpha_i \in \mathbb{Z}_+$, where we set $x_1^0 = 1$, form a K-linear basis of $A(X)/I$. We set

$$d(x_1^{\alpha_1} \cdots x_n^{\alpha_n}) = (\alpha_1, \ldots, \alpha_n) \in G = \mathbb{Z}_+^n.$$

Consider the commutation factor ε on G given by $\varepsilon(e_i, e_j) = q_{ij}$, $1 \le i, j \le n$, where $\{e_i \mid i = 1, \ldots, n\}$ is the standard set of generators of G. Then $K[X] = A(X)/I$ is the free ε-commutative associative algebra.

For more details on the algebra of quantum polynomials see, for example, the survey [13] by V. A. Artamonov.

Color Lie superalgebras

Definition 2.5
Let G be an Abelian semigroup, K a commutative associative ring with the unity element and without 2-torsion, K^* the group of invertible elements of K, $\varepsilon: G \times G \to K^*$ a commutation factor.

We say that a G-graded algebra

$$R = \bigoplus_{g \in G} R_g$$

over K is *a color Lie superalgebra* if

$$[a, b] = -\varepsilon(d(a), d(b))\,[b, a]$$

(*ε-anti-commutativity*),

$$\varepsilon(d(c), d(a))\big[a, [b, c]\big] + \varepsilon(d(a), d(b))\big[b, [c, a]\big] + \varepsilon(d(b), d(c))\big[c, [a, b]\big] = 0$$

(*ε-Jacobi identity*), and

$$\big[v, [v, v]\big] = 0 \qquad \text{with } d(v) \in G_-$$

for G-homogeneous elements $a, b, c, v \in R$.

A color Lie superalgebra L with the trivial multiplication ($[a, b] = 0$ for all $a, b \in L$) is called Abelian (ε-commutative).

Let p be a prime number, K a $\mathbb{Z}_p$-algebra. We say that a color Lie superalgebra R over K is *a color Lie p-superalgebra* if on homogeneous components R_g, $g \in G_+$, we have a mapping $a \to a^{[p]}$, $d(a^{[p]}) = pd(a)$, such that for all $\alpha \in K$, $a, b, c \in R$, $d(a) = d(b) \in G_+$, the following conditions are satisfied:

$$(\alpha a)^{[p]} = \alpha^p a^{[p]}; \qquad (\mathrm{ad}\,(a^{[p]}))(c) = [a^{[p]}, c] = (\mathrm{ad}\,a)^p(c);$$
$$(a + b)^{[p]} = a^{[p]} + b^{[p]} + \sum s_i(a, b),$$

where $js_j(a, b)$ is the coefficient on t^{j-1} in the polynomial $(\mathrm{ad}\,(ta+b))^{p-1}(a)$ and $(\mathrm{ad}\,a)(z) = [a, z]$.

Let

$$R_+ = \bigoplus_{g \in G_+} R_g, \qquad R_- = \bigoplus_{g \in G_-} R_g.$$

For convenience we use now the notation $\varepsilon(a, b)$ instead of $\varepsilon(d(a), d(b))$.

Note that sometimes it is more convenient to use the super Jacobi identity in the following form:

$$[a,[b,c]] = [[a,b],c] + \varepsilon(d(a),d(b))\,[b,[a,c]],$$

or

$$[[a,b],c] = [a,[b,c]] - \varepsilon(d(a),d(b))\,[b,[a,c]].$$

At the same time if we would like to have left normed monomials in the definition of color Lie p-superalgebra we may use operators $\operatorname{Ad} a$ where $(z)(\operatorname{Ad} a) = [z,a]$:

$$(c)(\operatorname{Ad} a^{[p]}) = [c, a^{[p]}] = (c)(\operatorname{Ad} a)^p.$$

We will consider only G-homogeneous subalgebras of color Lie superalgebras

$$A = \bigoplus_{g\in G} A_g \subseteq L = \bigoplus_{g\in G} L_g$$

such that $A_g \subseteq L_g$ for all $g \in G$. Also we will consider only homomorphisms of degree zero of color Lie superalgebras: if L, M are color Lie superalgebras over K (with the same G and ε), then a K-linear mapping $f\colon L \to M$ is a homomorphism of degree zero if $f([a,b]) = [f(a), f(b)]$ for all $a, b \in L$ and $f(L_g) \subseteq M_g$ for all $g \in G$.

Lemma 2.6
Let L be a color Lie superalgebra, x a homogeneous element of L, $d(x) \in G_-$, $y \in L$. Then

$$[[x,x],y] = 2[x,[x,y]], \qquad [y,[x,x]] = 2[[y,x],x].$$

PROOF Using super Jacobi identity, we get

$$[[x,x],y] = [x,[x,y]] - \varepsilon(x,x)\,[x,[x,y]] = 2[x,[x,y]].$$

Assuming y to be homogeneous and using super anti-commutativity we have

$$\begin{aligned}
[y,[x,x]] &= -\varepsilon(y,xx)\,[[x,x],y] = -2\varepsilon(y,x)^2\,[x,[x,y]] \\
&= 2\varepsilon(y,x)^2\varepsilon(x,y)\,[x,[y,x]] = -2\varepsilon(y,x)\,\varepsilon(x,yx)\,[[y,x],x] \\
&= -2\varepsilon(x,x)\,[[y,x],x] = 2[[y,x],x].
\end{aligned}$$

If y is not a G-homogeneous element, then it is sufficient to consider the latest formula for each its homogeneous component. ■

Remark 2.7
Not every color Lie superalgebra over a field of prime characteristic admits a structure of color Lie p-superalgebra. For example, if K is the finite field of p^n elements, L is a linear space over K with a basis $\{e_\alpha \mid \alpha \in K\}$ and the multiplication is given by $[e_\alpha, e_\beta] = (\alpha - \beta)e_{\alpha+\beta}$, then L is a Lie algebra which does not have any structure of Lie p-algebra.

However if for any element e_i, $d(e_i) \in G_+$, of a G-homogeneous basis E of a color Lie superalgebra L over a field K, $\operatorname{char} K = p > 2$, there exists an element $a_i \in L_{pd(e_i)}$ such that $(\operatorname{ad} e_i)^p = \operatorname{ad} a_i$, then there exists the unique mapping $a \to a^{[p]}$, $e_i^{[p]} = a_i$, which gives us a structure of a color Lie p-superalgebra on L.

Example 2.8
1) If ε is the trivial commutation factor, then a color Lie superalgebra is a G-graded Lie algebra. As an example one can consider the grading of a semisimple Lie algebra by the root system.

2) If $G = \mathbb{Z}_2$, $\varepsilon(f, g) = (-1)^{fg}$, then a color Lie superalgebra is a usual Lie superalgebra. Note that sometimes it is useful to consider $\mathbb{Z}$-graded Lie superalgebras such that these gradings are compatible with the $\mathbb{Z}_2$-grading:

$$L_0 = \bigoplus_{k\in\mathbb{Z}} L_{2k}, \qquad L_1 = \bigoplus_{k\in\mathbb{Z}} L_{2k-1}.$$

□

Example 2.9
Let

$$A = \bigoplus_{g\in G} A_g$$

be a G-graded associative K-algebra, ε a commutation factor on G. We consider a new multiplication $[\,,\,]$ on A given by

$$[a, b] = ab - \varepsilon(a, b)\, ba$$

for homogeneous elements a, b of A (for $G = \mathbb{Z}_2$ see Example 1.6). Then A with this new multiplication becomes a color Lie superalgebra denoted by $[A]$. Indeed,

$$[a, b] = ab - \varepsilon(a, b)\, ba = -\varepsilon(a, b)\,(ba - \varepsilon(b, a)\, ab) = -\varepsilon(a, b)\,[b, a],$$

and we obtain the anti-commutativity;

$$\begin{aligned} [a, [b, c]] &= [a, bc - \varepsilon(b, c)\, cb] \\ &= abc - \varepsilon(b, c)\, acb - \varepsilon(a, bc)\, bca + \varepsilon(b, c)\, \varepsilon(a, bc)\, cba; \\ [[a, b], c] &= [ab - \varepsilon(a, b)\, ba, c] \end{aligned}$$

$$
\begin{aligned}
&= abc - \varepsilon(a,b)\,bac - \varepsilon(ab,c)\,cab + \varepsilon(ab,c)\,\varepsilon(a,b)\,cba;\\
\varepsilon(a,b)\,[b,[a,c]] &= \varepsilon(a,b)\,[b, ac - \varepsilon(a,c)\,ca]\\
&= \varepsilon(a,b)(bac - \varepsilon(a,c)\,bca - \varepsilon(b,ac)\,acb\\
&\quad + \varepsilon(a,c)\,\varepsilon(b,ac)\,cab\\
&= \varepsilon(a,b)\,bac - \varepsilon(a,b)\,\varepsilon(a,c)\,bca - \varepsilon(b,c)\,acb\\
&\quad + \varepsilon(a,c)\,\varepsilon(b,c)\,cab,
\end{aligned}
$$

and we get the super Jacobi identity.

If p is a prime number, K is a $\mathbb{Z}_p$-algebra, A is a G-graded associative K-algebra, then A with the above defined new operation $[\,,\,]$ and the mapping $[p]\colon A_g \to A_{pg}$ for $g \in G_+$ given by $a^{[p]} = a^p$ for homogeneous elements of A with $d(a) \in G_+$, is a color Lie p-superalgebra (notation: $[A]^p$). Indeed, it is clear that $(\alpha a)^p = \alpha^p a^p$. To consider the other properties of this operation we need some notations.

Let a, b be homogeneous elements of $[A]$, l_a and r_a the operators of left and right multiplications respectively, $l_a(b) = ab = r_b(a)$. Suppose that $d(a) \in G_+$. Then $\varepsilon(a, a^k b) = \varepsilon(a,a)^k \varepsilon(a,b) = \varepsilon(a,b)$, and we get

$$(\operatorname{ad} a)^k(b) = (l_a - \varepsilon(a,b)\,r_a)^k(b).$$

It is clear that C_p^i is divided by p for $i = 1, \ldots, p-1$. Since $l_a r_a = r_a l_a$,

$$
\begin{aligned}
(l_a - \varepsilon(a,b)\,r_a)^p(b) &= l_a^p(b) - \varepsilon(a,b)^p r_a^p(b)\\
&\quad + \sum_{i=1}^{p} (-1)^i C_p^i l_a^i \varepsilon(a,b)^i r_a^{p-i}(b)\\
&= (l_a^p - \varepsilon(a,b)^p r_a^p)(b)\\
&= (l_{a^p} - \varepsilon(a^p,b) r_{a^p})(b) = (\operatorname{ad} a)^p(b).
\end{aligned}
$$

As there are no signs ε in this formula and bearing in mind that each element of A is a sum of its homogeneous components, we get the second identity.

It is well known that $C_p^i = C_{p-1}^i + C_{p-1}^{i-1}$. It follows by induction that $C_{p-1}^i = (-1)^i \pmod p$ for all $i = 0, 1, \ldots, p-1$. We consider the polynomial ring $A[t]$, $d(t) = 0$. If a, b be homogeneous elements of A, $d(a) = d(b) \in G_+$, then $\varepsilon(ta+b, (ta+b)^k b) = \varepsilon(ta+b, ta+b)^k \varepsilon(ta+b, b) = 1$ and

$$
\begin{aligned}
\operatorname{ad}(ta+b)^{p-1}(a) &= (l_{ta+b} - r_{ta+b})^{p-1}(a)\\
&= \sum_{i=0}^{p-1} (-1)^i C_{p-1}^i l_{ta+b}^i r_{ta+b}^{p-1-i}(a)\\
&= \sum_{i=0}^{p-1} l_{ta+b}^i r_{ta+b}^{p-1-i}(a).
\end{aligned}
$$

Now we consider the equation

$$(ta+b)^p = t^p a^p + b^p + \sum_{j=1}^{p-1} t^j s_j(a,b).$$

Taking the formal derivative on t we get

$$\sum_{i=0}^{p-1} (ta+b)^j a(ta+b)^{p-1-i} = \sum_{j=0}^{p-1} jt^{j-1} s_j(a,b).$$

Setting $t=1$ we obtain the third identity.

Note that above considered situation is typical, as Poincaré-Birkhoff-Witt theorems show, see Theorems 19.1, 19.2. □

Example 2.10
Let K be a field, $V = \bigoplus V_g$ a G-graded vector space over K. An element x of V is said to be homogeneous if $x \in V_g$ for some $g \in G$, in this case we set $d(x) = g$. By $\mathrm{END}_K(V)_g$, $g \in G$, we denote the set of all K-linear operators $f \in \mathrm{End}_K(V)$ such that the image of each homogeneous element $x \in V$ is a homogeneous element and $d(f(x)) = g + d(x)$. Let

$$\mathrm{END}_K(V) = \bigoplus_{g \in G} \mathrm{END}_K(V)_g.$$

It is clear that $\mathrm{END}_K(V)$ is a G-graded associative K-algebra with usual multiplications of operators (a K-subalgebra of $\mathrm{End}_K(V)$). The color Lie superalgebra $\mathrm{gl}(V) = [\mathrm{END}_K(V)_g]$ is called *the general linear color Lie superalgebra.* Remark that in the linear algebra of graded vector spaces instead of usual trace function one need to use the function grtr given by

$$\mathrm{grtr}\,(A) = \sum_{g \in G} \varepsilon(g,g)\,\mathrm{tr}(A_{gg})$$

where A_{gg} is the block on the diagonal of the matrix A corresponding to the subspace V_g, tr is the usual trace function. □

Example 2.11
Let

$$A = \bigoplus_{g \in G} A_g$$

be a G-graded algebra. A K-linear mapping $D\colon A \to A$ is *a derivation* of A of the degree $d(D) \in G$ if

$$D(ab) = D(a)b + \varepsilon(d(a), d(D))\, aD(b)$$

for all $b \in A$ and homogeneous $a \in A$. By $\mathrm{Der}(a)_g$, $g \in G$, we denote the set of all derivations of the degree g. It is easy to check that

$$\mathrm{Der}(A) = \bigoplus_{g \in G} \mathrm{Der}(A)_g$$

is a color Lie superalgebra relatively to the multiplication given by

$$[D_1, D_2] = D_1 D_2 - \varepsilon(D_1, D_2) D_2 D_1.$$

□

Example 2.12
Let $K = \mathbb{C}$, β a bilinear form on a G-graded vector space V, $\beta(x,y) = -\varepsilon(x,y)\,\beta(y,x)$ for G-homogeneous elements x, y of V. Let

$$H(V;\beta) = \bigoplus_{g \in G} H(V;\beta)_g,$$

$H(V;\beta)_0 = V_0 \oplus \mathbb{C} \cdot z$ and $H(V;\beta)_g = V_g$ for $g \in G$, $g \neq 0$,

$$[x,y] = \beta(x,y)\, z; \qquad [H(V;\beta), z] = [z, H(V;\beta)] = 0$$

for all homogeneous elements $x, y \in V$. Then $H(V;b)$ is a color Lie superalgebra (so called *ε-Heisenberg algebra*). □

Example 2.13
Let $G = \mathbb{R}^{2n}$, $z = (x,y) \in \mathbb{R}^{2n}$ with $x, y \in \mathbb{R}^n$, $I(z,z') = \langle x, y' \rangle - \langle y, x' \rangle$ a symplectic form in $\mathbb{R}^{2n}$, $\varepsilon(z,z') = \exp(\mathbf{i} \cdot I(z,z')/2)$. Let also L_z be the one-dimensional subspace generated by the operator

$$A(z) = \exp(\mathbf{i}xq) \cdot \exp\left(y \frac{\partial}{\partial q} \right)$$

on functions on variables $q_1, \ldots, q_n$,

$$L = \bigoplus_{z \in \mathbb{R}^{2n}} L_z.$$

The algebra L is an Abelian color Lie superalgebra. □

A subalgebra I of L is called *an ideal* of L if $[a,b] \in I$ for all $a \in I$ and $b \in L$. A color Lie superalgebra L is said to be *simple* if it has no proper ideals and its multiplication is not trivial.

Let A, B be subalgebras of a color Lie superalgebra L. By $[A,B]$ we denote the subalgebra of L generated by all elements $[a,b]$ with $a \in A$, $b \in B$.

For a color Lie superalgebra L we consider the subalgebras

$$L^1 = L \supseteq L^2 = [L^1, L] = [L, L] \supseteq \cdots \supseteq L^{n+1} = [L^n, L] \supseteq \cdots;$$
$$L^{(0)} = L \supseteq L^{(1)} = [L^{(0)}, L^{(0)}] \supseteq \cdots \supseteq L^{(n+1)} = [L^{(n)}, L^{(n)}] \supseteq \cdots.$$

A color Lie superalgebra L is said to be *nilpotent* of the degree c if $L^{c+1} = 0$ and $L^c \neq 0$. If $c = 2$, then L is an Abelian color super Lie superalgebra. If $L^{(l-1)} \neq 0$ and $L^{(l)} = 0$ for some $l \in \mathbb{N}$, then L is said to be *a solvable color Lie superalgebra* of the degree l.

Let V be a G-graded K-module. We consider the general linear color Lie superalgebra $\mathrm{gl}(V)$. *A representation* ρ of L in V is a homomorphism of color Lie superalgebras $\rho: L \to \mathrm{gl}(V)$. A representation ρ is said to be *faithful* if $\operatorname{Ker} \rho = 0$. A graded K-module V is an *L-module* if there exist a K-bilinear mapping $f: L \times V \to V$ such that $d(a(v)) = d(a) + d(v)$; $[a, b](v) = a(b(v)) - \varepsilon(a, b)\, b(a(v))$ for all homogeneous $a, b \in L$, $v \in V$. This notion is equivalent to the notion of a representation, it is sufficient to set $a(x) = \rho(a)(x)$. A K-linear mapping $D: L \to L$ is said to be *a derivation* of L of the degree $d(D) \in G$ if $D[a, b] = [D(a), b] + \varepsilon(d(D), d(a))\, [a, D(b)]$ for all homogeneous elements a, b of L. By $\mathrm{Der}_g(L)$ we denote the set of all derivations of L with the degree $g \in G$. By $\mathrm{Der}(L)$ we denote the color Lie superalgebra

$$\bigoplus_{g \in G} \mathrm{Der}_g(L)$$

with the multiplication $[D_1, D_2] = D_1 D_2 - \varepsilon(d(D_1), d(D_2)) D_2 D_1$.

For a homogeneous element $a \in L$ we consider mappings $\operatorname{ad} a, \operatorname{Ad} a: L \to L$ given by $(\operatorname{ad} a)(b) = [a, b]$; $(b)(\operatorname{Ad} a) = [b, a]$. This mappings are derivations of L of the degree $d(a)$. The mapping $\operatorname{ad}: L \to \mathrm{Der}L$ is a representation of L (it is called *the adjoint representation* of L). Indeed,

$$\begin{aligned}
(\operatorname{ad} a)([b, c]) &= [a, [b, c]] = [[a, b], c] + \varepsilon(a, b)\, [b, [a, c]] \\
&= [(\operatorname{ad} a)(b), c] + \varepsilon(a, b)[b, (\operatorname{ad} a)(c)]; \\
\operatorname{ad}([a, b])(c) &= [[a, b], c] = [a, [b, c]] - \varepsilon(a, b)\, [b, [a, c]] \\
&= (\operatorname{ad} a)((\operatorname{ad} b)(c)) - \varepsilon(a, b)\, (\operatorname{ad} b)\, (\operatorname{ad} a)(c) \\
&= [\operatorname{ad}(a), \operatorname{ad}(b)](c).
\end{aligned}$$

Using basic facts on representations of graded Lie algebras and Grassmann envelope one can prove the following analog of Ado-Iwasawa theorem. For the details in the case when $\operatorname{char} K = 0$ see [488].

Theorem 2.14
Let K be a field, L a finite dimensional color Lie superalgebra. Then L has a faithful finite dimensional representation.

Skew envelope of a color Lie superalgebra

Let L be a color Lie superalgebra. We consider the free ε-commutative associative K-algebra

$$K[X] = \bigoplus_{g\in G} K_g[X],$$

and the G-graded K-module

$$K[X](L) = \bigoplus_{g\in G} L_g \otimes_K K_g[X]$$

with the multiplication given by

$$[a \otimes a', b \otimes b'] = \varepsilon(b,a)[a,b] \otimes a'b'.$$

Then $K[X](L)$ is a G-graded Lie algebra which is called *the skew envelope* of L. Indeed, for homogeneous $a, b, c \in L$, $a', b', c' \in K[X]$, $d(a') = d(a)$, $d(b') = d(b)$, $d(c') = d(c)$ one can see that

$$\begin{aligned} [a \otimes a', b \otimes b'] &= \varepsilon(b,a)\,[a,b] \otimes a'b' = -\varepsilon(b,a)\,\varepsilon(a,b)\,[b,a] \otimes \varepsilon(a'b')\,b'a' \\ &= -\varepsilon(a,b)\,[b,a] \otimes b'a' = -[b \otimes b', a \otimes a']. \end{aligned}$$

Denoting $\xi = \varepsilon(b,a)\,\varepsilon(c,a)\,\varepsilon(c,b)$ we get

$$\begin{aligned} \big[a \otimes a', [b \otimes b', c \otimes c']\big] &= \xi \cdot \big[a, [b,c]\big] \otimes a'b'c', \\ \big[[a \otimes a', b \otimes b'], c \otimes c'\big] &= \xi \cdot \big[[a,b],c\big] \otimes a'b'c', \\ \big[b \otimes b', [a \otimes a', c \otimes c']\big] &= \xi \cdot \varepsilon(a,b)^2 \big[b, [a,c]\big] \otimes b'a'c' \\ &= \xi \cdot \varepsilon(a,b) \big[b, [a,c]\big] \otimes a'b'c', \end{aligned}$$

and it is clear that the Jacobi identity in $K[X](L)$ follows from the ε-Jacobi identity in L. Remark that in the case when G is an Abelian group it is also possible to consider

$$K[X](L)' = \bigoplus_{g\in G} L_g \otimes K_{-g}[X]$$

with the same multiplication. Since $\varepsilon(-g,-h) = \varepsilon(g,h)$, $K[X](L)'$ is also a Lie algebra.

On the other hand, if L is a G-graded Lie algebra,

$$K[X](L) = \bigoplus_{g\in G} L_g \otimes_K K_g[X]$$

with the multiplication given by

$$[a \otimes a', b \otimes b'] = [a,b] \otimes a'b'$$

for homogeneous $a, b \in L$, $a', b' \in K[X]$, $d(a') = d(a)$, $d(b') = d(b)$, then L is a color Lie superalgebra (which is called the skew envelope of L):

$$\begin{aligned}[a \otimes a', b \otimes b'] &= [a, b] \otimes a'b' = -[b, a] \otimes \varepsilon(a', b')\, b'a' \\ &= -\varepsilon(a, b)\, [b, a] \otimes b'a' = -\varepsilon(a, b)\, [b \otimes b', a \otimes a'],\end{aligned}$$

and we get the anti-commutativity;

$$\begin{aligned}\big[a \otimes a', [b \otimes b', c \otimes c']\big] &= \big[a, [b, c]\big] \otimes a'b'c, \\ \big[[a \otimes a', b \otimes b'], c \otimes c'\big] &= \big[[a, b], c\big] \otimes a'b'c', \\ \varepsilon(a, b)\big[b \otimes b', [a \otimes a', c \otimes c']\big] &= \varepsilon(a, b)\big[b, [a, c]\big] \otimes b'a'c' \\ &= \varepsilon(a, b)\big[b, [a, c]\big] \otimes \varepsilon(b', a')\, a'b'c' \\ &= \big[b, [a, c]\big] \otimes a'b'c',\end{aligned}$$

and we see that ε-Jacobi identity follows from the Jacobi identity in L.

Differential Lie superalgebras

Let G be an Abelian group,

$$R = \bigoplus_{g \in G} R_g$$

a G-graded associative ring with 1, $C(R)$ the center of the ring R, Φ a subring in R_0, $1 \in \Phi$, $\Phi \subseteq C(R)$, $U(\Phi)$ the group of invertible elements in Φ, $\varepsilon: G \times G \to U(\Phi)$ a commutation factor. We say that Φ-linear map $\mu: R \to R$, $\mu(\Phi) = 0$, is *an ε-derivation of the weight $d(\mu) \in G$* if $d(\mu(x)) = d(\mu) + d(x)$, $\mu(xy) = \mu(x)y + \varepsilon(\mu, x)\, x\, \mu(y)$ for all homogeneous elements $x, y \in R$. Denote the Φ-linear span of all ε-derivations of R by Der$R$. Then Der$R$ is a G-graded associative Φ-algebra, [Der$R$] is a color Lie superalgebra; in the case of prime characteristic $[\mathrm{Der}R]^p$ is a color Lie p-superalgebra over Φ.

Let $C = Z(R)$ be the color center of R, i.e. a G-graded subring of R which is a Φ-linear span of such G-homogeneous elements $x \in R_+$ that $xa = \varepsilon(x, a)\, ax$ for all homogeneous $a \in R$. It is clear that $\Phi \subseteq C_0$. Then Der$R$ is a left C-module: for $c \in C$, $\mu \in$ Der$R$, $x \in R$ we define $(c\mu)(x) = c\mu(x)$. The following identities are fulfilled in the C-module Der$R$:

$$[c\mu, \nu] = c[\mu, \nu] - \varepsilon(c, \nu)\, \varepsilon(\mu, \nu)\, \nu(c)\mu$$

for all homogeneous elements $c \in C$, $\mu, \nu \in$ Der$R$; in the case when p is a prime number, R is a $\mathbb{Z}_p$-algebra,

$$(c\delta)^p = T_{p,\delta}(c) + \varepsilon(\delta, c)^{p(p-1)/2} c^p \delta^p$$

where

$$T_{1,\delta}(c) = c, \qquad T_{k+1,\delta}(c) = c\delta(T_{k,\delta}(c))$$

for all homogeneous $c \in C$, $\delta \in (\mathrm{Der}R)_+$.

Let

$$R = \bigoplus_{g \in G} R_g$$

be a color Lie (p-)superalgebra over Φ,

$$F = \bigoplus_{g \in G} F_g$$

a G-graded super-commutative ring with 1 (i.e., $F = F_+$, $ab = \varepsilon(a,b)\, ba$ for all homogeneous elements $a, b \in F$), $\Phi \subseteq F_0$, $\mathrm{Der}F$ the color Lie (p-)superalgebra over Φ of all ε-derivations of F (in particular, we have $\mathrm{Der}F|_\Phi = 0$).

Definition 2.15

An algebra R is called *a (restricted) differential Lie F-superalgebra* if

(i) R is a G-graded left F-module;

(ii) there is a homomorphism $\partial: R \to \mathrm{Der}F$ of color Lie (p-)superalgebras over Φ such that $d(\partial x) = d(x)$ for all homogeneous elements $x \in R$, ∂ is a homomorphism of left F-modules and the following identities are fulfilled:

$$[c\mu, \nu] = c[\mu, \nu] - \varepsilon(c, \nu)\, \varepsilon(\mu, \nu)\, (\partial \nu)(c)\, \mu$$

for all homogeneous elements $c \in F$, $\mu, \nu \in R$;

(iii) in the case when $\mathrm{char}\, R = p$, p is a prime number,

$$(c\delta)^p = T_{p,\delta}(c) + \varepsilon(\delta, c)^{p(p-1)/2} c^p \delta^p$$

where

$$T_{1,\delta}(c) = c, \qquad T_{k+1,\delta}(c) = c\partial(\delta)(T_{k,\delta}(c))$$

for all homogeneous $c \in F$, $\delta \in R_+$.

3 Universal constructions

In this section we introduce free associative algebras and free color Lie superalgebras, (restricted) universal enveloping algebras of color Lie (p-)superalgebras and differential Lie superalgebras. Free products with the amalgamated subalgebra are also defined.

Free associative algebras

Let X be a G-graded set. Consider *the free semigroup* $S(X)$ on X. It consists of the unity element 1 (i.e., the empty word) and of all products of the form $x_{i_1} \cdots x_{i_n}$, $i_j \in I$. The multiplication is given by the rule

$$(x_{i_1} \cdots x_{i_n}) \cdot (x_{j_1} \cdots x_{j_m}) = x_{i_1} \cdots x_{i_n} \cdot x_{j_1} \cdots x_{j_m};$$
$$1 \cdot (x_{i_1} \cdots x_{i_n}) = (x_{i_1} \cdots x_{i_n}) \cdot 1 = x_{i_1} \cdots x_{i_n}.$$

By $S^+(X)$ we denote the free semigroup without unity element.

The grading function is as follows:

$$d(x_{i_1}\cdots x_{i_n})=\sum_{j=1}^{n} d(x_{i_j}),\qquad d(1)=0.$$

We set $S(X)_g=\{a\in S(X)\mid d(a)=g\}$.

Let K be a commutative associative ring with the unity element. *The free G-graded associative algebra $A(X)$ with the set X of free generators over K* is given by the following universal property. There exists a mapping $\xi: X\to A(X)$ of zero degree such that for any mapping $\varphi: X\to A$ of zero degree into an associative G-graded K-algebra A there exist the unique homomorphism ψ of zero degree of G-graded associative K-algebras such that $\varphi=\psi\xi$.

It is clear that $A(X)$ is uniquely determined by the universal property. This algebra is exactly the semigroup algebra of $S(X)$ over K. Its elements have the unique presentation in the form of finite sums

$$\sum_i \alpha_i a_i,\qquad 0\neq\alpha_i\in K,\ \ a_i\in S(X),\ \ a_j\neq a_s \text{ with } j\neq s.$$

Denoting by $A(X)_g$ the K-linear span of $S(X)_g$, we get the G-grading of $A(X)$:

$$A(X)=\bigoplus_{g\in G} A(X)_g.$$

Consider *a weight function* $\varphi: X\to \mathbb{N}$, let

$$\varphi(x_1\cdots x_n)=\sum_{i=1}^{n}\varphi(x_i)$$

for $x_1,\dots,x_n\in X$. If $\varphi(x)=1$ for all $x\in X$, then φ is just the usual length, $\varphi=\ell=\ell_X$. If $a\in A(X)$, $a=\sum\alpha_i a_i$, $0\neq\alpha_i\in K$, $a_i\in S(X)$, $a_j\neq a_s$ with $j\neq s$, then we set

$$\varphi(a)=\max_i \varphi(a_i).$$

By $\widehat{a}$ we denote *the leading part* of a:

$$\widehat{a}=\sum_{j,\varphi(a_j)=\varphi(a)}\alpha_j a_j.$$

It is clear that we have the following grading of $A(X)$:

$$A(X)=\bigoplus_{i=0}^{n} A(X)_i$$

where $A(X)_i$ is the K-linear span of all monomials from $S(X)$ with the value of φ (the length) equal to i.

Assuming X to be well ordered, by $\overline{a}$ we denote *the lexicographical greatest word* a_{j_0} in $\widehat{a}$. We set also $a^\circ = \alpha_{j_0} a_{j_0}$. If $a = \widehat{a}$, then a is said to be *φ-homogeneous*. By analogy we define the length and the homogeneity relatively to $x \in X$. If an element $a \in A(X)$ is ℓ-homogeneous relatively to all $x \in X$, then a is said to be *multihomogeneous*.

Note that $A(X)$ can be realized as a tensor algebra. Let V be the free K-module with the basis X

$$T(V) = \sum_{n=0}^{\infty} T^n(V),$$

where

$$T^n(V) = V \otimes_K \cdots \otimes_K V = \bigotimes^n V,$$

$n \in \mathbb{N}$, and $T^0(V) = K$. The K-linear multiplication is given by

$$(a_1 \otimes \cdots \otimes a_l) \cdot (a_{l+1} \otimes \cdots \otimes a_{l+m}) = a_1 \otimes \cdots \otimes a_{l+m}.$$

It is clear that $A(X) \cong T(V)$.

Consider the Abelian semigroup $H = (\mathbb{Z}_+)^{|X|}$ with the standard set of generators $\{h_x \mid x \in X\}$. *The multidegree function* $m: S(X) \to G$ is the semigroup homomorphism given by $m(x) = h_x$. For $a \in S(X)$ and $g \in G$ by $m_h(a)$ we denote the h-th component of $m(a)$. If an element $a \in A(X)$ is a linear combination of words of the same multidegree, then a is a multihomogeneous element. Let

$$|m(a)| = \sum_{h \in H} m_h(a).$$

Using the multidegree we get the following grading of $A(X)$ by multihomogeneous elements which is compatible with the G-grading and with the φ-grading:

$$A(X) = \bigoplus_{h \in H} A(X)_h,$$

where $A(X)_h$ is the K-linear span of the multihomogeneous elements of $A(X)$ with the multidegree h.

Let $x_1, \ldots, x_n \in X$ and $a = a(x_1, \ldots, x_n)$ an element of $A(X)$ which depends only on $x_1, \ldots, x_n$. If a is K-linear on each x_i, $i = 1, \ldots, n$, then a is said to be multilinear.

The free color Lie superalgebra

Let X be a G-graded set, $\Gamma(X)$ the free groupoid of nonassociative monomials in the alphabet X, $u \circ v = [u, v]$ for $u, v \in \Gamma(X)$, $\tilde{}\colon \Gamma(X) \to S(X)$

the bracket removing homomorphism, $[u]$ an arrangement of brackets on the word $u \in S(X)$. For $v \in \Gamma(X)$ we set $d(v) = d(\widetilde{v})$, $m(v) = m(\widetilde{v})$, $\ell(v) = \ell(\widetilde{v})$, and $\varphi(v) = d(\widetilde{v})$. By analogy, we define the length relative to a subset $Y \subseteq X$.

Let K be a commutative associative ring with 1, $F(X)$ the free nonassociative K-algebra. The algebra $F(X)$ is also *the free G-graded nonassociative algebra:*

$$F(X) = \bigoplus_{g \in G} F(X)_g$$

where $F(X)_g$ is the K-linear span of all elements w from $\Gamma(X)$ with $d(w) = g$. Let now I be the G-graded ideal of $F(X)$ generated by the homogeneous elements of the form

$$\begin{gathered} [a,b] + \varepsilon(a,b)\,[b,a], \\ [[a,b],c] - \big[a,[b,c]\big] + \varepsilon(a,b)\big[b,[a,c]\big], \\ \big[u,[u,u]\big], \end{gathered}$$

where a, b, c, u are homogeneous elements of $\Gamma(X)$, $d(u) \in G_-$. Then $L(X) = F(X)/I$ is *the free color Lie K-superalgebra* (i.e., every G-map φ of degree zero from X into any color Lie K-superalgebra R with the same group G and form ε $(d(\varphi(x)) = d(x),\ x \in X)$ can be uniquely extended to a color Lie superalgebra homomorphism $\varphi' \colon L(X) \to R$).

It is clear that this universal property defines the free color Lie superalgebra $L(X)$ uniquely up to a color Lie superalgebra isomorphism.

Remark 3.1
Free Lie superalgebras naturally appear in homotopy theory. Let $K = \mathbb{Q}$, and let C be a connected CW-complex, $\pi_i(C)$ the i-th homotopy group of C, $i \geq 1$. Then the linear space

$$\pi_*^{\mathbb{Q}}(C) = \bigoplus_{i=1}^{\infty} \pi_{i+1}(C) \otimes \mathbb{Q}$$

with the Whitehead product is a Lie superalgebra (see, for example, monographs [550] by H. Sze-Tsen and [216] by R. M. Hain, and review [427] by J. W. Milnor and J. C. Moore). If C is the union of spheres, then $\pi_*^{\mathbb{Q}}(C)$ is the free Lie superalgebra (see [231]).

Universal enveloping algebras

Definition 3.2
Let K be a commutative associative ring with 1, L a color Lie superalgebra, U a G-graded associative K-algebra with 1, $\delta \colon L \to [U]$ a homomorphism of color Lie superalgebras. We say that the algebra U (with δ) is *the universal*

enveloping algebra of L (we denote it $U(L)$) if for any homomorphism σ of the color Lie superalgebra L into a color Lie superalgebra $[R]$ (with the same ε and G) for some G-graded associative algebra R with 1 there exists a unique homomorphism $\theta: U \to R$ of G-graded associative algebras with 1 such that the diagram

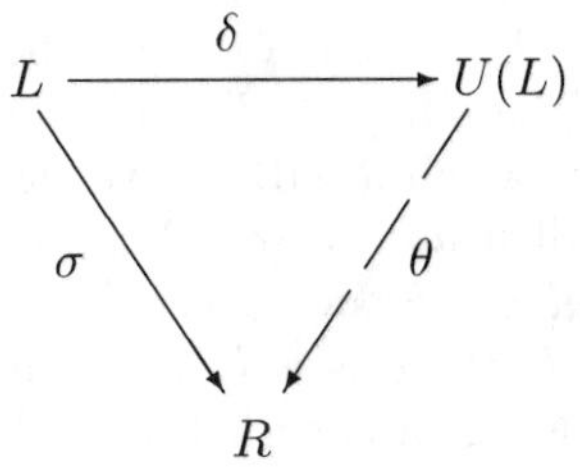

is commutative.

It is clear that the universal enveloping algebras are defined uniquely (up to isomorphisms of G-graded associative algebras with 1) and it is natural enough to consider the following question: is δ an embedding, or not? Using in Definition 3.2 only associative algebras without the unity element we obtain the definition of *the universal enveloping algebra without the unity element.* We denote it by $U^+(L)$.

Definition 3.3
Let K be a field, $\operatorname{char} K > 2$, L a color Lie p-superalgebra, U a G-graded associative K-algebra with 1, $\delta: L \to [U]^p$ a homomorphism of color Lie p-superalgebras. We say that U (with δ) is *the restricted universal enveloping algebra* of the color Lie p-superalgebra L (notation $u(L)$) if for any homomorphism $\sigma: L \to [R]^p$ of color Lie p-superalgebras, where R is a G-graded associative K-algebra with 1 (with the same ε and G), there exists a unique homomorphism $\theta: U \to R$ of G-graded associative K-algebras with 1 such that $\sigma = \theta\delta$.

It is clear that the restricted universal enveloping algebra is defined uniquely (up to an isomorphism of G-graded associative K-algebras). As above, by $u^+(L)$ we denote *the restricted universal enveloping algebra of L without the unity element.*

Let A be a G-graded associative algebra with 1 over a commutative ring Φ with 1. Suppose that A is a faithful left module over a G-graded supercommutative ring F ($\Phi \subseteq F_0$). Let also L be a left F-submodule in A such that L is a color Lie (p-)subsuperalgebra over Φ in $[A]$ (in $[A]^p$, respectively) and

$$a(c) = a(c \cdot 1) - \varepsilon(a, c)\, ca \in F \cdot 1$$

for all homogeneous $c \in F$, $l \in L$. Then L is a (restricted) differential Lie F-superalgebra.

Definition 3.4
We say that a G-graded associative algebra A over Φ is *a (restricted) associative envelope of (restricted) differential Lie F-superalgebra D* if there are a (restricted) differential Lie F-superalgebra $L \subseteq A$ and a F-linear homomorphism $\xi_A: D \to L$ of color Lie (p-)superalgebras over Φ such that $\partial f = \xi_A(f) \in \mathrm{Der}F$ for all $f \in D$ and the set $\xi_A(D)$ generates A as a G-graded associative algebra with 1 over Φ. A (restricted) associative envelope U of D is called *a (restricted) universal enveloping algebra of a (restricted) differential Lie F-superalgebra D* if for any (restricted) associative envelope A of D there is a F-linear homomorphism $\psi: U \to A$ of G-graded associative Φ-algebras with 1 such that $\psi\xi_U = \xi_A$.

Free products

Definition 3.5
Let K be a field, T a set, ε a commutation factor on G, and H^0, H_α, $\alpha \in T$, color Lie K-superalgebras (with the same group G and the commutation factor ε), and let H_α^0 be a G-homogeneous subalgebra of H_α, $\delta_\alpha: H^0 \to H_\alpha^0$ an isomorphism of color Lie superalgebras. We say that a color Lie K-superalgebra Q (with the group G and the commutation factor ε) is *the free product of color Lie superalgebras H_α, $\alpha \in T$, with the amalgamated homogeneous subalgebra H^0* (notation: $Q = \prod^*_{\substack{H_0\\ \alpha\in T}} H_\alpha$) if the following conditions are satisfied:

(i) there exist homogeneous subalgebras H'_α in Q and isomorphisms $\sigma_\alpha: H_\alpha \to H'_\alpha$ of color Lie superalgebras such that $\sigma_\alpha\delta_\alpha(h) = \sigma_\beta\delta_\beta(h)$ for all $\alpha, \beta \in T$ and $h \in H^0$;

(ii) if L is a color Lie K-superalgebra (with the same group G and the commutation factor ε), $\gamma_\alpha: H_\alpha \to L$ is a homomorphism of color Lie superalgebras, $\gamma_\alpha\delta_\alpha(h) = \gamma_\beta\delta_\beta(h)$ for all $\alpha, \beta \in T$ and $h \in H^0$, then there is one and only one homomorphism θ: $\prod^*_{\substack{H_0\\ \eta\in T}} H_\eta \to L$ of color Lie superalgebras such that $\theta\delta_\alpha = \gamma_\alpha$ for all $\alpha \in T$.

It is clear that the color Lie superalgebra $\prod^*_{\substack{H_0\\ \alpha\in T}} H_\alpha$ is defined uniquely (up to an isomorphism of color Lie superalgebras).

If, in Definition 3.5, in place of color Lie superalgebras we consider color Lie p-superalgebras then we get the definition of *the free product of color Lie p-superalgebras with the amalgamated homogeneous subalgebra*. By analogy, if we consider associative algebras (without the unity element), then we obtain the definition of the free product with the amalgamated subalgebra of associative algebras.

4 Group actions on free Lie superalgebras

In this section we discuss free Lie superalgebras as modules under the actions of the symmetric and the general linear groups (i.e., modules over group algebras of corresponding groups).

In all what follows in this section S_n will denote *the symmetric group of permutations* of the set $\{1,2,\ldots,n\}$, and $(i_1,\ldots,i_k)\in S_n$ will denote the cyclic permutation of $i_1,\ldots,i_k$.

S_n-modules and GL(V)-modules

Let K be a commutative ring, $X=\{x_1,x_2,\ldots\}$, $V=Kx_1\oplus Kx_2\oplus\cdots$ the free K-module generated by the set X, and $A(X)$ the free associative algebra. Consider the K-submodules $A_n(X)\subseteq A(X)$ of all monomials of degree n, and the K-submodules $A_n\subseteq A_n(X)$ of all multilinear polynomials on variables $x_1,\ldots,x_n$.

Any automorphism $g\in \mathrm{GL}(V)$ of the K-module V can be uniquely extended to an automorphism of K-algebra $A(X)$. Hence $A(X)$ can be considered as a left $\mathrm{GL}(V)$-module. It is clear that $A_n(X)$ are $\mathrm{GL}(V)$-submodules of $A(X)$. If $X=\{x_1,\ldots,x_m\}$ then the symmetric group S_m is naturally embedded into $\mathrm{GL}(V)$, and $S_m(A_m)\subseteq A_m$. In this case for any $\sigma\in S_m$ and $x_{i_1}\cdots x_{i_m}\in A_m$ we have

$$\sigma(x_{i_1}\cdots x_{i_m})=x_{\sigma(i_1)}\cdots x_{\sigma(i_m)}.$$

Consider the free Lie K-algebra $L(X)\subseteq A(X)$, and let

$$L_n(X)=A_n(X)\cap L(X),\qquad L_n=A_n\cap L(X).$$

Since $g(L(X))\subseteq L(X)$ for any $g\in \mathrm{GL}(V)$, $L_n(X)$ are left $\mathrm{GL}(V)$-modules. If $|X|=m$, then $L_n(X)$ and L_n are left S_m-modules.

We introduce also the right action of S_n on $A_n(X)$, $n\in\mathbb{N}$, by the following way:

$$(x_{i_1}\cdots x_{i_n})\sigma=x_{i_{\sigma(1)}}\cdots x_{i_{\sigma(n)}}$$

for any $\sigma\in S_n$ and any word $x_{i_1}\cdots x_{i_n}\in A_n(X)$ (extending the action of S_n on $A_n(X)$ by linearity).

The K-module $L_n(X)$ is not invariant with respect to the right action of S_n. For example, for $\sigma=(2,3)$ and $a=\big[[x_1,x_2],x_3\big]$ one can see that

$$\begin{aligned} a\sigma &= (x_1x_2x_3)\sigma-(x_2x_1x_3)\sigma-(x_3x_1x_2)\sigma+(x_3x_2x_1)\sigma\\ &= x_1x_3x_2-x_2x_3x_1-x_3x_2x_1+x_3x_1x_2\notin L(X). \end{aligned}$$

Lemma 4.1
The left action of GL(V) *and the right action of* S_n *on* $A_n(X)$ *are commuting.*

For the free Lie K-superalgebra $L(X,Y)$ with the even set of generators $X = \{x_1, x_2, \ldots\}$ and the odd set of generators $Y = \{y_1, y_2, \ldots\}$ we can naturally define the left action of the group $\mathrm{GL}(V) \times \mathrm{GL}(W)$, where $V = Kx_1 \oplus Kx_2 \oplus \cdots$, $W = Ky_1 \oplus Ky_2 \oplus \cdots$. If $X = \{x_1, \ldots, x_n\}$, $Y = \{y_1, \ldots, y_m\}$, then $S_n \times S_m \subseteq \mathrm{GL}(V) \times \mathrm{GL}(W)$, and $L(X,Y)$ can be considered as a left $(S_n \times S_m)$-module.

For a field K of characteristic zero the theory of representations of S_n could be found, for example, in [139], [245]. We mention only the following classical results.

Let $n \in \mathbb{N}$. The sequence $\lambda = (\lambda_1, \ldots, \lambda_k)$ of natural numbers is called *a partition* of n if $\lambda_1 \geq \cdots \geq \lambda_k > 0$ and $\lambda_1 + \cdots + \lambda_k = n$. The set of points

$$\{(i,j) \mid j \in \{1, 2, \ldots, k\}, i \in \{1, 2, \ldots, \lambda_{k+1-j}\}\}$$

is called *the Young diagram of the shape* $\lambda = (\lambda_1, \ldots, \lambda_k)$. A Young diagram of the shape $\lambda = (\lambda_1, \ldots, \lambda_k)$ with a one-to-one correspondence between the set of points of the Young diagram and the set $\{1, 2, \ldots, n\}$, $n = \lambda_1 + \cdots + \lambda_k$, is called *a Young tableau.* We will say that a number $r \in \{1, 2, \ldots, n\}$ is placed in the i-th row and in the j-th column if r corresponds to the point $(j, k+1-i)$. A Young tableau is called *regular* if for any $i, j \in \{1, 2, \ldots, n\}$ such that j is placed not leftward and not higher than i we have $j \geq i$.

Theorem 4.2
There exists a natural one-to-one correspondence between the set of all pairwise non-isomorphic irreducible S_n*-modules and the set of all partitions of* n.

Let T be a Young tableau of the shape $\lambda = (\lambda_1, \ldots, \lambda_k)$, $\lambda_1 + \cdots + \lambda_k = n$. Denote by $C(T)$ (by $R(T)$) the subgroup of S_n transposing numbers inside of the columns of T (inside of the rows of T, respectively), and let

$$e_T = \sum_{\sigma \in C(T)} \sum_{\tau \in R(T)} (-1)^\sigma \sigma \tau$$

be an element of the group algebra $K[S_n]$.

Theorem 4.3
(i) *For any Young tableau* T *of the shape* λ *the left* S_n*-module* $M_T = K[S_n]e_T$ *is an irreducible* S_n*-module of type* λ.

(ii) *Let* $\{T_1, \ldots, T_r\}$ *be the set of all regular Young tableaux of the shape* λ. *Then*

$$I_\lambda = M_{T_1} \oplus \cdots \oplus M_{T_r}$$

is a two-sided ideal of $K[S_n]$ *isomorphic to the full matrix algebra over* K.

(iii) *Let* $\{\lambda_1, \ldots, \lambda_s\}$ *be the set of all partitions of* n. *Then*

$$K[S_n] = I_{\lambda_1} \oplus \cdots \oplus I_{\lambda_s}$$

and

$$I_{\lambda_i} I_{\lambda_j} = 0$$

for all $i \neq j$.

Lie idempotents

Let K be an arbitrary commutative ring with unity and without zero divisors, $X = \{x_1, x_2, \ldots\}$ an infinite set of generators. An element $e \in K[S_n]$ is called *a Lie idempotent* if $e^2 = e$ and $A_n(X) \cdot e = L_n(X)$.

For any $k = 2, 3, \ldots, n$ we consider the element

$$w_n = (1 - \omega_2)(1 - \omega_3) \cdots (1 - \omega_{n-1})(1 - \omega_n),$$

where ω_k is the cyclic permutation $(k, k-1, \ldots, 2, 1)$. It is easy to see that w_n acts on any word of $A_n(X)$ as the operator of left normed arrangement of Lie brackets:

$$(x_{i_1} x_{i_2} \cdots x_{i_n}) \cdot w_n = [[\ldots[[x_{i_1}, x_{i_2}], x_{i_3}], \ldots], x_{i_n}].$$

Hence $A_n(X) \cdot w_n \subseteq L_n(X)$. By Theorem 9.1,

$$u \cdot w_n = n \cdot u$$

for all $u \in L_n(X)$, and since all coefficients in w_n are integers, we get

$$w_n^2 = n \cdot w_n.$$

If $1/n \in K$ then the element

$$v_n = \frac{w_n}{n}$$

is a Lie idempotent. This element was introduced by E. B. Dynkin [171], [172], W. Specht [536] and F. Wever [600], see also [73], [189].

Let $\sigma \in S_n$, $i \in \{1, \ldots, n\}$. The number i is called a descent of σ if $\sigma(i) > \sigma(i+1)$ (or $\sigma(n) > \sigma(1)$ for $i = n$). The set of all descents of σ is denoted by $D(\sigma)$. Then the element v_n can be rewritten in the following form:

$$v_n = \frac{1}{n} \sum_{k=0}^{n-1} (-1)^k \sum_{\substack{\sigma \in S_n \\ D(\sigma) = \{1, \ldots, k\}}} \sigma$$

(see [73], [189], [481]).

The well known Lie idempotent

$$s_n = \frac{1}{n}\sum_{\sigma\in S_n}\frac{(-1)^{|D(\sigma)|}}{C_{n-1}^{|D(\sigma)|}}\sigma = \frac{1}{n}\sum_{S\subseteq\{1,\ldots,n-1\}}\frac{(-1)^{|S|}}{C_{n-1}^{|S|}}\sum_{\substack{\sigma\in S_n\\ D(\sigma)=S}}\sigma$$

$$= \sum_{S\subseteq\{1,\ldots,n-1\}}\frac{(-1)^{|S|}}{|S|+1}\sum_{\substack{\sigma\in S_n\\ D(\sigma)\subseteq S}}\sigma$$

was introduced by L. Solomon [533], [534], see also [478]. This element is well defined only if all numbers $2, 3, \ldots, n$ are multiplicative invertible in K.

The next important Lie idempotent considered here is the Klyachko's idempotent l_n, see [272]. For any $\sigma \in S_n$ denote

$$\operatorname{ind}(\sigma) = \sum_{i\in D(\sigma)} i.$$

Let K contain $1/n$ and a primitive n-th root ε of unity. Then the element

$$l_n = \frac{1}{n}\sum_{\sigma\in S_n}\varepsilon^{\operatorname{ind}(\sigma)}\sigma$$

is a Lie idempotent.

The following lemma can be used to recognize whether an element $e \in K[S_n]$ is a Lie idempotent, or not.

Lemma 4.4
Let $1/n \in K$, $e \in K[S_n]$. The element e is a Lie idempotent if and only if $v_n e = v_n$ and $e v_n = e$.

Corollary 4.5
For any two Lie idempotents $e, f \in K[S_n]$

$$ef = e.$$

Corollary 4.6
Let $e \in K[S_n]$ be a Lie idempotent. Then an element $f \in K[S_n]$ is a Lie idempotent if and only if $ef = e$ and $fe = f$.

Definition 4.7
The Möbius function is given by $\mu(1) = 1$ and

$$\mu(n) = \begin{cases} (-1)^r & \text{if } n \text{ is the product of } r \text{ distinct prime numbers,}\\ 0 & \text{if } n \text{ is divisible by square of a prime number.}\end{cases}$$

It is clear that for $n = p_1^{s_1} \cdots p_k^{s_k}$ with p_i prime numbers we get

$$\sum_{d \mid n} \mu(d) = \sum_{t=0}^{k} C_k^t (-1)^t = 0.$$

Lemma 4.8 (A. M. Garsia [189])
Let $1/n \in K$, and let

$$e = \sum_{\sigma \in S_n} c_\sigma \sigma \in K[S_n]$$

be a Lie idempotent, $c_\sigma \in K$, and C a conjugate class in S_n. Then

$$\sum_{\sigma \in C} c_\sigma = \begin{cases} \mu(d)/n & \text{if } d \mid n \text{ and } (1, 2, \ldots, n)^{n/d} \in C \\ 0 & \text{otherwise.} \end{cases}$$

The theory of Lie idempotents for Lie superalgebras takes the first steps. Thus, I. A. Zavorotchenko considered Lie superalgebra $L'(X)$ with the set of odd generators X. In [622] he constructed the analog of idempotent v_n for Lie superalgebra $L'(X)$:

$$v'_n = \frac{1}{n}(1 + \omega_2)(1 - \omega_3) \cdots (1 + (-1)^{n-1}\omega_{n-1})(1 + (-1)^n \omega_n)$$

and proved that

$$(v'_n)^2 = v'_n, \qquad A_n(X) \cdot v'_n = L'_n(X).$$

Klyachko's theorem

Let K be an arbitrary commutative ring with the unity element and without zero divisors, V a K-module. Let

$$T(V) = \bigoplus_{n=0}^{\infty} T^n(V)$$

be the tensor algebra of the K-module V,

$$L(V) = \bigoplus_{n=0}^{\infty} L_n(V)$$

its Lie subalgebra generated by V. It is clear that if V is a free K-module, then $L(V)$ is the free Lie K-algebra with the set V of free generators, and $T(V)$ is the free associative K-algebra with the same set of generators.

Let $n \in \mathbb{N}$. Suppose that the ring K contains $1/n$ and n-th primitive root ε of unity. Consider the cyclic group Γ of order n with the generator γ. We define left and right actions of Γ on K as follows:

$$\gamma k = \varepsilon^{-1} k, \qquad k\gamma = \varepsilon k,$$

where $k \in K$. The right action of Γ on $T^n(V)$ is given by

$$(a_1 \otimes a_2 \otimes \cdots \otimes a_{n-1} \otimes a_n) \cdot \gamma = a_2 \otimes a_3 \otimes \cdots \otimes a_n \otimes a_1,$$

for all $a_i \in V$, i.e. $\gamma = (1, 2, \ldots, n-1, n) \in S_n$. By K_ε we denote the ring K with the structure of Γ-bimodule. Let

$$C_n(V) = \mathrm{Hom}_{K\Gamma}(K_\varepsilon, T^n(V)) \cong T^n(V) \otimes_{K\Gamma} K_\varepsilon.$$

Theorem 4.9 (A. A. Klyachko [272])
Left $\mathrm{GL}(V)$*-modules* $C_n(V)$ *and* $L_n(V)$ *are isomorphic.*

PROOF We present here only a sketch of proof. Remind that $D(\sigma)$ denotes the set of all descents of $\sigma \in S_n$.

One can directly verify that for any $\sigma \in S_n$

$$\begin{aligned} |D(\sigma\gamma)| &= |D(\sigma)|, \\ \mathrm{ind}(\sigma\gamma^i) &= \mathrm{ind}(\sigma) - i|D(\sigma)|, \\ \mathrm{ind}(\gamma^i\sigma) &= \mathrm{ind}(\sigma) - i, \end{aligned}$$

and that $|D(\sigma)| = 1$ if and only if $\sigma = \gamma^i$ for some i.

Consider the element

$$c_n = \frac{1}{n}\sum_{i=0}^{n-1} \varepsilon^{-i}\gamma^i$$

of the group algebra $K[S_n]$ and the Klyachko's Lie idempotent l_n. It follows from the above mentioned relations, that c_n is an idempotent of $K[S_n]$ and

$$c_n l_n = l_n, \qquad l_n c_n = c_n.$$

Therefore left $\mathrm{GL}(V)$-modules Mc_n and Ml_n are isomorphic for any left $\mathrm{GL}(V)$-module M. It is easy to show that $T^n(V)c_n = C_n(V)$, $T^n(V)l_n = L_n(V)$ because l_n is a Lie idempotent. ■

Remark 4.10
The condition $1/n \in K$ is essential. For example, if $K = \mathbb{Z}[\varepsilon]$ then the assertion of Theorem 4.9 is not valid even for the free K-module V.

For the free K-module V the Theorem 4.9 can be written also in the following form.

Theorem 4.11
Let K be a commutative ring with unity and without zero divisors, $n \in \mathbb{N}$. Suppose that K contains $1/n$ and n-th primitive root ε of unity, $X = \{x_1, x_2, \ldots\}$, $V = Kx_1 \oplus Kx_2 \oplus \cdots$. Then $\mathrm{GL}(V)$*-modules $L_n(X)$ and $C_n(V)$ are isomorphic.*

Let G_0 be a subgroup of a group G, and M a G_0-module. The G-module $M^G = G \otimes_{G_0} M$ is called *the induced module* of M with respect to G_0.

As a corollary of Theorem 4.11 we have the following

Theorem 4.12 (A. A. Klyachko [272])
Let M be a one-dimensional faithful Γ-module over K. Then $L_n(X)$ is isomorphic to the induced module M^{S_n}.

Since the dimension of the homogeneous component of free Lie algebra is exactly the number of orbits of length n of the subgroup $\Gamma \subseteq S_n$ with respect to its left action on the set of left cosets of the corresponding Young subgroup of S_n, all above presented propositions are connected very closely with Witt's dimension formula, because the dimension of homogeneous component of free Lie algebra is exactly the number of orbits of length n of the subgroup $\Gamma \subseteq S_n$ with respect to its left action on the set of left cosets of the corresponding Young subgroup of S_n. The connections of Lemma 4.8, Theorem 4.11, and Theorem 4.12 with Witt's formula were considered in [71], [72]. The equivalence between Theorem 4.12 and Witt's formula for dimensions of homogeneous components of Lie superalgebra was pointed out in [409].

Problem 4.13
Is the analog of Theorem 4.11 valid for Lie superalgebra $L(X, Y)$?

Characters and multiplicities of representations

In all what follows in this subsection let K be a field of zero characteristic.

Let G be a group, V a finite dimensional G-module. The function

$$\mathrm{ch}_V : G \to K,$$

given by

$$\mathrm{ch}_V(\sigma) = \mathrm{tr}(\sigma)$$

for all $\sigma \in G$, is called *the character* of G-module V, where $\mathrm{tr}(\sigma)$ is the trace of the operator $\sigma \in \mathrm{End}_K V$. It is clear that if V is a direct sum of G-submodules $V_1, \ldots, V_k$ then

$$\mathrm{ch}_V = \mathrm{ch}_{V_1} + \cdots + \mathrm{ch}_{V_k},$$

and that the G-module V can be uniquely determined by its character.

We denote by L_n the left S_n-module of all multilinear Lie polynomials of degree n on variables $x_1, \ldots, x_n$. The character of this S_n-module will be denoted by χ_n.

Theorem 4.14 (A. Brandt [98])
If $\sigma \in S_n$ has n/d cycles of the length d for some $d \mid n$ then

$$\chi_n(\sigma) = \frac{1}{n} d^{n/d}(n/d)!\mu(d).$$

Otherwise $\chi_n(\sigma) = 0$.

Problem 4.15
Give a formula for the characters of the spaces of multilinear components of the free Lie superalgebra considered as left modules over the group $S_n \times S_m$.

The following theorems describe the structure of S_n-module L_n.

Theorem 4.16 (A. A. Klyachko [272])
Let V_λ be an irreducible S_n-module corresponding to a partition λ of n, χ the character of V_λ, $\gamma = (1, 2, \ldots, n) \in S_n$. Then for the multiplicity of V_λ in L_n we have:

$$m(V_\lambda, L_n) = \frac{1}{n} \sum_{d \mid n} \mu(d) \chi(\gamma^{n/d}).$$

Theorem 4.17 (A. A. Klyachko [272])
Let λ be a partition of n. Then the irreducible S_n-module corresponding to λ appears in the S_n-module L_n if and only if λ is not of the following form:
(i) $\lambda = (n)$ *with* $n \geq 2$;
(ii) $\lambda = (1, 1, \ldots, 1)$ *with* $n \geq 3$;
(iii) $\lambda = (2, 2)$ *with* $n = 4$;
(iv) $\lambda = (2, 2, 2)$ *with* $n = 6$.

For any $\alpha_1, \ldots, \alpha_k \in \mathbb{N}$ we denote

$$W(\alpha_1, \ldots, \alpha_k) = \frac{1}{n} \sum_{\substack{d,\mid \alpha_i \\ i=1,\ldots,k}} \mu(d) \frac{(|\alpha|/d)!}{(\alpha_1/d)! \cdots (\alpha_k/d)!},$$

where $|\alpha| = \alpha_1 + \cdots + \alpha_k$ (see (2.3)).

Theorem 4.18 (V. M. Zhuravlev [630])
Let $\lambda = (\lambda_1, \ldots, \lambda_k)$ be a partition of n, V_λ the irreducible S_n-module corresponding to the partition λ. Then

$$m(V_\lambda, L_n) = \sum_{\sigma \in S_n} (-1)^\sigma W(\lambda_1 - 1 + \sigma(1), \ldots, \lambda_k - k + \sigma(k)).$$

We say that partitions $\lambda = (\lambda_1, \ldots, \lambda_k)$ and $\nu = (\nu_1, \ldots, \nu_m)$ of n are *conjugated* if $m = \lambda_1$ and $\nu_i = \max\{j \mid \lambda_j \geq i\}$, $i = 1, \ldots, m$.

Theorem 4.19 (V. M. Zhuravlev [630])
Let λ and ν be conjugated partitions of n, $\gamma = (1, 2, \ldots, n) \in S_n$, V_λ and V_ν the irreducible S_n-modules corresponding to partitions λ and ν respectively, χ_λ the character of the module V_λ.

(i) *If* $n \neq 2 \pmod 4$, *then*

$$m(V_\nu, L_n) = m(V_\lambda, L_n).$$

(ii) *If* $n = 2 \pmod 4$, *then*

$$m(V_\nu, L_n) = m(V_\lambda, L_n) + \frac{2}{n} \sum_{d\,|\,(n/2)} \mu(d)\chi_\lambda(\gamma^{n/2d}).$$

For a standard tableau T of shape λ, where λ is a partition of n, a number $i \in \{1, 2, \ldots, n-1\}$ is called *a descent* of T if $i + 1$ is located in a lower row that i in T. The set of all descents of T is denoted by $D(T)$, and the number

$$\operatorname{ind}(T) = \sum_{i \in D(T)} i$$

is called *the index* of T.

Theorem 4.20 (W. Kraskiewich, J. Weyman [284]; see also [481])
Let V_λ be the irreducible S_n-module corresponding to a partition λ of n, $i \in \{1, \ldots, n\}$ a number such that i and n are relatively prime. Then the multiplicity of V_λ in L_n is equal to the number of standard tableaux of shape λ and of index congruent to $i \pmod n$.

Problem 4.21
Describe the multiplicity of irreducible components of a space of multilinear component of the free Lie superalgebra considered as a left $(S_n \times S_m)$-module.

A. I. Molev and L. M. Tsalenko in [431] considered characters of infinite dimensional graded S_n-modules.

Let

$$V = \bigoplus_{k \in \mathbb{Z}_+^n} V^{(k)}$$

be a $\mathbb{Z}_+^n$-graded space, $k = (k_1, \ldots, k_n)$, $\sigma \in S_n$. We define the action of S_n on $\mathbb{Z}_+^n$ by relations

$$\sigma(k_1, \ldots, k_n) = (k_{\sigma^{-1}(1)}, \ldots, k_{\sigma^{-1}(n)}),$$

where $(k_1, \ldots, k_n) \in \mathbb{Z}_+^n$ and $\sigma \in S_n$. Assume that V has a structure of left S_n-module and

$$\sigma(V^{(k)}) \subseteq V^{(\sigma(k))}$$

for all $k \in \mathbb{Z}_+^n$. The power series

$$\mathrm{ch}_V(\sigma,t) = \sum_{\substack{k\in\mathbb{Z}_+^n\\ \sigma(k)=k}} \left(\mathrm{tr}(\sigma)|_{V^{(k)}}\right) t^k$$

with $t = (t_1,\dots,t_n)$, $t^k = t_1^{k_1}\cdots t_n^{k_n}$ is called *the character of the graded S_n-module V.*

Let $X = \{x_1,\dots,x_n\}$, and let $L(X)$ be the free Lie algebra on X, $L'(X)$ the free Lie superalgebra with the set of odd generators X. We consider the natural left action of S_n on $L(X)$ and $L'(X)$. The algebras $L(X)$ and $L'(X)$ are naturally $\mathbb{Z}_+^n$-graded.

Theorem 4.22 (A. I. Molev and L. M. Tsalenko [431]) *The character $\mathrm{ch}_{L(X)}(\sigma,t)$ of the Lie algebra $L(X)$ is equal to*

$$\sum_{d\in\mathbb{N}}\sum_{k\in\mathbb{Z}_+^n} \frac{\mu(d\cdot|\sigma|)}{d\cdot|\sigma|}\frac{(|k|-1)!}{k_1!\cdots k_n!} t^{d(k+\sigma(k)+\cdots+\sigma^{|\sigma|-1}(k))},$$

where $|\sigma|$ denotes the order of $\sigma\in S_n$, $|k| = k_1+\cdots+k_n$.

Theorem 4.23 (A. I. Molev and L. M. Tsalenko [431]) *The character $\mathrm{ch}_{L'(X)}(\sigma,t)$ of the Lie superalgebra $L'(X)$ is given by*

$$\sum_{d\in\mathbb{N}}\sum_{k\in\mathbb{Z}_+^n} (-1)^{|k|}\frac{\mu(d\cdot|\sigma|)}{d\cdot|\sigma|}\frac{(|k|-1)!}{k_1!\cdots k_n!}(-t)^{d(k+\sigma(k)+\cdots+\sigma^{|\sigma|-1}(k))},$$

where $|\sigma|$ denotes the order of $\sigma\in S_n$, $|k| = k_1+\cdots+k_n$.

Comments

Taking some attention to the history of considered problems we could see that Lie superalgebras have arisen in nature in many different ways proving the remarkable versatility of supersymmetry in mathematics. We indicate only some sources of the theory of Lie superalgebras.

Exposition of basic facts of the theory of supermanifolds could be found, for example, in reviews and books of B. Kostant [278], F. A. Berezin [55] and D. Leites [325], [326]. Mathematical aspects of the theory of supersymmetry were included in many books and surveys: [22], [51], [89], [137], [149], [179], [210], [237], [251], [260], [271], [358], [362], [441], [546], [598].

Another source of Lie superalgebras was the theory of deformation of algebraic structures. We mention here only papers [186], [449], [450], [451] by A. Frölicher, A. Nijenhuis and R. W. Richardson.

On the other hand in algebraic topology, in particular in homotopy theory, Lie superalgebras were appearing under consideration of homotopy groups with Whitehead product, see monographs [216], [550], [601]. Skew graded Lie type structures were used frequently in homological algebra, see, for instance, [17], [108], [194], [330], [343], [427], [467].

The systematic study of Lie superalgebras was started by V. G. Kac in [247], [249] (see also M. Scheunert's monograph [489] and D. Leites review [327]). In our book we don't touch seriously the theory of representations of Lie superalgebras, one of the most important branches of the general theory, in which, especially in the finite dimensional case, one can see now an extremely active period of research. We concentrated our attention primarily on combinatorial aspects of Lie superalgebras, mostly infinite dimensional. We would like to mention first of all the following papers in this field: [20], [25], [27], [33], [47], [52], [57], [58], [97], [141], [193], [204], [223], [234], [248], [256], [257], [323], [324], [365], [366], [383], [385], [386], [388], [397], [399], [452], [456], [461], [464], [491], [492], [507], [527], [532], [539], [576], [591], [592], [613].

Remark that nonhomogeneous subalgebras in Lie superalgebras were touched in [36], [328], [499], [591].

Generally speaking, superalgebras (associative, Lie, alternative, Jordan and others) are now becoming to be a rather useful tool in the study of varieties of (even nongraded) algebras. For example, A. R. Kemer [265], [266] invented so called a "supertrick" to work with $\mathbb{Z}_2$-graded algebras, E. I. Zelmanov used effectively the techniques of superalgebras and Grassmann envelopes in his articles on solutions of some Burnside type problems ([623], [624], [625], [626]). We mention also in this line some important papers [258], [259] by I. Kaplansky, [429] by S. P. Mishchenko, [505], [506] by I. P. Shestakov, [530] by O. N. Smirnov, [542] by V. V. Stovba, [593] by A. Wais, [614], [615], [616], [617], [618] by M. V. Zaicev, [509] by E. I. Zelmanov and I. P. Shestakov.

Identities of Grassmann envelopes of superalgebras with additional structures were considered in [471], [472], [474] by Yu. P. Razmyslov and in [633], [636], [637] by A. A. Zolotykh.

Generalized Lie algebras were introduced by R. Ree in 1960 ([476]). In physics color Lie superalgebras appeared as generalizations of Lie superalgebras in papers [336], [439], [483], [484], [488] by W. Nahm, V. Rittenberg, M. Scheunert, and D. Wyler. Later they were studied systematically in series of articles: [6], [27], [30], [33], [37], [56], [209], [222], [260], [273], [304], [305], [306], [307], [308], [363], [383], [385], [388], [393], [394], [395], [396], [397], [414], [418], [419], [430], [438], [490], [491], [579], [580], [581].

In recent works [181], [276], [607], [608] some connections with other colored structures could be found.

Differential Lie algebras appeared in papers of R. Baer [21] and N. Jacobson [238] devoted to derivations of associative rings. V. K. Kharchenko used them in [267] to study differential identities of semiprime rings. As for general theory of differential Lie algebras see monographs [50] by K. I. Beidar, W. S. Martindale and A. V. Mikhalev and [268] by V. K. Kharchenko. Differential (color) Lie superalgebras were introduced in [399] by A. A. Mikhalev.

One of the main objects of our book is the theory of free (color) Lie superalgebras. Remark that $\mathbb{Z}$-graded free Lie superalgebras have naturally appeared in homological algebra, see for example [16], [17], [231], [330]. The systematic development of free Lie superalgebras was started in articles [476] by R. Ree, [18] by I. K. Babenko, [255] by I. L. Kantor, [379]–[405] by A. A. Mikhalev, [410]–[419] by A. A. Mikhalev and A. A. Zolotykh, [527] by A. S. Shtern, [431] by A. I. Molev and L. M. Tsalenko, [369] by G. Melançon, [372] by G. Melançon and Ch. Reutenauer, [621], [622] by I. A. Zavorotchenko, [148] by J. Désarménien, G. Duchamp, D. Krob, and G. Melançon, [165] by G. Duchamp.

In particular, in papers [148], [372], [409], [431], [621], [622], the structure of free Lie superalgebras as S_n- and $\mathrm{GL}(V)$-modules is under consideration. It's only the beginning of the story. We mention only that in Lie algebras we have a nice theory of group actions, see [38], [39], [40], [61], [62], [70], [71], [72], [73], [75], [98], [104], [105], [245], [272], [480], [555], [630]. The recent monograph [481] by Ch. Reutenauer is devoted to this subject.

In papers [470], [471], [472] by Yu. P. Razmyslov and [633], [635], [636] by A. A. Zolotykh nonstandard actions of symmetric groups on free algebras were under consideration.

Chapter 2

Linear Structure

In this chapter we consider the linear structure of free Lie superalgebras. Section 5 is devoted to combinatorics on Lyndon-Shirshov words. In Section 6 we consider properties of regular monomials. In Section 7 we discuss induced arrangements of brackets on regular monomials. Various bases of free Lie superalgebras are constructed in Section 8. PBW-bases of the free associative algebra and various types of super Witt formula are also presented. In Section 9 the Dynkin-Specht-Wever criterion for an element of the free associative algebra to be a super Lie element is proved. In Section 10 we touch equations in free Lie superalgebras. Section 11 is devoted to algorithms and examples.

5 Regular words

This section collects some combinatorics on Lyndon-Shirshov words.

Suppose that the set X is G-graded and totally ordered, and the set $S(X)$ is ordered lexicographically, i.e. for $u = x_1 \cdots x_r$ and $v = y_1 \cdots y_m$ where $x_i, y_j \in X$ we have $u > v$ if either $x_i = y_i$ for $i = 0, 1, \ldots, t-1$ and $x_t > y_t$ or $x_i = y_i$ for $i = 1, 2, \ldots, m$ and $r > m$. Then $S(X)$ is linearly ordered. Consider the order relation on $\Gamma(X)$ defined by $a > b$ for $a, b \in \Gamma(X)$ if either $\widetilde{a} > \widetilde{b}$ in $S(X)$ or $\widetilde{a} = \widetilde{b}$, $a = [u, v]$ and $b = [c, d]$ with $\ell(u) < \ell(c)$ or $\ell(u) = \ell(c)$, but $u > c$, or $u = c$ and $v > d$. It is clear that this ordering is total. If in the product $[u, v]$ we have $u > v$ then we say that v is *the least factor* (analogously, if $u \leq v$ then we say that u is the least factor).

Definition 5.1
An word $u \in S^+(X)$ is said to be *regular* if for any decomposition $u = ab$, where $a, b \in S^+(X)$, we have $u > ba$.

A word $w \in S^+(X)$ is said to be *s-regular* if either w is a regular word or $w = vv$ with v a regular word and $d(v) \in G_-$.

Let p be a prime number. A word $w \in S^+(X)$ is said to be *ps-regular* if either w is an s-regular word, or $w = v^{p^k}$, $k \in \mathbb{N}$, v is an s-regular word and $d(v) \in G_+$.

New letters

Lemma 5.2
Let $x \in X$, $t \in \mathbb{N}$, $t \geq 2$. Consider the following subsets of $S(X)$:

$$\begin{aligned} V = V(x) &= \{y, yx^r \mid y \in X \setminus \{x\}, r \in \mathbb{N}\}, \\ V_t = V_t(x) &= \{y, yx^s, x^t \mid y \in X \setminus \{x\}, 1 \leq s < t\}. \end{aligned}$$

Then V, V_t are free generating sets of the submonoids which they generate.

PROOF It is sufficient to prove that values of different words on elements from V (from V_t, respectively) are again different words. Let $a_i \in V$ ($a_i \in V_t$) and

$$a_{i_1} \cdots a_{i_k} = a_{j_1} \cdots a_{j_l}.$$

We show that $k = l$, $i_q = j_q$ for $1 \leq q \leq k$. The proof will be given by induction on the length $\ell_X(a_{i_1} \cdots a_{i_k}) = N$. It is clear that our assertion is valid for $N = 1$. Let now $N = n > 1$.

The first case: in $a_{i_1} \cdots a_{i_k} = a_{j_1} \cdots a_{j_l}$ we have a subword xy, where $y \in X \setminus \{x\}$. Consider the first such subword from the left. By the definition of elements of V (V_t, respectively) it is clear that a_i does not contain xy, $y \in X \setminus \{x\}$ for any i. Hence,

$$\begin{aligned} a_{i_1} \cdots a_{i_p} &= a_{j_1} \cdots a_{j_r}, \\ a_{i_{p+1}} \cdots a_{i_k} &= a_{j_{r+1}} \cdots a_{j_l} \end{aligned}$$

for some p and r. By the induction hypothesis, we have our statement verified.

The second case: in $a_{i_1} \cdots a_{i_k} = a_{j_1} \cdots a_{j_l}$ there is no subword xy with $y \in X \setminus \{x\}$. If $a_{i_1} \cdots a_{i_k}$ contains $y \in X \setminus \{x\}$, then

$$a_{i_1} \cdots a_{i_k} = y_1 \cdots y_r x^m,$$

$m \in \mathbb{Z}_+$. For V we have $a_{i_1} \cdots a_{i_k} = y_1 \cdots y_{r-1} y_r x^m$ and it is clear that such presentation is unique. Therefore $k = l$, $a_{i_s} = a_{j_s}$. For V_t we get

$$y_1 \cdots y_r x^m = \begin{cases} y_1 \cdots y_{r-1} w_1 w^s, & \text{where } m = st + r,\ 1 \leq r < t, \\ & w_1 = y_r x^r,\ w = x^t; \\ y_1 \cdots y_r w^s, & \text{where } m = ts,\ w = x^t. \end{cases}$$

It is obvious that this presentation is also unique, and therefore $k = l$ and $a_{i_s} = a_{j_s}$.

If $a_1, \ldots, a_k$ does not contain any $y \in X \setminus \{x\}$, then $a_{i_1} \cdots a_{i_k} = x^m$. For V it is impossible. For V_t it follows that $m = ts$. Since the presentation $x^m = w^s$, $w = x^t$, is unique, we get $k = l = t$, $i_1 = \cdots = i_k = j_1 = \cdots = j_l$. ■

Lemma 5.3
Let x be the least letter in X, $a, b \in S(X)$, $a, b \neq x$, and such that it is possible to rewrite a and b in letters $V = V(x)$. Then $a > b$ as words in letters from X if and only if $a > b$ as words in letters from V, where $x_i > x_i x^k > x_i x^l > x_j$ with $k < l$, $x_i > x_j$.

PROOF Let $a > b$ in letters from V. If $b \neq ab'$, then $a > b$ by the first different letter (we may assume that this letter is the first one), for example $x_i x_k > x_j x_l > x^r$. If $x_i > x_j$, then it is clear that $a > b$ in $S(X)$. If $x_j = x_i$, then $k < l$. Since $b \neq ab'$, $\ell_V(a) > 2$, $a = x_i x^k x_j a'$, and therefore $a > b$ in letters from X because of $x_j \neq x$ and x is the least letter in X.

If now $b = ab'$, then $b = ab'$ in letters from X, hence $a > b$.

Let now $a > b$ in letters from X. By Lemma 5.2 there exists the unique presentation of a and b in letters from V. Since the ordering on V is compatible with the lexicographical ordering on X, $a > b$ in letters from V. ■

Corollary 5.4
Under the conditions of Lemma 5.3 a is a regular word in letters V if and only if a is a regular word in letters X.

Remark 5.5
The assertion of Lemma 5.3 does not take place relatively to the letters V_t. Indeed, let $t = 2$, $X = \{x, y, z\}$, $y > z > x$. Consider the words $a = yx^2$ and $b = yxz$. Then we have the following letters from V_2 in presentations of a and b: $y > yx > z > x^2$. We see that $a < b$ in letters from X, but at the same time $a = y \cdot x^2 > yx \cdot z = b$ in letters from V_2.

Elementary properties

Lemma 5.6
Let $u, v \in S^+(X)$, u be a regular word and $u < v$. Then $uv < vu$.

PROOF If $u < v$ and $uv \geq vu$, then $u = vz$ and therefore $vzv \geq vvz$. Hence $zv \geq vz$, and we get a contradiction. ■

Lemma 5.7
If $u = vw$ is a regular word, $v, w \in S^+(X)$, then $u > w$.

PROOF Since u is a regular word, $u = vw > wv$. If now $u = vw \leq w$, then $u = wz$, $z \in S^+(X)$, and therefore it follows from $u = wz > wv$

we have $z > v$. As u is a regular word, $u = wz > zw$. Since $z > v$, $zw > vw = u$, and we have come to a contradiction. ■

Lemma 5.8
A word $u \in S^+(X)$ is regular if and only if it follows from $u = vw$, $v, w \in S^+(X)$, that $u > w$.

PROOF If u is a regular word, then by Lemma 5.7 $u > w$. Otherwise, it is clear that $u = vw > w > vw$, and the word u is regular. ■

Lemma 5.9
Let x be the least letter of X, $v \in S(X)$, $u = xv$ is a regular word. Then $v = 1$ and $u = x$.

PROOF Assume to the contrary that $u = xv$, $v \in S^+(X)$. Since $u = xv$ is a regular word, $x > v$ and $x > w$ where w is any ending of v. Bearing in mind that x is the least letter we obtain that $v = x^l$, $l \geq 2$, and we have come to a contradiction with the regularity of v. ■

Lemma 5.10
Let u, v be regular words and $u > v$. Then $u^k > v$ for any $k \in \mathbb{N}$.

PROOF By the way of contradiction assume that $u^k \leq v$. Then $v = ua$, $a \in S^+(X)$, and $a \geq u$. Since v is a regular word, $v > a$, hence $u > ua = v > a$, and we have come to a contradiction. ■

Lemma 5.11
Let u be a regular word, $u > v$. Suppose that $u > v_0$ for all endings v_0 of v. Then $uv > vu$.

PROOF We use the induction on the length $\ell(v)$ of v. The basis of induction: $v \in X$. Then from $u > x$ it follows that $ux > xu$. Suppose that our assertion is true for all v with $\ell(v) < n$. Let now $\ell(v) = n$. If $v \neq uv_0$ then $uv > vu$ (comparing first letters). Let $v = uv_0$. Since v_0 is an ending of v, $u > v_0$. By the induction hypothesis $uv_0 > v_0u$ (observe that any ending of v_0 is also an ending of v). Therefore $uv = uuv_0 > uv_0u = vu$. ■

Lemma 5.12
Let a and b be regular words and $a > b$. Then ab is a regular word.

PROOF Consider possible decompositions of the word ab:

(i) $ab > ba$ by Lemma 5.6;

(ii) $a_1a_2b > a_2a_1b > a_2ba_1$ by regularity of a and Lemma 5.6;

(iii) $ab_1b_2 > ab_2b_1$ by regularity of b. Since $a > b$, by Lemma 5.7 we have $a > b_0$ for any ending b_0 of b, therefore $a > b_2$ and $a > b_0'$ for any end b_0' of b_2. By Lemma 5.6 $ab_2 > b_2a$ and therefore $ab_2b_1 > b_2ab_1$. Thus we have $ab_1b_2 > b_2ab_1$. ■

Corollary 5.13
Let x be the greatest letter in X and u a regular word, $u \neq x$. Then xu is a regular word.

Remark 5.14
A. D. Chanyshev and V. A. Ufnarovsky in [116], [117], [561] proved that for any integers r and m there exists an integer $N(r,m)$ such that in each word of the length $N(r,m)$ on $X = \{x_1, \ldots, x_r\}$ there exists either a subword ax_ia with a regular word, or a subword $x_jb_1x_j \ldots x_jb_mx_j$ with b_k regular words. As a corollary of this result a combinatorial proof of Kostrikin-Zelmanov theorem [283] that every Lie (super) ring generated by a finite number of sandwiches is nilpotent was given.

In [115] A. D. Chanyshev also proves that there exists an infinite word w_∞ on two ordered letters such that w_∞ does not contain subwords of the form u^3, $u_1u_2u_3$, $v_1v_2v_3v_4$, where u, u_i, v_j are regular words and $u_1 > u_2 > u_3$, $v_1 \geq v_2 \geq v_3 \geq v_4$.

Remark 5.15
In [559] V. A. Ufnarovsky suggested the following ordering $\rhd$ on words: firstly lexicographically (up to the first different letter), and then $a \rhd b$ for $a, b \in S^+(X)$ if and only if there exist integers k and l such that $f^k > g^l$ lexicographically (equivalent condition: $fg > gf$). Then a word a is regular if and only if for any presentation $a = bc$ with $b, c \in S^+(X)$, we get $a \rhd c$ (equivalent conditions: $a \lhd b$; $b \rhd c$). The lexicographical ordering and the ordering $\rhd$ coincide on the set of regular words.

Cyclic conjugates

Now we compute the number Ψ_n of regular words of the fixed length n in the alphabet $X = \{x_1, \ldots, x_q\}$. Consider a word u with $\ell(u) = n$ and all its cyclic conjugates, i.e. the words of the form wv where $u = vw$, $v, w \in S^+(X)$. We consider possible cases.

Case 1. All cyclic conjugates of u are distinct and therefore in the conjugate class of u we have exactly n words. We may take as a representative of this class the lexicographically greatest word which is a regular word.

Case 2. Some cyclic conjugates may coincide. To consider this situation we need the following.

Lemma 5.16
Let $a, b \in S^+(X)$ and $ab = ba$. Then $a = v^k$, $b = v^l$, $v \in S(X)$, $k, l \in \mathbb{N}$, and all cyclic conjugates of v are distinct.

PROOF We give the proof using the induction on $\ell(a) + \ell(b)$. The basis of induction: $\ell(a) = 1$, $\ell(b) = 1$, $a, b \in X$. Then from $ab = ba$ it follows that $a = b$. Suppose that our assertion holds for all $a, b \in S(X)$ with $\ell(a) + \ell(b) < n$ and let $\ell(a) + \ell(b) = n$.

If $a = b$ and all cyclic conjugates of a are distinct, then our statement is obvious. If $u = u_1u_2 = u_2u_1$ for some cyclic conjugate u of a then $\ell(u_1) + \ell(u_2) = \ell(a) < 2\ell(a)$ and by the induction hypothesis $u_1 = v_1^k$, $u_2 = v_1^l$, and all cyclic conjugates of v_1 are distinct. Therefore $a = v^{k+l}$ where v is a cyclic conjugate of v_1.

If $a \neq b$ and $\ell(a) < \ell(b)$, then from $ab = ba$ it follows that $b = aw$. Hence $aaw = awa$ if and only if $aw = wa$. But $\ell(a) + \ell(w) < \ell(a) + \ell(b)$. By induction we get $a = v^k$, $w = v^l$, $b = aw = v^{k+l}$ and all cyclic conjugates of v are distinct. The proof in the case $\ell(a) \geq \ell(b)$ is analogous. ■

Now by Lemma 5.16 for Case 2 we have that some cyclic conjugate of u has the form v_1^k where all cyclic conjugates of v_1 are distinct. Therefore, $u = v^k$ where v is a cyclic conjugate of v_1 and all cyclic conjugates of v are distinct.

One can see now that

$$q^n = \sum_{l \mid n} l\Psi_l, \tag{2.1}$$

where $q = |X|$, Ψ_l is the number of all regular words of the length l.

The following Möbius Inversion Formula is well known (see for example, [26], [131], [219]).

Theorem 5.17
For any functions f, g on $\mathbb{N}$ such that

$$f(n) = \sum_{m \mid n} g(m),$$

we have

$$g(n) = \sum_{m \mid n} \mu(m) f(n/m).$$

Theorem 5.18
If Ψ_n is the number of regular words of the length n in the alphabet $X = \{x_1, \ldots, x_q\}$ then

$$\Psi_n = \frac{1}{n} \sum_{m \mid n} \mu(m) q^{n/m}.$$

PROOF The assertion of the theorem follows immediately from (2.1) and Theorem 5.17. ■

For the multidegree function we get the following result.

Theorem 5.19
If $\Psi(\alpha)$ is the number of regular words in the alphabet $X = \{x_1, \ldots, x_q\}$ of

the multidegree $\alpha = (\alpha_1, \ldots, \alpha_q)$, *then*

$$\Psi(\alpha) = \frac{1}{|\alpha|} \sum_{\substack{l \mid \alpha_i \\ i=1,\ldots,q}} \mu(l) \frac{(|\alpha|/l)!}{(\alpha_1/l)! \cdots (\alpha_q/l)!},$$

where $|\alpha| = \alpha_i + \cdots + \alpha_q$. *In particular,*

$$\sum_{|\alpha|=n} \Psi(\alpha) = \Psi_n.$$

Factorization

We show now that each associative word can be presented as a product of regular words.

Lemma 5.20
Let $a \in S^+(X)$. *Then there exist the unique factorization of* a *in the form* $a = b_1 \cdots b_k$ *where* $b_1, \ldots, b_k$ *are regular words and* $b_1 \leq \cdots \leq b_k$.

PROOF Consider the longest regular beginning subword of a and denote it by b_1. If $a = b_1$, then we get the desirable factorization. At the same time if we have another factorization $a = b_1' \cdots b_k'$, then $b_1' \leq b_k'$. Since $a = b_1$ is a regular word, $b_1' > a > b_k'$, and we have come to a contradiction.

If $a = b_1 a'$, then we continue this procedure starting with a'. By the way of contradiction suppose that $b_1 > b_2$. Then by Lemma 5.12 the word $b_1 b_2$ is regular. But $\ell(b_1 b_2) > \ell(b_1)$, and we get a contradiction.

Assume now that we have two different factorizations of u:

$$u = b_1 \cdots b_k = b_1' \cdots b_l'.$$

One may suppose that $b_1' = b_1 u$, $u \in S^+(X)$. Then we see that $b_2 = uv$, $v \in S(X)$, $b_1 > u > uv = b_2$, and we again come to a contradiction. ∎

Lemma 5.21
Let $u, v, w, z \in S^+(X)$, $u = vw$, $u = wz$, $\ell(u) = k\ell(v)$, $k \in \mathbb{N}$, $k \geq 2$. *Assume that it follows from* $u = v'w'$, *where* $v', w' \in S^+(X)$ *that* $u > w'$. *Then* v *is a regular word and* $u = v^k$.

PROOF Since $\ell(u) = k\ell(v)$, $\ell(w) \geq \ell(v)$, it follows from $u = vw = wz$ that $w = vt$, $t \in S(X)$. Now $u = vvt = vtz$ implies that either $t = 1$ (and $w = v$, $u = v^2$), or $t = vt'$ (because of $\ell(t) = (k-2)\ell(v)$). Continuing this procedure we get $u = v^k$.

On the other hand, suppose that $v = v_1 v_2$, $v_1, v_2 \in S^+(X)$. By our assumption $v_1 v_2 w > v_2 w$. Bearing in mind that $w = v^{k-1}$, we get $v_1 v_2 v^{k-1} > v_2 v_1 v_2 v^{k-2}$. Since $\ell(v_1^k) > \ell(v_2 v_1^{k-1})$, it follows that $v_1 v_2 > v_2 v_1$. Thus v is a regular word. ∎

Note that if v is a regular word, $u \in S^+(X)$, $vu = uv$, then $u = v^k$ for some $k \in \mathbb{N}$. We mentioned in connection with this fact G. M. Bergman's Centralizer Theorem (see [130, Chapter 6]) that the centralizer of any non-scalar element of the free associative algebra is a polynomial ring in one variable over the same field.

Ambiguities

Lemma 5.22 (A. I. Shirshov [516])
Let $e_1, e_2, e_3 \in S^+(X)$ and e_1e_2, e_2e_3 be regular words. Then the word $e_1e_2e_3$ is regular.

PROOF Let $e_1e_2e_3 = w_1w_2$. If w_1 is a subword of e_1e_2 then $e_1e_2 = w_1v$, $v \in S^+(X)$. Since e_1e_2 is a regular word, by Lemma 5.7 $e_1e_2 > v$. It follows from $\ell(v) < \ell(e_1e_2)$, that $w_1w_2 = e_1e_2e_3 > ve_3w_1 = w_2w_1$.

If w_2 is a subword of e_3 (possibly, $w_2 = e_3$) then $e_3 = uw_2, u \in S(X)$. Since e_1e_2 is a regular word, $e_1e_2 > e_2$. It is clear that $\ell(e_1e_2) > \ell(e_2)$. As e_2e_3 is a regular word we have $e_1e_2e_3 > e_2e_3 = e_2uw_2 > w_2$. On the other hand, $w_2 > w_2w_1$. Thus $w_1w_2 > w_2w_1$. ∎

Lemma 5.23 (A. I. Shirshov [516])
Let w be a regular word. Then there are no $e_1, e_2, e_3 \in S^+(X)$ such that $w = e_1e_2 = e_2e_3$.

PROOF By the way of contradiction assume that $w = e_1e_2 = e_2e_1$. Since w is a regular word, $e_2e_3 = e_1e_2 > e_2e_1$ and therefore $e_3 > e_1$. On the other hand, $e_1e_2 = e_2e_3 > e_3e_2$. Hence $e_1 > e_3$, and we have come to a contradiction. ∎

Lemma 5.24 ([387], [388], [397])
Let u, v be regular words, $u^k = e_1e_2$, $v^l = e_2e_3$, $e_1, e_2, e_3 \in S^+(X)$, $u \neq v$, $k, l \in \mathbb{N}$. Then $e_1e_2e_3$ is a regular word and either $e_2 = u$ and $v = ue$, or $e_2 = v$ and $u = fv$ with $e, f \in S^+(X)$, or $u = u'e_2$, $v = e_2v'$, where $u', v' \in S^+(X)$.

If $u = v$, then $e_2 = u^m$, $m \in \mathbb{N}$.

PROOF Let $u \neq v$, $u = u_1u_2$, $v = v_1v_2$, $u_2, v_1 \in S(X)$, $e_2 = u_2u^{l'} = v^{k'}v_1$, $l', k' \in \mathbb{N}$.

Case 1. $v = u_2u^{l_1}a$ where $l_1 \geq 0$, $u = ab$. If $u_2 \neq 1$ and $a \neq 1$ then taking into account that v is a regular word we get $v > a$ and therefore $u_2 > a$. Since u is a regular word, $u > u_2$. It follows from $u = ab$ that $a > u$, and we have a contradiction. If $a = 1$ and $u_2 \neq 1$, then $v = u_2u^{l_1}$. In the case $l_1 > 0$, since v is a regular word, $v = u_2u^{l_1} > u$. But u is also a regular word, therefore $u > u_2$, $u > u_2 > u_2u^{l_1} > v$, and we get a contradiction. If $a \neq 1$, $u_2 = 1$ and $l_1 > 0$ then $v = u^{l_1}a$, $v > a$ (v is a regular word), $a > u$

and $u > u^{l_1}a = v$. Again we have come to a contradiction. If $u_2 = 1 = a$, then $v = u^{l_1}$ is not a regular word.

Case 2. $u = dv^{k_1}v_1$ where $k_1 \geq 0$ and $v = cd$. If $d \neq 1$ and $v_1 \neq 1$, then $u > v_1$, $v > d > dv^{k_1}v_1 = v$ and thus we have a contradiction. If $d \neq 1$, $v_1 = 1$ and $k_1 > 0$ then $u = dv^{k_1} > v > d > dv^{k_1}$ and again we have a contradiction. If $d = 1$, $v_1 \neq 1$ and $k_1 > 0$ then $u = v^{k_1}v_1 > v_1 > v > v^{k_1}v_1$. If $d = 1$ and $v = 1$, then the word $u = v^{k_1}$ is not a regular word.

Thus for $e_2 = u$ we have $v = ue$, $c \in S^+(X)$, $u > ue$. Since the words u and v are regular, by Lemma 5.12 the word uue is regular. It follows that $uue > ue$, therefore the word $u^{k-1}v^l = e_1e_2e_3$ is regular. If $e_2 = v$ then $u = fv$, $f \in S^+(X)$, $fv > v$ and therefore fvv is a regular word, $u = fv > fvv$, the word u^kv^{l-1} is regular. If $u = u'e_2$, $v = e_2v'$, u', $v' \in S^+(X)$, then it follows from Lemma 5.22 that the word $u'e_2v$ is regular. Since $u = u'e_2 > u'e_2v' > e_2v'$, using again Lemma 5.22 we get that the word $e_1e_2e_3$ is regular.

In the case $u = v$ the contrary assumption follows that $u = u_2u_1 = u_1u_2$ where $u_1, u_2 \in S^+(X)$, and we get a contradiction. ■

Using the same arguments we obtain the following result.

Lemma 5.25 ([387], [388], [397])
Let u and v be regular words, $u \neq v$, $k, l \in \mathbb{N}$, v^k is a subword of u^l. Then v^k is a subword of u.

6 Regular monomials

In this section we analyse connections between regular monomials and regular words.

Definition 6.1
A monomial $u \in \Gamma(X)$ is said to be *regular* if either $u \in X$ or:

(i) from $u = [u_1, u_2]$ it follows that u_1, u_2 are regular monomials with $\widetilde{u_1} > \widetilde{u_2}$;

(ii) from $u = \big[[u_1, u_2], u_3\big]$ it follows that $\widetilde{u_2} \leq \widetilde{u_3}$.

A monomial $u \in \Gamma(X)$ is said to be *s-regular* if either u is a regular monomial or $u = [v, v]$ with v a regular monomial and $d(v) \in G_-$.

Observe that if $x \in X$, v is a regular monomial and $x > \widetilde{v}$ then $[x, v]$ is a regular monomial.

Lemma 6.2
If u is a regular monomial then $\widetilde{u}$ is a regular word. Conversely, for any regular word v there is a unique arrangement of brackets $[v]$ on v such that $[v]$ is a regular monomial.

PROOF We use induction on $\ell(u)$. If $\ell(u) = 1, 2$ then our assertion is evident. If $\ell(u) > 2$ then $u = [u_1, u_2]$ where u_1 and u_2 are regular monomials with $\widetilde{u_1} > \widetilde{u_2}$. By the induction hypothesis and Lemma 5.12 $\widetilde{u} = \widetilde{u_1}\widetilde{u_2}$ is a regular word.

To prove the second statement we use the induction on $\ell(v)$. The starting point $\ell(v) = 1$ is obvious. Let $X = \{x_1, \ldots, x_n\}$ and $x = x_1$ be the least letter from X. We may suppose that u contains x_1. Now we show that the letter x_1 does not come as the first letter to any regular word distinct from x_1. Indeed, if $w = x_1 v$ is a regular word, then $x_1 v > v x_1$ and therefore $v = x_1 v_1$, in other words $w = x_1^l$ and w is not a regular word. Thus we may rewrite our word using the follows as new letters $V = V(x)$ (see Lemma 5.2). Since x_1 occurs to u, $\ell_V(u) < \ell_X(u)$. By Corollary 5.4, the induction hypothesis and by the uniqueness of the regular arrangement of brackets on $x_i x_1^k$: $[x_i x_1^k] = (x_i)(\mathrm{Ad}\, x_1)^k$ we get our assertion. ∎

As a corollary of Lemma 6.2 we have the following statement.

Lemma 6.3
Let u be an s-regular monomial. Then $\widetilde{u}$ is an s-regular word. Conversely, if u is an s-regular word, then there is a unique arrangement of brackets $[u]$ on u such that $[u]$ is an s-regular monomial.

Lemma 6.4
Let x be the least letter in X and $u = [u_1, u_2]$ a regular monomial, $u_2 = x$. Then $u = (y)(\mathrm{Ad}\, x)^l$ for some $y \in X \setminus \{x\}$ and $l \in \mathbb{N}$.

Let z be the greatest letter in X and v a regular monomial, $v \neq z$. Then $[z, v]$ is a regular monomial.

PROOF Since $[u_1, x]$ is a regular monomial, u_1 is a regular monomial and $\widetilde{u_1} > x$; if $u_1 = [u_{11}, u_{12}]$, then u_{12} is a regular monomial and $\widetilde{u_2} \leq x$. If $u_1 \in X$, then $u_1 \neq x$ and $u = (u_1)(\mathrm{Ad}\, x)$. If $u = [u_{11}, u_{12}]$, then, by Lemma 6.2, $\widetilde{u_{12}}$ is a regular word. Since $\widetilde{u_{12}} \leq x$, by Lemma 5.9, $u_{12} = x$. Continuing this procedure we complete the proof. ∎

Lemma 6.5
Let u be a regular word, $\ell(u) > 1$. Then the external brackets in the regular arrangement of brackets $[u]$ on u take the place $u = [a, b]$, where b is the longest proper ending of u such that b is a regular word.

PROOF We use the induction on $\ell(u)$. The basis of induction is the case $\ell(u) = 2$, $u = yx$, $y, x \in X$, $y > x$. In this case $a = y$, $b = x$, $[u] = [a, b]$.

Let now $\ell(u) > 2$ and x be the least letter from X in u. If $w = xw'$, $w' \in S(X)$, then $b = x$ by Lemmas 5.9, 6.2. Since $[u]$ is a regular monomial, we see that $v = yx^k$ for some $y \in X \setminus \{x\}$ and $k \in \mathbb{N}$. It is clear that in this situation b is the longest proper regular ending of u. If $b \neq x$, then it is possible to rewrite a and b in new letters V, $\ell_V(u) < \ell_X(u)$, and, by Corollary 5.4, u is a regular word in letters from V. Using the inductive assumption we complete the proof. ■

Lemma 6.6 ([383])
If $[X]$ is a subalgebra of the color Lie superalgebra $[A(X)]$ generated by X and $\pi\colon L(X) \to [X]$ is the homomorphism induced by the identity mapping of X, then $\pi(u) = \widetilde{u} + u^*$ for any regular monomial u and $\pi(u) = 2\widetilde{u} + u^*$ for $u = [v, v]$, where v is a regular monomial with $d(u) \in G_-$. Moreover, in both cases u^* is a linear combination of some words which are less than $\widetilde{u}$ and have multidegree equal to $m(u)$.

PROOF We use the induction on the length of a regular monomial. The basis of induction: $\ell(u) = 1$, $u \in X$; here $\pi(u) = u = \widetilde{u}$. Suppose that $\ell(u) \geq 2$ and our assertion be valid for all regular monomials v such that $\ell(v) < \ell(u)$. Then $u = [u_1, u_2]$, where u_1, u_2 are regular monomials. By the induction assumption, $\pi(u_1) = \widetilde{u_1} + u_1^*$, $\pi(u_2) = \widetilde{u_2} + u_2^*$ and therefore

$$\begin{aligned}\pi(u) = [\pi(u_1), \pi(u_2)] &= \widetilde{u_1}\widetilde{u_2} + u_1^*\widetilde{u_2} + \widetilde{u_1}u_2^* + u_1^*u_2^* \\ &\quad - \varepsilon(u_1, u_2)(\widetilde{u_2}\widetilde{u_1} + \widetilde{u_2}u_1^* + u_2^*\widetilde{u_1} + u_2^*u_1^*).\end{aligned}$$

It is obvious that $\widetilde{u_1}\widetilde{u_2}$ is the greatest word among the words occurring in the first four summands and $\widetilde{u_2}\widetilde{u_1}$ is the greatest word among words occurring in the last four summands. Since $\widetilde{u} = \widetilde{u_1}\widetilde{u_2}$ is a regular word, $\widetilde{u_1}\widetilde{u_2} > \widetilde{u_2}\widetilde{u_1}$.

Let now $u = [v, v]$ where v is a regular monomial, $d(v) \in G_-$. As we saw above $\pi(v) = \widetilde{v} + v^*$. Since $\varepsilon(v, v) = -1$, we get

$$\pi(u) = [\pi(v), \pi(v)] = 2\widetilde{v}\widetilde{v} + 2(\widetilde{v}v^* + v^*\widetilde{v} + v^*v^*).$$

Thus $\pi(u) = 2\widetilde{u} + u^*$. ■

7 Arrangements of brackets

In this section we consider arrangements of brackets induced by the regular arrangements of brackets on regular subwords.

Lemma 7.1 ([387], [388], [397])
Let u, v be regular words. Then

(i) *if $u = avb$ $(a, b \in S(X))$, then on u there is an arrangement of brackets $[u] = \big[a[v]b\big]$ such that $[v]$ is a regular monomial, $[u]^\circ = u$ in $[A(X)]$;*

(ii) *if $u = avvb$, $d(v) \in G_-$, $a, b \in S(X)$, then on u there is an arrangement of brackets $[u]$ such that $[u] = \big[a[v, v]b\big]$ where $[v, v]$ is an s-regular monomial, $[u]^\circ = 2u$ in $[A(X)]$;*

(iii) *if Let* $u = av^{p^t}b$, $d(v) \in G_+$. *Then on* u *there is an arrangement of brackets* $[u]$ *such that* $[u] = [a[v]^{p^t}b]$ *where* $[v]$ *is a regular monomial,* $[u]^\circ = u$ *in* $[A(X)]^p$;

(iv) *if* $u = av^{2p^t}b$, $d(v) \in G_-$, *then there is an arrangement of brackets on* $[u]$ *such that* $[u] = [a[v,v]^{p^t}b]$ *where* $[v,v]$ *is an s-regular monomial,* $[u]^\circ = 2^{p^t}u$ *in* $[A(X)]^p$.

PROOF To prove the first statement we use the induction on the length $\ell(u)$ of u. The basis of induction when $u = v$ is obvious. If $\ell(v) = 1$ then $[u]$ is the arrangement of brackets on u such that $[u]$ is a regular monomial and by Lemma 6.6 our statement is true. Suppose that we have proved this statement for all u with $\ell(u) < n$. Let $\ell(u) = n > 1$ and x be the least possible letter from X which occurs in u. Since u, v are regular words and $\ell(u) > 1$, $\ell(v) > 1$, the initial letters of the words u and v are different from x. Writing down u in new letters from $V = V(x)$ (see Lemmas 5.2, 5.3) we get $\ell_V(u) < \ell_X(u)$.

Let v' be a subword of u in new letters V of the form $v' = vx^l$ where $l \geq 0$ is such that either to the right of v' in u we have a letter which is different from x or v' is the end of u. Since v is a regular word and $v > x$, by Lemmas 5.11, 5.12 we have that v' is a regular word. By the inductive hypothesis we have the desirable statement for the subword v' of u in letters from V.

We consider arrangement of brackets on x_ix^k of the form: $(x_i)(\operatorname{Ad} x)^k$. Since $(x_i)(\operatorname{Ad} x)^k$ is a regular monomial, $((x_i)(\operatorname{Ad} x)^k)^\circ = x_ix^k$. Let $[v']$ be the arrangement of brackets on v' such that $[v']$ is a regular monomial in letters from V and on x_ix^k we have the following arrangement of brackets: $(x_i)(\operatorname{Ad} x)^k$. Now we consider another arrangement of brackets on v': $[v']' = ([v])(\operatorname{Ad} x)^l$ where $[v]$ is a regular monomial. Since v is a regular word and $v > x$, $([v']')^\circ = v' = ([v'])^\circ$. Now we use the following arrangements of brackets on u: on v' as $[v']'$; all other brackets in accordance with inductive hypothesis; on x_ix^k as $(x_i)(\operatorname{Ad} x)^k$. It proves the first statement.

By analogy using the induction on $\ell(u)$ one can prove other statements of the lemma. ■

Remark 7.2

Using Lemma 7.1 one can show that a Lie superalgebra of at most countable rank can be isomorphically embedded into a Lie superalgebra with two generators over the same field. It is sufficient to consider $X_0 = \{x\}$, $X_1 = \{y\}$, and the subset of elements

$$D = \{[[x,(x)(\operatorname{Ad} y)^k],[x,y]] \mid k \in \mathbb{N}\}.$$

These elements form a set of free generators of the subalgebra B which they generate. At the same time, if J is an ideal of B and J_1 is the ideal of $L(X)$ generated by J, then $B \cap J_1 = J$ (to prove this it is necessary to use Lemma 7.1). For any Lie superalgebra L we may consider the following presentation: $L = B/J$. Since in $A(x, y)$ we have that $J_1 \cap B = J$, there is a subalgebra in $L(X)$ which is isomorphic to L. Such theorem takes also place for a color Lie superalgebra in the case when the grading group G is 2-generated.

For Lie algebras this result was obtained by A. I. Shirshov. For Lie superalgebras see [390].

Problem 7.3
G. P. Kukin in [292] proved that a solvable Lie algebra of degree n of countable rank can be embedded into a solvable Lie algebra with two generators of degree $n+2$. Is this statement true for Lie superalgebras?

Lemma 7.4
Let w be a regular word such that $w = b_1 \cdots b_k$ where $b_1, \ldots, b_k$ are s-regular (ps-regular) words. Then there exists an arrangement of brackets $[w]$ on w such that $[w] = [[b_1] \ldots [b_k]]$ where $[b_1], \ldots, [b_k]$ are s-regular (ps-regular) monomials and $[w]^\circ = \alpha w$ with $\alpha \in \mathbb{N}$.

PROOF We use the induction on

$$\sum_{i=1}^{k} \ell_X(b_i).$$

The basis of induction is the case when $\ell(b_1) = \cdots = \ell(b_k) = 1$. In this case $[w]$ is the regular arrangement of brackets. Let now $\ell(b_{i_0}) > 1$ for some i_0 and x the least letter from X in w. By Lemma 5.2 and Corollary 5.4, w is a regular word in the letters $V(x)$. If $b_i \neq x^k$, then by the same arguments b_i is a s-regular (ps-regular) word in letters from $V(x)$. Otherwise we consider the subword b_{i-1}. If $b_{i-1} = x^s$, then we consider b_{i-2}, and so on. If $b_{i-1} \neq x^s$, then we consider the regular word $b = b_{i-1}b_i$, and set $k' = k - 1$. After possible rewriting of w in letters from $V(x)$ we get $\ell_V(w) < \ell_X(w)$. Using the inductive hypothesis and the arrangement of brackets $[b]$ on b such that $[b] = [[b_{i-1}], [b_i]]$ where $[b_{i-1}], [b_i]$ are s-regular (ps-regular) monomials, we complete the proof. ■

8 Bases of free color Lie superalgebras

In this section we construct linear bases of the free color Lie superalgebra. We present some dimension formulas for the subspace of elements of given length (multidegree) in the free color Lie superalgebra.

In all what follows in this section we assume that the ring K has no 2-torsion.

Lyndon-Shirshov bases

Lemma 8.1 ([385])
Let u, v be s-regular monomials. Then:

$$[u, v] = \sum_k \alpha_k w_k$$

in $L(X)$ where w_k is an s-regular monomial for all k, $m(w_k) = m(uv)$. If $\widetilde{u} \neq \widetilde{v}$ then $\widetilde{w_k} > \min(\widetilde{u}, \widetilde{v})$.

PROOF One can prove the statement of the theorem by induction on $n = \ell(u) + \ell(v)$ using properties of s-regular words and monomials and the following "procedure of increasing the least factor": if u_1, u_2, v are regular monomials, $\widetilde{u_1} > \widetilde{u_2} > \widetilde{v}$, then

$$\big[[u_1, u_2], v\big] = -\varepsilon(u_1, u_2 v)\big[[u_2, v], u_1\big] + \varepsilon(u_2, v)\big[[u_1, v], u_2\big], \tag{2.2}$$

and the summands at the right side has the least factor greater than v. ■

Theorem 8.2 ([379], [381], [383], [385])
Let K be a commutative associative ring with 1 and without 2-torsion, and let $L(X) = F(X)/I$ be the free color Lie superalgebra. Then the cosets in $F(X)/I$ whose representatives are s-regular monomials give us a basis of $L(X)$ as a free G-graded K-module.

PROOF First of all we prove that for any $w \in \Gamma(X)$ the coset with representative w is a linear combination of cosets whose representatives are s-regular monomials. We use induction on the length $\ell(w)$. The basis of induction is $\ell(w) = 1$. Here, obviously, w is a regular monomial. Let $\ell(w) > 1$ and suppose that our statement holds for all $w' \in \Gamma(X)$ with $\ell(w') < \ell(w)$. Then $w = [u, v]$, $u, v \in \Gamma(X)$, $\ell(u) < \ell(w)$, $\ell(v) < \ell(w)$. By the induction hypothesis, the cosets with representatives u and v are linear combinations of cosets such that their representatives are s-regular monomials. Now it is sufficient to apply Lemma 8.1.

Secondly we show that the cosets in $L(X) = F(X)/I$ whose representatives are s-regular monomials are linearly independent. Indeed, by the way of contradiction assume that

$$\sum_i \alpha_i u_i = 0$$

where u_i are pairwise distinct s-regular monomials, $0 \neq \alpha_i \in K$. We choose the greatest s-regular monomial u_k. Since

$$\pi\left(\sum \alpha_i u_i\right) = 0,$$

by Lemmas 6.2, 6.3, 6.6 we get either $\alpha_k \widetilde{u_k} = 0$ or $2\alpha_k \widetilde{u_k} = 0$ (and then since 2 is not a ziro divisor in K, $\alpha_k = 0$). Thus we have come to a contradiction.

Now it is sufficient to mention only that $\{z + I \mid z \in \Gamma(X)\}$ is a set of generators of the K-module $L(X)$. ■

Corollary 8.3
The homomorphism $\pi: L(X) \to [X]$ is an isomorphism of color Lie superalgebras; $A(X)$ is the universal enveloping algebra of $L(X)$.

Remark 8.4
As we have seen above, the universal enveloping algebra of the free color Lie superalgebra is the free associative algebra. But the reverse assertion does not take place as Theorem 28.10 shows. However it would be interesting to consider the following open questions.

Problem 8.5
Let K be a field.

(i) *Let* $\operatorname{char} K = 0$ *and L be a Lie algebra such that $U(L)$ is a free associative algebra of finite rank. Is L a free Lie algebra? The same question for Lie color superalgebras.*

(ii) *Let* $\operatorname{char} K = p > 2$ *and L be a Lie p-algebra such that $u(L)$ is a free associative algebra of finite rank. Is L a free Lie p-algebra? The same question for Lie color p-superalgebras.*

In some special cases we have

Proposition 8.6
Let K be a field and L a color Lie (p-)superalgebra generated by n G-homogeneous elements such that its (restricted) universal enveloping algebra $U(L)$ ($u(L)$, respectively) is the free associative algebra of the rank n. Then L is the free color Lie (p-)superalgebra of the rank n.

PROOF The algebra $U^+(L)$ is the free associative algebra without unit. Let $X = \{x_1, \ldots, x_n\}$ be a free generating set of $U^+(L)$ and $y_1, \ldots, y_n$ generators of L. Then $Y = \{y_1, \ldots, y_n\}$ generates $U^+(L) = A(X)$. Since $x_1, \ldots, x_n$ are associative elements on $\{y_1, \ldots, y_n\}$, the linear parts of elements $y_1, \ldots, y_n$ relatively to X are linear independent. Hence Y is a set of free generators of $A(X)$. Thus Y is a set of free generators of L. ■

Remark 8.7
In Theorem 8.2, instead of regular monomials we can use the arrangement of brackets $[u]'$ on each regular word u such that the greatest associative word in the expression of $\pi([u])$ is equal to u. Examples of such arrangements of brackets which are different from the canonical arrangement of brackets on a regular monomial could be found in Lemma 7.1.

Definition 8.8
The length, the function φ*, the multidegree* of an element $a \in L(X)$ are defined as for the element $\pi(a) \in A(X)$. Usually we in the same way identify elements a and $\pi(a)$. It is clear that the length of $a \in L(X)$ is equal to the maximal length of s-regular monomials in the s-regular presentation of a.

The cardinality of X is called *the rank* of $L(X)$ (notation: $\operatorname{rank}(L(X))$). It is clear that the rank does not depend on the choice of free generators and

$$\begin{aligned}\operatorname{rank}(L(X)) &= \sum_{g\in G} \operatorname{rank}_g(L(X)),\\ \operatorname{rank}_g(L(X)) &= \dim L_g(X)/[L(X), L(X)]_g.\end{aligned}$$

Other bases

Definition 8.9
A subset $R \in \Gamma(X)$ with a linear ordering $<$ is called *a basic family* if the following conditions are satisfied:

1. $X \subseteq R$;
2. $w = [u, v] \in R$ if and only if $u, v \in R$, and
   (a) $u > v$;
   (b) if $u = [u_1, u_2]$, then $u_2 \leq v$;
   (c) $w > v$.

Theorem 8.10 (A. I. Shirshov [515])
Any basic family R of $\Gamma(X)$ forms a basis of the free Lie algebra $L(X)$ as a free K-module.

In the case of the lexicographical ordering (with the additional condition $u > uv$ for any $u, v \in S^+(X)$) we get Lyndon-Shirshov basis; if we consider an ordering when from $\ell(u) > \ell(v)$ it follows that $u > v$, then we get Hall basis.

Theorem 8.11
Let K have no 2-torsion, and let X be a G-graded set, R a basic family of $\Gamma(X)$,

$$R' = \{a, [b, b] \mid a, b \in R,\ d(b) \in G_-\}.$$

Then R' forms a basis of the free color Lie superalgebra $L(X)$.

PROOF It follows from Theorem 8.10 that the number of elements of any basic family with given multidegree is the same as the number of regular words with this multidegree (and this number is given by Witt's formula, see Theorems 5.18, 5.19). On the other hand, it is clear that each element of $L(X)$ can be presented as a linear combination of elements from R' (one need to consider "the procedure of increasing of the least multiplier" like in (2.2)). Thus, by Theorem 8.2, R' forms a linear basis of $L(X)$. ■

Another way to construct bases of $L(X)$ is as follows. For $x \in X$, $d(x) \in G_+$ by Corollary 13.5 and Theorem 13.7 we get

$$L(X) = Kx \oplus T$$

where T the subalgebra of $L(X)$ with the set $Z(x)$ of free generators. One can consider a basis of $L(X)$ in the following form: x and any basis of T.

Analogously, if $x \in X$, $d(x) \in G_-$, then we have that either

$$L(X) = Kx \oplus K[x,x] \oplus T,$$

or

$$L(X) = Kx \oplus R$$

where R is the subalgebra of $L(X)$ with the set $W(x)$ of free generators (see Section 13).

Moreover, it is possible to consider a finite number of such iterations, and then to construct a basis. For example, if $X = \{x,y\}$, $d(x) \in G_+$, $d(y) \in G_-$, then

$$\begin{aligned} L(X) &= Kx \oplus L(Z_X(x)) = Kx \oplus Ky \oplus K[y,y] \oplus L\big(Z_{Z_X(x)}(y)\big) \\ &= Kx \oplus Ky \oplus K[y,y] \oplus K[x,y] \oplus L\big(W_{Z_{Z(x)}}([x,y])\big), \end{aligned}$$

where $W_{Z_{Z(x)}}([x,y])$ is the set

$$\big\{ [[x,y],[x,y]];\ (y)(\operatorname{Ad} x)^k(\operatorname{Ad} y)^l;\ (y)(\operatorname{Ad} x)^k(\operatorname{Ad} y)^l(\operatorname{Ad}[x,y]) \mid k,l \in \mathbb{N}\big\}.$$

Remark 8.12

For any basic family R and its corresponding set R' we have the following factorization of associative words: if $a \in S^+(X)$, then a has the unique presentation in the form $a = u_1 \cdots u_k$, where $u_1, \dots, u_k$ have the form $\widetilde{w_i}$ with w_i basic monomials, $w_1 \leq \cdots \leq w_k$ (for R' we assume that w_i are s-basic monomials and $w_i \neq w_j$ with $d(w_i) \in G_-$). Moreover, in number of cases it is possible to construct a basis of the free Lie superalgebra using a factorization of the free monoid $S^+(X)$. For the details in the case of free Lie algebras see [590].

Problem 8.13

Construct a basis of the free Lie algebra (of the free Lie superalgebra) which consists of left normed monomials.

This problem is open even in the case of the free Lie algebra of rank two. In [295] it was claimed that such basis was constructed, but there were serious incorrections in the article (the suggested set of monomials, in fact, is not a basis of the free Lie algebra). Note that in [73] D. Blessehohl and H. Laue constructed a basis of the free Lie algebra which consists of sums of left normed monomials.

Remark 8.14
It would be desirable to have suitable linear bases of relatively free algebras to solve combinatorial problems and to study ideals and subalgebras of such algebras. It seems that generally it is not easy to construct such linear bases. As an example we mention only the relatively free Lie algebra $L = L(X)/J$ where $L(X)$ is the free Lie algebra on X, and J is the ideal of $L(X)$ generated by all elements which are obtained by substitutions of elements from $L(X)$ in the standard identity of the degree 4:

$$T_4(x_1, x_2, x_3, x_4, y) = \sum_{\sigma \in S_4} (-1)^\sigma (\mathrm{ad}\, x_{\sigma(1)})(\mathrm{ad}\, x_{\sigma(2)})(\mathrm{ad}\, x_{\sigma(3)})(\mathrm{ad}\, x_{\sigma(4)})(y).$$

This algebra L is important from the point of view of the problem to describe all identities of Lie algebras of vector fields. Up to now we don't have bases of L which consist of monomials.

Let $k \geq 1$, $n_1 \geq 2$, $n_2 \geq 2$, $\ldots, n_k \geq 2$, $L(X)$ the free Lie algebra over a field K, $L(X)/L(X)^{(n)}$ and $L(X)/(\ldots(L(X)^{n_1})^{n_2}\ldots)^{n_k}$ *the free solvable Lie algebra* and *the free polynilpotent Lie algebra* respectively.

In [78] L. A. Bokut' constructed bases of free solvable and of free polynilpotent Lie algebras.

Problem 8.15
Construct linear bases of free solvable and free polynilpotent Lie superalgebras.

If in a color Lie superalgebra L any two commutators commute, i.e. $\big[[x, y], [z, t]\big] = 0$ for any $x, y, z, t \in L$, we say that L is *a metabelian color Lie superalgebra.* The free metabelian color Lie superalgebra with the set X of free generators is the algebra $M(X) = L(X)/L(X)^{(2)}$. Linear bases for free color metabelian Lie superalgebras were constructed by Yu. A. Bahturin and V. Drensky in [27].

Bases of the free color Lie p-superalgebra

In all what follows in this subsection we assume that p is a prime number and K is a $\mathbb{Z}_p$-algebra.

Definition 8.16
Let X be a G-graded set and $A(X)$ the free G-graded associative algebra. We say that an element $w \in A(X)$ is *an as-regular monomial* relatively to the alphabet X if $w = \pi(u)$ where u is an s-regular monomial relatively to X, and, respectively, *a ps-regular monomial*, if either w is an as-regular monomial or $w = (\pi(v))^{p^k}$, where $k \in \mathbb{N}$, and v is an s-regular monomial with $d(v) \in G_+$.

By $L^p(X)$ we denote the subalgebra of $[A(X)]^p$ generated by X.

Lemma 8.17
The set of ps-regular monomials gives us a linear basis of the color Lie p-superalgebra $L^p(X)$.

PROOF By Theorem 8.2, Corollary 8.3 and the definition of the operation $a \to a^{[p]}$, we set that any element of $L^p(X)$ is a linear combination of ps-regular monomials in the elements of X. By Lemmas 5.25 and 6.6 the set of ps-regular monomials is K-linear independent. ■

Proposition 8.18
The color Lie p-superalgebra $L^p(X)$ constructed above is a free color Lie p-superalgebra with the set X of free generators (i.e., X generates the algebra $L^p(X)$ and any mapping $\varphi\colon X \to R$ where R is a color Lie p-superalgebra over K with the same group G and commutative factor ε, can be extended to a homomorphism $\varphi'\colon L^p(X) \to R$ of color Lie p-superalgebras).

PROOF Any element a of $L^p(X)$ can be written as

$$a = \sum_i \lambda_i \big(\pi(w_i)\big)^{p^{t_i}} + \sum_j \beta_j \pi(v_j),$$

where $\lambda_i, \beta_j \in K$, w_i, v_j are s-regular monomials, $d(w_i) \in G_+$, $t_i \in \mathbb{N}$ for all i, j. Since $[X]$ is the free color Lie superalgebra, the mapping φ can be extended to a homomorphism $\varphi_1\colon [X] \to R$ of color Lie superalgebras. Now we set

$$\varphi'(a) = \sum_i \lambda_i \big(\varphi_1(\pi(w_i))\big)^{[p]^{t_i}} + \sum_j \beta_j \varphi_1(\pi(v_j)).$$

■

Definition 8.19
The number $\operatorname{rank}(L^p(X)) = |X|$ is called *the rank of the free color Lie p-superalgebra.* It is clear that the rank of $L^p(X)$ does not depend on the choice of a set of free generators.

The length and the multidegree of an element $a \in L^p(X)$ can be defined in the sme way using the fact that $a \in A(X)$.

PBW-basis of the free associative algebra

Each element of the free associative algebra has the unique presentation as a sum of associative words. At the same time it is possible to consider another basis of $A(X)$ which is compatible with the structure of $A(X)$ as the universal enveloping algebra of the free color Lie superalgebra $L(X)$ (as the restricted universal enveloping algebra of $L^p(X)$, respectively).

Theorem 8.20
Let K be a field, $\operatorname{char} K \neq 2$. *Then the unity element and the monomials $\pi(w_1)\cdots\pi(w_k)$ where w_i are s-regular monomials, $\widetilde{w_1} \leq \cdots \leq \widetilde{w_k}$, $\widetilde{w_i} \neq \widetilde{w_{i+1}}$ with $d(w_i) \in G_-$ for all $i = 1, \ldots, k-1$, form a linear basis of $A(X)$. Moreover,*

$$\left(\pi(w_1)\cdots\pi(w_k)\right)^{\circ} = 2^l\, \widetilde{w_1}\cdots\widetilde{w_k}$$

with $l \in \mathbb{Z}_+$.

If $\operatorname{char} K = p > 2$, *then the unity element and the monomials*

$$(\pi(w_1))^{\lambda_1}\cdots(\pi(w_k))^{\lambda_k}$$

where w_i are ps-regular monomials, $\widetilde{w_1} < \cdots < \widetilde{w_k}$, $0 \leq \lambda_i \leq p-1$ with $d(w_i) \in G_+$, $\lambda_j = 0, 1$ for $d(w_j) \in G_-$, form a linear basis of $A(X)$. Moreover,

$$\left((\pi(w_1))^{\lambda_1}\cdots(\pi(w_k))^{\lambda_k}\right)^{\circ} = 2^l\, (\widetilde{w_1})^{\lambda_1}\cdots(\widetilde{w_k})^{\lambda_k}$$

with $l \in \mathbb{Z}_+$.

PROOF The assertions of the theorem immediately follow from Lemmas 5.20, 7.4, and Theorem 8.2. Note that they are particular cases of Poincaré-Birkhoff-Witt theorems, see Theorems 19.1, 19.2. ∎

Remark 8.21
Consider the set of all associative words which appear in associative presentations of elements from $L(X)$ (from $L^p(X)$). This set is called *the associative support* of $L(X)$ (of $L^p(X)$, respectively). In [168] G. Duchamp and J.-Y. Thibon described the associative support of the free Lie algebra. Analogous description of the associative support of the color Lie p-superalgebra was obtained by A. A. Mikhalev in [402].

Let $K = \mathbb{Q}$, $L(X)$ be the free Lie algebra, $\langle\, ,\, \rangle$ the bilinear form on $A(X)$ such that $S(X)$ is an orthogonal basis of $A(X)$. In [164] G. Duchamp proves that there exists the orthogonal projection of $A(X)$ onto $L(X)$: $A(X) = L(X) \oplus L(X)^{\perp}$. He described this projection using Pólya action.

Witt formulas

We start this section with the classical formula for the dimension of the component of the given multidegree of the free Lie ring.

Theorem 8.22 ([385])
Let $X = \{x_1, \dots, x_n\}$, $L(X)$ the free Lie ring. Then the rank W_k of the free module of elements of length k is equal to

$$\frac{1}{k}\sum_{d\,|\,k} \mu(d) n^{k/d},$$

and the rank $W(\alpha_1, \dots, \alpha_n)$ of the free module of elements of multidegree $(\alpha_1, \dots, \alpha_n)$ is as follows

$$W(\alpha_1, \dots, \alpha_n) = \frac{1}{|\alpha|} \sum_{\substack{d\,|\,\alpha_i \\ i=1,\dots,n}} \mu(d) \frac{(|\alpha|/d)!}{(\alpha_1/d)! \cdots (\alpha_n/d)!}, \tag{2.3}$$

where $|\alpha| = \alpha_1 + \cdots + \alpha_n$, μ is the Möbius function.

Theorem 8.23 (R. Ree [476])
Let K be a field, $\operatorname{char} K = 0$, $|X_+| = q$, $|X_-| = r$, $k, l \in \mathbb{N}$, k is odd, $L(X)$ the free color Lie superalgebra with the grading on the length

$$L(X) = \bigoplus_{i=1}^{\infty} L(X)_i.$$

Then

$$\begin{aligned}
\dim L(X)_k &= \frac{1}{k}\sum_{d\,|\,k} \mu(d)(q+r)^{k/d}, \\
\dim L(X)_{2^l k} &= \frac{1}{2^l k} \sum_{d\,|\,(2^l k)} \mu(d)(q+r)^{2^l k/d} \\
&\quad + \frac{1}{2^l k} \sum_{d\,|\,k} \mu(d) \Big((q+r)^{2^{l-1} k/d} - (q-r)^{2^{l-1} k/d} \Big).
\end{aligned}$$

Theorem 8.24
Let K have no 2-torsion, $X_+ = \{x_1, \dots, x_t\}$, $X_- = \{x_{t+1}, \dots, x_{t+s}\}$. Then the rank $SW(\alpha_1, \dots, \alpha_{t+s})$ of the free K-module of elements of multidegree $\alpha = (\alpha_1, \dots, \alpha_{t+s})$ in the free color Lie superalgebra $L(X)$ of the rank $t+s$ is as follows

$$SW(\alpha_1, \dots, \alpha_{t+s}) = W(\alpha_1, \dots, \alpha_{t+s}) + \beta W\Big(\frac{\alpha_1}{2}, \dots, \frac{\alpha_{t+s}}{2}\Big),$$

where

$$\beta = \begin{cases} 0, & \text{if there exists an } i \text{ such that } \alpha_i \text{ is odd;} \\ 0, & \text{if } (\alpha_{t+1}/2) d(x_{t+1}) + \cdots + (\alpha_{t+s}/2) d(x_{t+s}) \in G_-; \\ 1, & \text{otherwise.} \end{cases}$$

Now we discuss another forms of Witt formulas for Lie superalgebras. Let $K = \mathbb{C}$. For a polynomial

$$q(t) = \sum_{i=0}^{n} \alpha_i z^i,$$

with $\alpha_i \in \mathbb{C}$ denote by $N_k(q(t))$ the k-th *Newton's polynomial* on roots of the reciprocal with respect to $q(t)$, in other words, if $\xi_1, \ldots, \xi_n$ are all roots of the polynomial

$$\sum_{i=0}^{n} \alpha_{n-i} z^i,$$

then

$$N_k(q(t)) = \sum_{i=1}^{n} \xi_i^k.$$

Theorem 8.25 (I. K. Babenko [18])
Let $K = \mathbb{C}$, $X = \{x_1, \ldots, x_n\}$, $\varphi(x_i) = k_i \in \mathbb{N}$, $i = 1, \ldots n$, $L(X)$ the free Lie superalgebra, L_j the component of $L(X)$ of the weight j relatively to φ

$$q(t) = 1 - \sum_{i=1}^{n} t^{k_i}.$$

Then

$$\dim L_j = \frac{(-1)^j}{j} \sum_{d\,|\,j} (-1)^d \mu(j/d) N_d(q(t)).$$

Theorem 8.26 (I. L. Kantor [255])
Let $|X_-| = m$, $|X_+| = n$, $r = m + n$, $s = m - n$, $L(X)$ the free Lie superalgebra, $M_k = \dim((L_k)_-)$, $N_k = \dim((L_k)_+)$. Then

$$\begin{aligned} M_k &= \frac{1}{2k} \sum_{\substack{d\,|\,k \\ d \text{ odd}}} \mu(d)\left(r^{k/d} - s^{k/d}\right); \\ N_k &= \frac{1}{2k} \sum_{\substack{d\,|\,k \\ d \text{ odd}}} \mu(d)\left(r^{k/d} + s^{k/d}\right) + \frac{1}{k} \sum_{\substack{d\,|\,k \\ d \text{ even}}} \mu(d) s^{k/d}. \end{aligned}$$

In particular, if $m = n$, then

$$M_l = N_l = \frac{1}{2l} \sum_{\substack{d\,|\,l \\ d \text{ odd}}} \mu(d)(2n)^{l/d};$$

if $n = 0$, *then* $M_{2i} = 0$, $N_{2i+1} = 0$,

$$M_{2i+1} = \frac{1}{2i+1} \sum_{d \mid (2i+1)} \mu(d) m^{(2i+1)/d};$$

$$N_{2i} = \frac{1}{2i} \sum_{d \mid (2i)} \mu(d)(-m)^{(2i)/d}$$

For the multidegree $\alpha = (\alpha_1, \ldots, \alpha_{m+n})$ *we have*

$$SW(\alpha_1, \ldots, \alpha_{m+n}) = \frac{1}{|\alpha|} \sum_{\substack{d \mid \alpha_i \\ i=1,\ldots,m+n}} \mu(d) \frac{(-1)^{|\alpha|'(d-1)/d}(|\alpha|/d)!}{(\alpha_1/d)! \cdots (\alpha_{m+n}/d)!}$$

where $|\alpha| = \alpha_1 + \cdots + \alpha_{m+n}$, $|\alpha|' = \alpha_1 + \cdots + \alpha_m$.

Theorem 8.27 (A. I. Molev and L. M. Tsalenko [431])
Let $L(X)$ *be the free Lie superalgebra with the odd set of generators* $X = \{x_1, \ldots, x_s\}$, K *a field,* $\operatorname{char} K = 0$. *Then*

$$SW(\alpha_1, \ldots, \alpha_s) = \frac{(-1)^{|\alpha|}}{|\alpha|} \sum_{\substack{d \mid \alpha_i \\ i=1,\ldots,s}} \mu(d) \frac{(-1)^{|\alpha|/d}(|\alpha|/d)!}{(\alpha_1/d)! \cdots (\alpha_s/d)!}$$

where $|\alpha| = \alpha_1 + \cdots + \alpha_s$.

Remark 8.28
In [157] D. Dokovič considered the situation when $Z = X \cup Y$, $X \cap Y = \emptyset$, and $L(X, Y) = L(Z)/I$ where I is the ideal of the free Lie algebra $L(Z)$ generated by all elements $[x, x']$ with $x, x' \in X$. He constructed a basis of $L(X, Y)$ and proved the following analog of Witt formula.

Theorem 8.29 (D. Dokovič [157])
Let $X = \{x_1, \ldots, x_m\}$, $Y = \{y_1, \ldots, y_n\}$, *and let* $c(\alpha, \beta)$ *be the dimension of the component with the multidegree* (α, β) *of* $L(X, Y)$, *where* $\alpha = (\alpha_1, \ldots, \alpha_m)$, $\beta = (\beta_1, \ldots, \beta_n)$.. *If* $|\alpha| + |\beta| \geq 1$, *then*

$$c(\alpha, \beta) = \left(\frac{1}{|\beta|} \prod_{i=1}^{m} \frac{|\beta|}{|\beta| + \alpha_i} \right) \sum_{\substack{d \mid \alpha_i \\ d \mid \beta_j}} \mu(d) \frac{(|\beta|/d)!}{(\beta_1/d)! \cdots (\beta_n/d)!} \prod_{i=1}^{m} \frac{((|\beta| + \alpha_i)/d)!}{(|\beta|/d)!(\alpha_i/d)!}$$

where $|\alpha| = \alpha_1 + \cdots + \alpha_m$, $|\beta| = \beta_1 + \cdots + \beta_n$, *and the term*

$$\frac{1}{|\beta|} \prod_{i=1}^{m} \frac{|\beta|}{|\beta| + \alpha_i}$$

in the case $\beta = 0$ *should be interpreted as* 0 *unless there is an* i_0 *such that* $\alpha_i = 0$ *for all* $i \neq i_0$, *in which case this term should be interpreted as* $1/\alpha_i$.

It would be interesting to obtain an analog of this theorem for color Lie (p-)superalgebras.

Remark 8.30
In [167] G. Duchamp and D. Krob studied the free partially commutative Lie algebra. Let A be the alphabet and θ is a partial commutation relation on A, i.e. a symmetric subset of $A \times A$ which does not contain any element of the diagonal, $M(A,\theta)$ the free partially commutative monoid on A associated to θ, $M(A,\theta) = S(A)/\theta$. Let I_θ be the ideal of the free Lie algebra $L(A)$ generated by the elements $\{[a,b] \mid (a,b) \in \theta\}$. *The free partial commutative Lie algebra* is the algebra $L(A,\theta) = L(A)/I_\theta$. In [167] a basis of $L(A,\theta)$ was constructed and analogs of Witt formulas were obtained.

Identical relations among left normed monomials

Let $\delta, \delta' \colon A(X) \to [X]/I$ be linear mappings such that for all $x_1, \dots, x_n \in X$, $\delta(x_1, \dots, x_n) = [x_1 \cdots x_n]$ is a left normed monomial and $\delta'(x_1, \dots, x_n) = [x_1 \cdots x_n]'$ is a right normed monomial.

It is clear that any element from $L(X)$ is a linear combination of left (right) normed monomials.

P. M. Cohn gave the following description of identical relations among left normed monomials in free Lie rings.

Theorem 8.31 (P. M. Cohn [121], Theorem 10.1)
Let $L(X)$ be a free Lie ring. Then $\delta(a) = 0$ if and only if

$$a = \sum_i u_i\delta(u_i) + \sum_j v_j\delta(v_j)w_j,$$

where $u_i, v_j, w_j \in A(X)$.

For free Lie superalgebras we get the following result.

Theorem 8.32
For mappings δ, δ' we have:

(i) $\delta(a) = 0$ *if and only if*

$$a = \sum_i (a_i\delta(b_i) + \varepsilon(a_i, b_i)d_i\delta(a_i))w_i;$$

(ii) $\delta'(a) = 0$ *if and only if*

$$a = \sum_i w_i(\delta'(a_i)b_i + \varepsilon(a_i, b_i)\delta'(b_i)a_i),$$

where a_i, b_i are homogeneous elements from $A(X)$, $w_i \in A(X)$.

Images of inner derivations

Let K be a commutative associative ring without 2-torsion, $L(X)$ the free color Lie superalgebra over K. For $u \in L(X)$ we consider *the inner derivation* $\operatorname{ad} u$. We use the notation

$$\operatorname{Im}(\operatorname{ad} u) = \{a \in L(X) \mid a = [u, w], w \in L(X)\}.$$

Theorem 8.33 ([400])
Let $X = X_+$, $u \in L(X)$, $\alpha u^\circ \in S(X)$ for some $\alpha \in K^*$. Then the K-submodule $\operatorname{Im}(\operatorname{ad} u)$ is free and a direct summand of the free K-module $L(X)$.

PROOF By the assumption of our theorem we obtain

$$\alpha u = v + \alpha_1 v_1 + \cdots + \alpha_k v_k$$

where v and v_i are regular monomials, $\alpha_i \in K$, and either $\ell(v) > \ell(v_i)$, or $\ell(v) = \ell(v_i)$ and $\widetilde{v} > \widetilde{v_i}$ for all i. It is clear that any element of $\operatorname{Im}(\operatorname{ad} u)$ is a linear combination with coefficients from K of elements $[u, w_j]$ with w_j regular monomials. If $\widetilde{w_j} > \widetilde{v}$, then

$$\alpha[w_j, u]^\circ = \widetilde{w_j}\widetilde{v}$$

and $\widetilde{w_j}\widetilde{v}$ is a regular word. If $\widetilde{w_j} < \widetilde{v}$ then

$$\alpha[w_j, u]^\circ = \widetilde{v}\widetilde{w_j}$$

and $\widetilde{v}\widetilde{w_j}$ is a regular word. If $\widetilde{w_j} = \widetilde{v}$, then $w_j = v$ and we get

$$\begin{aligned} \alpha[u, w_j] &= [v + \alpha_1 v_1 + \cdots + \alpha_k v_k, v] \\ &= [v + \alpha_1 v_1 + \cdots + \alpha_k v_k, -\alpha_1 v_1 - \cdots - \alpha_k v_k] \\ &= -\alpha_1[\alpha u, v_1] - \cdots - \alpha_k[\alpha u, v_k], \end{aligned}$$

$v_i \neq v$ for all $i = 1, \ldots k$.

Now we change the set W of all regular monomials by the following way. If $w \in W$ and the word $\widetilde{w}$ is either $\widetilde{v}a$ no $b\widetilde{v}$ where a and b are regular words, then firstly we show that these equalities are not valid at the same time. Indeed, let $\widetilde{w} = \widetilde{v}a = b\widetilde{v}$. Since $\widetilde{w}$ is a regular word we have a contradiction: $b\widetilde{v} > \widetilde{v} > \widetilde{v}a = b\widetilde{v}$. Let $\widetilde{w} = \widetilde{v}a$ where a is a regular word. Then we replace w by $w' = \big[\alpha u, [a]\big]$ with $[a]$ a regular monomial. If $\widetilde{w} = b\widetilde{v}$, b is a regular word, then we replace w by $w'' = \big[[b], \alpha u\big]$ where $[b]$ is a regular monomial. On this way we come to the set W' which is a basis of free K-module $L(X)$. On the other hand, the subset $W'' = W' \setminus W$ forms a basis of the free K-module $\operatorname{Im}(\operatorname{ad} u)$. ■

Note that in the case when $L(X)$ is the free Lie algebra and $u \in X$ the assertion of Theorem 8.1 is due to D. Dokovič [158].

Remark 8.34
The statement of Theorem 8.33 is not valid for free color Lie (p-)superalgebras. Indeed, let $\mathbb{Q}[t]$ be the ring of polynomials in one variable over the field $\mathbb{Q}$, $G = \mathbb{Z}_2$, $X_+ = \{x\}$, $X_- = \{y\}$, $L(X)$ the free Lie superalgebra, $u = tx + [y, y]$. Then $[x, y] \notin \mathrm{Im}\,(\mathrm{ad}\, u)$, and at the same time $t[x, y] \in \mathrm{Im}\,(\mathrm{ad}\, u)$. Hence the K-module $\mathrm{Im}\,(\mathrm{ad}\, u)$ is not a direct summand of the free K-module $L(X)$.

If $K = \mathbb{Z}_p[t]$ is the ring of polynomials in one variable over the field $\mathbb{Z}_p$, $X = \{x, y\}$, $L^p(X)$ is the free Lie p-algebra over K, $u = x^{[p]} + ty$, then $[x, u] = t[x, y]$, and thus the K-module $\mathrm{Im}\,(\mathrm{ad}\, u)$ is not a direct summand of the free K-module $L^p(X)$.

In the general case we have the following theorems.

Theorem 8.35 ([400])
Let K have no 2-torsion, $L(X)$ the free color Lie superalgebra over K (in the case when K is a $\mathbb{Z}_p$-algebra, p is a prime number, let also $L^p(X)$ be the free color Lie p-superalgebra). If a G-homogeneous element u belongs to $L(X)$ ($u \in L^p(X)$), $\alpha u^\circ \in S(X)$ for some $\alpha \in K^*$, αu° is a regular word, then the K-submodule $\mathrm{Im}\,(\mathrm{ad}\, u)$ *is free and a direct summand of the free K-module $L(X)$ (of $L^p(X)$).*

PROOF If $u \in L(X)$, then $\alpha u = v + \alpha_1 v_1 + \cdots + \alpha_k v_k$ where v is a regular monomial, v_i are s-regular monomials, and either $\ell(v) > \ell(v_i)$, or $\ell(v) = \ell(v_i)$ and $\widetilde{v} > \widetilde{v}_i$. Using the same arguments as in the proof of Theorem 8.33 we get that the K-module $\mathrm{Im}\,(\mathrm{ad}\, u)$ is free,

$$L(X) = \mathrm{Im}\,(\mathrm{ad}\, u) \oplus M$$

where M is the free K-module which basis form s-regular monomials w_j such that $\widetilde{w_j}$ are not equal to $a\widetilde{v}$ and $\widetilde{v}b$ where a and b are regular words, $a > \widetilde{v} > b$ (in the case when $d(u) \in G_-$ we suppose also that

$$\widetilde{w_j} \neq \widetilde{v}\widetilde{v}, cc\widetilde{v}, \widetilde{v}ff$$

where c and f are regular words, $d(c), d(e) \in G_-$, $c > \widetilde{v} > f$.

If $u \in L^p(X)$, then $d(v) = d(u) \in G_+$ and elements v_i are ps-regular monomials such that either $\ell(v) > \ell(v_i)$, or $\ell(v) = \ell(v_i)$ and $\widetilde{v} > \widetilde{v}_i$. In this situation we see that $\mathrm{Im}\,(\mathrm{ad}\, u)$ is a free K-module,

$$L^p(X) = \mathrm{Im}\,(\mathrm{ad}\, u) \oplus V$$

where V is the free K-module with the free basis consisting of ps-regular monomials w_j such that

$$\widetilde{w_j} \neq a\widetilde{v}, \widetilde{v}b, c^{p^k}\widetilde{v}, \widetilde{v}f^{p^l}$$

where a and b are s-regular words, $k, l \in \mathbb{N}$, c, f are s-regular words, $d(c), d(f) \in G_+$. ∎

Theorem 8.36 ([400])

(i) *Let K have no 2-torsion, and $L(X)$ be the free color Lie superalgebra over K, v a G-homogeneous element of $L(X)$, $d(v) \in G_-$, $u = [v, v]$, $\alpha u^\circ \in S(X)$ for some $\alpha \in K^*$. Then the K-module* Im (ad u) *is free and a direct summand of the free K-module $L(X)$.*

(ii) *Let K be a $\mathbb{Z}_p$-algebra, p a prime number, and $L^p(X)$ the free color Lie p-superalgebra over K. If $v \in L^p(X)$, $d(v) \in G_+$, $u = v^{[p]^k}$, $k \in \mathbb{N}$, $\alpha u^\circ \in S(X)$ for some $\alpha \in K^*$, then the K-module* Im (ad u) *is free and a direct summand of the free K-module $L^p(X)$. The same statement is also valid in the case when $d(v) \in G_-$, $l \in \mathbb{N}$, $u = ([v, v])^{[p]^l}$, $\alpha u^\circ \in S(X)$ for some $\alpha \in K^*$.*

PROOF One can easily see that

$$L = \mathrm{Im}\,(\mathrm{ad}\, u) \oplus V$$

where L is either $L(X)$, or $L^p(X)$, V is a free K-module with the basis consisting of s-regular (ps-regular) monomials w_j such that $\widetilde{w_j}$ are not equal to $\tilde{a}\tilde{u}$ and $\tilde{u}\tilde{b}\tilde{u}$ where a and b are s-regular (ps-regular) monomials. ■

Problem 8.37 (D. Dokovič [158])

Let $L(X)$ be the free Lie ring, and $x, y \in X$. Is it true that the module Im (ad x) + Im (ad y) *is free and a direct summand of the free $\mathbb{Z}$-module $L(X)$?*

Note that Problem 8.37 has some relations with Problem 8.13.

9 Dynkin-Specht-Wever criterion

In this section we prove a super analog of Dynkin-Specht-Wever criterion for an element of the free associative algebra to be a Lie element.

Any associative derivation δ with $d(\delta) \in G$ of a G-graded associative algebra Q gives us the derivation of the color Lie superalgebra $[Q]$:

$$\begin{aligned}
\delta([a,b]) &= \delta(ab - \varepsilon(a,b)\,ba) \\
&= \delta(a)\,b + \varepsilon(d(\delta), d(a))\,a\,\delta(b) \\
&\quad - \varepsilon(a,b)(\delta(b)\,a + \varepsilon(d(\delta), d(b))\,b\,\delta(a)) \\
&= (\delta(a)\,b - \varepsilon(d(\delta(a)), d(b))\,b\,\delta(a)) \\
&\quad + \varepsilon(d(\delta), d(a))(a\,\delta(b) - \varepsilon(d(a), d(\delta(b))\,\delta(b)\,a) \\
&= [\delta(a), b] + \varepsilon(d(\delta), d(a))\,[a, \delta(b)].
\end{aligned}$$

It is easy to see by induction on n that all derivations δ with $d(\delta) \in G$ of the free G-graded associative algebra $A(X)$ have the following form:

$$\begin{aligned}\delta(x_1\cdots x_n) &= \delta(x_1)\,x_2\cdots x_n+\varepsilon(d(\delta),d(x_1))\,x_1\,\delta(x_2)\,x_3\cdots x_n\\ &\quad+\cdots+\varepsilon\left(d(\delta),\sum_{i=1}^{n-1}d(x_i)\right)x_1\cdots x_{n-1}\,\delta(x_n)\end{aligned}$$

where $x_i\in X$.

Since $[X]$ is the free color Lie superalgebra generated by X, any mapping $\delta\colon X\to[X]$ with $d(\delta)\in G$ (i.e., $d(\delta(x))=d(x)+d(\delta)$) has the unique extension to a derivation of the free color Lie superalgebra $[X]$.

Let $x_i\in X$ and $[x_1]=x_1,\ldots,[x_1x_2\cdots x_n]=\big[x_1,[x_2\cdots x_n]\big]$ be left normed monomials. Consider a linear mapping $\delta\colon A(X)\to[X]$ such that $\delta(x_1\cdots x_n)=[x_1\cdots x_n]$ for all $x_i\in X$, $n\in\mathbb{N}$. Observe that $d(\delta(u))=d(u)$ for all $u\in S(X)$.

Since $A(X)$ is the universal enveloping algebra of $[X]$, there exists a homomorphism of G-graded associative algebras $\theta\colon A(X)\to\operatorname{End}[X]$ such that $\theta|_{[X]}=\operatorname{Ad}$.

We may write

$$\delta(x_{i_1}\cdots x_{i_n}\cdot x_{j_1}\cdots x_{j_m})=\delta(x_{i_1}\cdots x_{i_n})\,\theta(x_{j_1}\cdots x_{j_m}).$$

For homogeneous elements $a,b\in[X]$ we get

$$\begin{aligned}\delta([a,b]) &= \delta(ab-\varepsilon(a,b)\,ba)=\delta(a)\,\theta(b)-\varepsilon(a,b)\,\delta(b)\,\theta(a)\\ &= [\delta(a),b]+[a,\delta(b)]\end{aligned}$$

because of $d(\delta)=0$, i.e. δ is a derivation with $d(\delta)=0$ of the color Lie superalgebra $[X]$.

Theorem 9.1 (R. Ree [476])
Let K be a field, $\operatorname{char}K=0$. *Then an ℓ-homogeneous element $a\in A(X)$ with $\ell(a)=m$ belongs to $[X]$ if and only if $\delta(a)=ma$.*

PROOF If $\delta(a)=ma$, then $a=\delta(a)/m\in[X]$. We show by induction on the length that any element $a\in[X]$ is a linear combination of monomials of the form $[x_{i_1}\cdots x_{i_k}]$ where $x_{i_j}\in X$. Clearly, it is necessary to prove this only for ℓ-homogeneous elements. The basis of induction: $\ell(a)=1$, then a has the required form. Let now

$$a=\sum_i\gamma_i[u_i,v_i]$$

where $\gamma_i\in K$ and $[u_i,v_i]$ are s-regular monomials, $\ell(u_iv_i)=\ell(a)$. It is sufficient to consider only one summand $[u_j,v_j]$. By the induction hypothesis,

$$u_j=\sum_i\alpha_ia_i,\qquad \ell(a_i)=\ell(u_j),$$

$$v_j=\sum_k\beta_kb_k,\qquad \ell(b_k)=\ell(v_j),$$

where $\alpha_i, \beta_k \in K$ and a_i, b_k have the desired form for all i, k. Then

$$[u_j, v_j] = \sum_{i,k} \alpha_i \beta_k [a_i, b_k]$$

and it is sufficient to consider the elements $\big[[x_{i_1} \cdots x_{i_n}], [x_{j_1} \cdots x_{j_m}]\big]$. By the Jacobi identity we get

$$\begin{aligned}\big[[x_{i_1} \cdots x_{i_n}], [x_{j_1} \cdots x_{j_m}]\big] &= \alpha\big[\big[[x_{i_1} \cdots x_{i_n}], [x_{j_1} \cdots x_{j_{m-1}}]\big], x_{j_m}\big] \\ &\quad + \beta\big[[x_{i_1} \cdots x_{i_n} x_{j_m}], [x_{j_1} \cdots x_{j_{m-1}}]\big]\end{aligned}$$

where $\alpha, \beta \in K$, $x_{i_j} \in X$. The first factor in the first summand is, by induction, a linear combination of monomials having the wanted form. The second summand has the same form, but with first factor of shorter length. The second factor in the second summand is, by induction, a linear combination of monomials having the wanted form. To complete the proof we use the induction by the length of the second factor.

Thus, it is sufficient to prove the theorem only for left normed monomials $[x_{i_1} \cdots x_{i_n}]$. We use the induction on n. If $n = 1$ then $\delta(x_i) = x_i$. Furthermore, since δ is a derivation with $d(\delta) = 0$,

$$\begin{aligned}\delta([x_{i_1} \cdots x_{i_{n-1}}], x_{i_n}) &= [\delta([x_{i_1} \cdots x_{i_{n-1}}]), d(x_{i_n})] \\ &\quad + [[x_{i_1} \cdots x_{i_{n-1}}], \delta(x_{i_n})] \\ &= ((n-1)+1)[x_{i_1} \cdots x_{i_n}] = n[x_{i_1} \cdots x_{i_n}].\end{aligned}$$

■

10 Equations in free Lie superalgebras

In this section we discuss equations in free Lie superalgebras.

Lemma 10.1
Let K be a field, $\operatorname{char} K \neq 2$, $L(X)$ the free color Lie superalgebra, v a nonzero G-homogeneous element of $L(X)$, $d(v) \in G_-$, $w = [v, v]$. If z is a G-homogeneous element of $L(X)$, $d(z) \in G_-$, $[w, z] = 0$, then $z = \alpha v$, $0 \neq a \in K$.

PROOF If $z = \alpha v$, $\alpha \in K$ then $[w, z] = 0$. If now $z \neq \alpha v$, $\alpha \in K$, then $z = \beta v + u$, $\beta \in K$, and the leading word $\overline{u}$ of u is different from $\overline{v}$. Since the associative leading word of $[w, z]$ is equal either to $\gamma \overline{v}\,\overline{v}\,\overline{u}$ or $\gamma' \beta \overline{u}\,\overline{v}\,\overline{v}$, $0 \neq \gamma, \gamma' \in K$, we have $[w, z] = [w, u] \neq 0$. ■

Remark 10.2
The analog of Lemma 10.1 does not take place in the case of free color Lie

p-superalgebras for the extracting procedure of p-th root. Indeed, let K be a field, $\operatorname{char} K = p > 2$, $L = L^p(X)$ the free Lie p-algebra over K, $v \in L$, $w = v^{[p]^2}$, $z = v^{[p]} + v$. Then one can see that $[w, z] = 0$ and at the same time $z^{[p]} = w + v^{[p]} \neq w$.

J. P. Labute in [310] considered free Lie algebras as modules over their universal enveloping algebras. Let K be a commutative associative ring with unit, $X = \{x_1, \dots x_n\}$, $L(X)$ the free Lie algebra over K, $U = U(L(X)) = A(X)$ the universal enveloping algebra of $L(X)$. Consider $L(X)$ as a left U-module relatively to the adjoint action ad. Let also $\{e_1, \dots, e_n\}$ be the standard basis of the left U-module U^n, $e_i = (\delta_{i1}, \dots, \delta_{in})$ where δ_{ij} is the Kronecker delta. For $u, v \in U$, $i, j = 1, \dots, n$ denote

$$\begin{aligned} f_i(u) &= ((\operatorname{ad} u)x_i)ue_i, \\ f_{ij}(u, v) &= ((\operatorname{ad} v)x_j)ue_i + ((\operatorname{ad} u)x_i)ve_j. \end{aligned}$$

Let F be the U-submodule of U^n generated by the elements $f_i(u)$ and $f_{ij}(u, v)$ with $u, v \in U$, $i, j = 1, \dots, n$.

Theorem 10.3 (J. P. Labute [310])

$$(\operatorname{ad} u_1)x_1 + \cdots + (\operatorname{ad} u_n)x_n = 0$$

if and only if $u_1, \dots, u_n \in F$.

If K is a field and x is a nonzero homogeneous element of $L(X)$, then $(\operatorname{ad} u)(x) = 0$ *if and only if*

$$u = \sum_{v \in U} w_v((\operatorname{ad} v)x)v$$

with $w_v \in U$.

Problem 10.4
Give a description of the set annihilators of an element of the free color Lie (p-)superalgebra.

Note that the super case is more complicated, for example for a G-homogeneous element x and

$$u' = \sum_{v \in U} w_v((\operatorname{ad} v)x)((\operatorname{ad} v)x)v$$

with G-homogeneous v and $w_v \in U$ we have that $(\operatorname{ad} u)(x) = 0$. At the same time if $d(v) + d(x) \in G_-$, then $(((\operatorname{ad} v)x)v)x \neq 0$. As in Freiheitssatz (see Remark 20.6), the main problem is related with the case when at some step we get an even element which is not a full square, but its leading part is a full square.

Problem 10.5
Let a,b be G-homogeneous elements of $L(X)$. Describe G-homogeneous $v, w \in L(X)$ such that $[w, a] = [w, b]$.

Problem 10.6
Develop the theory of systems of equations over the free color Lie (p-)superalgebra.

For the case of Lie algebras see also [42], [43], [44], [77], [121], [310], [459], [460], [503], [504].

11 Algorithms and examples

In this section we give the following series of algorithms: to decide whether a word is regular or not (Algorithm 11.1); to find a factorization of the given word into a product of regular words (Algorithm 11.2); to construct the canonical arrangement of brackets on a regular word and the arrangement of brackets induced by the canonical arrangement of brackets on a regular subword (Algorithms 11.3, 11.4); to obtain the s-regular (ps-regular) form of an element (Algorithms 11.5, 11.7); to decide whether an associative element is a super Lie element (a super Lie p-element), or not (Algorithm 11.8); to get the PBW-presentation of an associative element (Algorithm 11.9); to decide whether an element is a full power, or not (Algorithm 11.10).

We present also some concrete examples.

Algorithms

Algorithm 11.1
Let the set X be well ordered. The problem is to decide whether a word $u \in S(X)$ is regular, or not.

This algorithm is based on Lemma 5.8. Let $u = x_{i_1} \cdots x_{i_k}$, $x_{i_j} \in X$. It is sufficient to check that u is greater than the words

$$x_{i_2} \cdots x_{i_k}, \quad x_{i_3} \cdots x_{i_k}, \quad \ldots, \quad x_{i_{k-1}} x_{i_k}, \quad x_{i_k}.$$

Algorithm 11.2
Let the set X be well ordered. The problem is to find a factorization of the given word $u \in S(X)$ into a product of regular words (see Theorem 5.20).

Using the proof of Theorem 5.20 we see that it is necessary on each step to find the longest regular beginning subword of the word.

Algorithm 11.3
Let u be a regular word. The problem is to construct a (unique) arrangement of brackets $[u]$ on u such that $[u]$ is a regular monomial.

(i) External way. Using Lemma 6.5 we see that to construct the desirable arrangement of brackets it is necessary on each step to find the longest proper regular ending of the word. In other words, it is sufficient starting with the left side to consider all endings, and to choose the first proper regular ending.

(ii) Internal way. Let x be the least letter in u. Starting with the right side of u we rewrite u in letters $V(x)$. By Corollary 5.4, u is a regular word relatively to $V(x)$. We construct the desirable arrangement of brackets inductively: in the arrangement of brackets $[u]$ such that $[u]$ is a regular monomial relatively to $V(x)$ we replace yx^k by $(y)(\mathrm{Ad}\,x)^k$.

Algorithm 11.4
Let u, v be s-regular (ps-regular) words, v a subword of u. The problem is to find an arrangement of brackets $[u]$ on u such that $[v]$ is a part of $[u]$ (where $[v]$ is a s-regular (ps-regular) monomial), $[u]^\circ = \alpha u$, $0 \neq \alpha \in K$. The same factorization problem for u is also under consideration.

In fact, such algorithms were given in the proofs of Lemmas 7.1, 7.4.

Algorithm 11.5
Let $w \in L(X)$. The problem is to give the s-regular presentation of w.

We assume that w is a sum of monomials. Each monomial of the length greater than one has two multipliers. First of all we find the s-regular presentations for these multipliers. Then it is sufficient to consider products of two s-regular monomials. If this product is not an s-regular monomial, then we use the procedure of increasing the least multiplier (2.2). For the details see the proof of Lemma 8.1.

Note that one can obtain s-regular presentation also by the following way. Firstly to rewrite w in associative form. Then to find the leading word of this form, and to consider the element $w - \alpha u$, where u is the corresponding s-regular monomial, $0 \neq \alpha \in K$ is such that $w^\circ = \alpha u^\circ$. After a finite number of steps we get the desirable presentations.

Algorithm 11.6
Let $w \in L(X)$. The problem is to give the presentation of w using a super Hall basis.

It is sufficient to act as in Algorithm 11.5, but with the following ordering on words: firstly by the length, and then lexicographically.

Algorithm 11.7
Let $w \in L^p(X)$. The problem is to write down the ps-regular presentation of w.

We assume that w is a sum of monomials and p^k-powers of such sums. First of all using Algorithm 11.5 we construct s-regular presentations of these monomials. Then using the properties of the p-operation we get terms of p^k-powers, and reduce Lie monomials of these powers to the s-regular form using Algorithm 11.5.

One can reduce w to the ps-regular form by the following way. To rewrite w in the associative form, and then starting with the leading associative word a in w to consider the element $w - \alpha[a]$, where $[a]$ is a ps-regular monomial and $\alpha \in K$ is such that $w^\circ = \alpha[a]^\circ$. After a finite number of steps we obtain the ps-regular form of w.

Algorithm 11.8
Let $w \in A(X)$. The problem is to decide whether $w \in L(X)$ ($w \in L^p(X)$), or not.

We find the leading associative word a in w. If a is not an s-regular (a ps-regular) word, then a is not an element of $L(X)$ (of $L^p(X)$, respectively), here we use Algorithms 11.1, 11.2. Otherwise we consider the element $w-\alpha[a]$, where $[a]$ is an s-regular (a ps-regular) monomial and $\alpha \in K$ is such that $w^\circ = \alpha[a]^\circ$. It is clear that after a finite number of steps we get the solution of the problem.

If K is a field, $\operatorname{char} K = 0$, then to solve the problem for $L(X)$ it is possible to use Friedrichs criterion (Theorem 19.21). Analogously it is possible to use Theorem 19.22 for $L^p(X)$ in the case when $\operatorname{char} K = p > 2$.

For an ℓ-homogeneous element w in the case when $\operatorname{char} K = 0$ it is possible to use Theorem 9.1.

Algorithm 11.9
Let $w \in A(X)$. The problem is to get the presentation of w using a PBW-basis (see Theorem 8.20). In the case when $\operatorname{char} K = p > 2$ we have to find also the PBW-presentation using the structure of $A(X)$ as the restricted universal enveloping algebra of $L^p(X)$.

Let a be the leading word in w. Using Algorithm 11.2 we construct the factorization $a = u_1 \cdots u_l$ where u_i are regular words. Using this factorization we get the unique factorization $a = v_1^{l_1} \cdots v_r^{l_r}$, where v_i are s-regular words, $v_1 < \cdots < v_r$ (if $d(v_i) \in G_-$, then $l_i = 1$).

Consider the element $a = [v_1]^{l_1} \cdots [v_r]^{l_r}$, where $[v_i]$ are s-regular monomials, and denote $a' = a - \alpha b$ with $\alpha \in K$ such that $a^\circ = \alpha b^\circ$. Then we consider the element a', and so on. After a finite number of steps we find the desirable presentation.

In the case when $\operatorname{char} K = p > 2$ we use the presentation of a in the form $a = w_1^{m_1} \cdots w_s^{m_s}$, where w_j are s-regular words, (if $d(w_j) \in G_+$, then $m_j \leq p-1$, and if $d(w_j) \in G_-$, then $m_j = 1$).

Algorithm 11.10
Let w be a G-homogeneous element of $L(X)$, $d(w) \in G_+$. The problem is to decide whether $w = [u,u]$ for some $u \in L(X)$, $d(u) \in G_-$, or not.

We consider the leading monomial a in the s-regular presentation of w (firstly by the length, and then lexicographycally): $w = \alpha a + w^*$, $0 \neq \alpha \in K$. If either $a \neq [b,b]$ where b is a regular monomial, or $\sqrt{\alpha} \notin K$, then w is not a full square. Otherwise if $w = [u,u]$, then the leading term in u is $\sqrt{\alpha}\, b$. We find u in the form $u = \sqrt{\alpha}\, b + u^*$, where u^* is a linear combination of s-regular monomials of the grading weight $d(b)$ and less than b. We consider the system of linear equations (relatively to the coefficients with s-regular monomials) $[w, u] = 0$. By Lemma 10.1 if there are no solutions of this system, then w is not a full square. Otherwise we get the unique solution (up to a coefficient from K), and we have two square roots from w: $\sqrt{\alpha}\, b + u^*$ and $-\sqrt{\alpha}\, b - u^*$.

Examples

In all what follows in this subsection $X = \{x, y, z\}$, $x > y > z$.

Example 11.11
Let $u = y^3zxy^3xyz^2xy^2xy^2xy$. The problem is to obtain the canonical factorization of u into a product of regular words.

We find the longest regular beginning subword of u: y, y^2, y^3, y^3z are regular, and y^3zx can not be a beginning of a regular word, thus the wanted beginning is y^3z. Continuing this procedure we get the following factorization of u:

$$u = y^3z \cdot xy^3xyz^2 \cdot xyy \cdot xyy \cdot xy;$$
$$y^3z < xy^3xyz^2 < xyy = xyy < xy.$$

□

Example 11.12
The problem is to construct the regular arrangement of brackets on a regular word $u = xzy^2zy^2zxz^2yz^3$.

Applying the program **CALS** we find this arrangement

$$[u] = [[[[x, z], [y, [y, z]]], [y, [y, z]]], [[[x, z], z], [[[y, z], z], z]]].$$

□

Example 11.13
Let $d(x), d(z) \in G_+$, $d(y) \in G_-$,

$$u = xyxyzxyz^4xyzxz^2, \qquad v = xyzxyz.$$

The problem is to construct an arrangement of brackets $[u]$ on u induced by the canonical arrangements of brackets $[v]$ on v.

Using Algorithm 11.4 we get the following arrangement of brackets:

$$[u] = [[[x, y], [[[[[x, y], z], [[x, y], z]], z], z], z]], [[[x, y], z], [[x, z], z]]],$$

$[u]^\circ = 2u$. We note that the regular arrangement of brackets $[u]'$ is as follows:

$$[u]' = [[x, y], [[[x, [y, z]], [x, [[[[y, z], z], z], z]]], [[x, [y, z]], [[x, z], z]]]].$$

□

Example 11.14
Let $K = \mathbb{Z}$, $G = \mathbb{Z}_2$, $d(x) = 0$, $d(y) = d(z) = 1$, $L(X)$ be the free Lie superalgebra,

$$\begin{aligned} u &= 20yxz^2x + 2z^2xyx - 20xz^2xy + 10x^2z^2y - 7xz^2 + 9z^2x^2y \\ &\quad + 3z^2y^2x + x^2yz^2 - 9yx^2z^2 - 3z^2xy^2 - 3y^2xz^2 \\ &\quad - 2xyxz^2 - z^2yx^2 - 10yz^2x^2 + 3xy^2z^2 + 7z^2x \end{aligned}$$

The problem is to decide whether $u \in L(X)$, or not.
Applying the program **CALS** we see that

$$\begin{aligned} u &= [x,[x,[[y,z],z]]] + 11[x,[[[x,z],z],y]] + 9[[x,y],[[x,z],z]] \\ &\quad + 3[x,[y,[[y,z],z]]] + 3[[[[x,z],z],y],y] - 7[[x,z],z]. \end{aligned}$$

Thus $u \in L(X)$. □

Comments

Two-way communications between group theory and the theory of Lie algebras proved for many years to be very fruitful. As for the theory of free bases it goes back to pioneering papers of P. Hall [220] (1933, Hall basic commutators), W. Magnus [345], [346] (1935, 1937), and E. Witt [603] (1937) devoted to the study of free groups. The first construction of linear bases of free Lie algebras has been obtained by M. Hall [217] (1950, so called Hall bases). R. Lyndon [339], [340] (1954–55), introduced another type of basic group commutators (Lyndon words).

Bases of free Lie algebras consisting of regular monomials were constructed by A. I. Shirshov in [512] (1958) and by K. T. Chen, R. H. Fox, and R. C. Lyndon in [119] (1958). Remark that these versions of bases are exactly symmetric. For our exposition we prefer to use the Shirshov version. Other variations of linear bases of free Lie algebras could be found, for example, in [74], [111], [183], [202], [368], [377], [495], [496], [497], [515], [537], [586], [587], [588], [589], [590], [595].

We mention also some books and reviews in which these topics were touched: [26], [169], [170], [282], [303], [342], [351], [371], [517].

Note that many combinatorial facts on associative words could be found in [54], [335], [508].

Bases of free color Lie superalgebras (consisting of s-regular monomials) over commutative rings were constructed by A. A. Mikhalev [379], [381], [383], [384]. Independently, such bases for free Lie superalgebras over fields were constructed by A. S. Shtern [527]. An exposition of a super version of Hall bases were given by G. Melançon [370] (see also [148] where some other bases were also considered).

Chapter 3

Free Subalgebras

This chapter deals with the structure of subalgebras of the free color Lie superalgebra and of the free color Lie p-superalgebra. In Section 12 we present P. M. Cohn's weak algorithm which is rather useful for recognizing free subalgebras of the free associative algebra. Section 13 is devoted to elementary properties of reduced subsets. In Section 14 we prove the main theorem of this chapter (Theorems 14.1, 14.2) that any reduced subset of $L(X)$ (of $L^p(X)$, respectively) is an independent subset (Nielsen property). As a corollary we prove that any G-homogeneous subalgebra of $L(X)$ (of $L^p(X)$) is free (Theorems 14.5, 14.9, Schreier property). Theorem 14.21 states that the automorphism group of the free color Lie (p-)superalgebra of a finite rank is generated by elementary automorphisms.

In Section 15 subalgebras of finite rank are under consideration. In particular, we present analogs of Schreier formula, prove that in the case when the main field has prime characteristic the intersection of two subalgebras of finite rank is again a subalgebra of finite rank. We consider also odd components of subalgebras of finite rank.

Section 16 examines properties of subalgebras of infinite rank. A number of sufficient conditions for a subalgebra to be of infinite rank can be found here. For example, almost always the nontrivial even component of $L(X)$ is a subalgebra of infinite rank. Using this fact we prove that fixed points of $L(X)$ under the action of a finite group of linear automorphisms form a subalgebra of infinite rank.

In Section 17 we present algorithms and examples. We mention only Algorithm 17.4 to solve the entry problem, and Algorithm 17.6 to decide whether an endomorphism of the free color Lie (p-)superalgebra of finite rank is an automorphism, or not.

12 Weak algorithm

In this section we discuss the generalization of Euclidean algorithm in the free associative algebra. We also recall P. M. Cohn's technique with weak algorithm and analyze with free subalgebras of the free associative algebra.

Definition 12.1
The family $\{a_i \mid i \in I\}$ of elements of $A(X)$ is called *left ℓ-dependent* if there exist elements $b_i \in A(X)$, almost all zero, such that

$$\ell\left(\sum_i b_i a_i\right) < \max\{\ell(a_i) + \ell(b_i)\},$$

or if some a_i is equal to zero.

Otherwise the family $\{a_i\}$ is called *left ℓ-independent.*

An element $a \in A(X)$ is said to be *left ℓ-dependent on a family* $\{a_i\}$ if either $a_i = 0$, or there exist $b_i \in A(X)$, almost all zero, such that

$$\ell\left(a - \sum_i b_i a_i\right) < \ell(a), \qquad \ell(a_i) + \ell(b_i) \leq \ell(a)$$

for all i. Otherwise a is said to be *left ℓ-independent on a family* $\{a_i\}$.

In the free associative algebra over a field K the following analog of the Euclidean algorithm takes place:

Theorem 12.2 (Weak algorithm, P. M. Cohn [124])
Let K be a field, $n \in \mathbb{N}$, $a_1, \ldots, a_n \in A(X)$, $\ell(a_1) \leq \cdots \leq \ell(a_n)$. If the family $\{a_1, \ldots, a_n\}$ is left ℓ-dependent, then there exists i, $1 \leq i \leq n$, such that the element a_i is left dependent on the family $\{a_1, \ldots, a_{i-1}\}$.

Note that the right version of this algorithm also takes place in $A(X)$.

Since the weak algorithm takes place in $A(X)$, the algebra $A(X)$ is *a free ideal ring (a fir)*, i.e. all left ideals of $A(X)$ are left $A(X)$-modules of unique rank (see [130]). Note that a very elegant proof of this fact is given by P. M. Cohn's theorem that the free product of fir's is again fir. For a left ideal M of $A(X)$ we denote by rank(M) *the rank* of the free left $A(X)$-module M.

In [332] J. Lewin with the use of Schreier technique gave a direct proof of the fact that $A(X)$ is a fir. In particular he obtained the following analog of Schreier's formula.

Theorem 12.3 (J. Lewin [332])
Let $|X| = n$ and let I be a left ideal of $A(X)$ such that

$$\dim_K(A/I) = k < \infty.$$

Then I is a free left $A(X)$-module, rank$(I) = k(n-1)+1$.

Subalgebras of the free associative algebra are not necessary free. For example, if we consider elements x^2 and x^3 in $A(x)$, then $x^2x^3 = x^3x^2$. Hence the subalgebra generated by x^2 and x^3 is not free. Criteria for the subalgebras of free associative algebra to be free were given by P. M. Cohn, for the details see his monograph [130]. For our considerations we need the following sufficient condition.

Let B be a subalgebra of $A(X)$. We consider $A(X)$ as a left B-module relatively to usual multiplication. A system of elements $S = \{s_i \mid i \in I\}$ of $A(X)$ is said to be B-independent if for almost all zero elements a_i

$$\ell\left(\sum_i v_i s_i\right) = \max\{\ell(v_i) + \ell(s_i)\}.$$

Theorem 12.4 (P. M. Cohn [127])
If B is any subalgebra of $A(X)$ such that $A(X)$ is free as left $A(X)$-modulc, with a B-independent basis, then B is a free associative algebra over K. In particular, If B is a subalgebra of $A(X)$ such that $A(X)$ satisfies the weak algorithm as B-module, then B is a free subalgebra of $A(X)$.

At the end of this section we give some examples of nonfree subalgebras of $A(X)$.

Example 12.5 (P. M. Cohn)
If I is any nonzero ideal of $A(X)$ such that the subalgebra B generated by I is different from $A(X)$, then B is not free. □

Example 12.6 (G. M. Bergman)
Let $X = \{x, y, z\}$, δ be the derivation of $A(X)$ given by

$$\delta(x) = xyx + x, \qquad \delta(y) = -yxy - y, \qquad \delta(z) = -x.$$

Then $\operatorname{Ker}\delta$ coincides with the subalgebra generated by

$$a = xyz + x + z, \quad b = yx + 1, \quad c = xy + 1, \quad d = zyx + x + z.$$

Since $ab = cd$, $\operatorname{Ker}\delta$ is not a free subalgebra. □

Example 12.7 (G. M. Bergman)
Let $X = \{x, y_1, y_2, y_3, z\}$, $Y = \{x, y_1, y_3, y_2z, z\}$, $Z = \{x, xy_1, y_1y_2 - y_3, y_2, z\}$, A and B subalgebras of $A(X)$ generated by Y and Z, respectively.

Then A and B are free subalgebras of $A(X)$, but the subalgebra $A \cap B$ is not free. □

13 Reduced subsets

In this section we define reduced subsets of $L(X)$ (of $L^p(X)$) and consider their elementary properties. The main attention is paid for the subsets $Z(x)$ and $W(z)$ of $L(X)$ and the subset $ZP(x)$ of $L^p(X)$. We prove that these subsets are independent subsets. These subsets are important tools in our further proofs with the aim to rewrite elements to get lesser length.

In all what follows in this section let K be a field, $\operatorname{char} K \neq 2$.

Subsets $Z(x)$ and $W(z)$

Definition 13.1
We say that a subset M of G-homogeneous elements of $L(X)$ (of $L^p(X)$) is *independent* if M is a set of free generators of the subalgebra, generated by M in $L(X)$ (in $L^p(X)$, respectively).

Definition 13.2
Let $x \in X_+$, $z \in X_-$. The following notation is used systematically:

$$\begin{aligned} Z(x) &= \{y,\ y(\operatorname{Ad} x)^n \mid y \in X \setminus \{x\},\ n \in \mathbb{N}\} \subseteq L(X); \\ W(z) &= \{y,\ [y,z],\ [z,z] \mid y \in X \setminus \{x\}\} \subseteq L(X). \end{aligned}$$

We consider a well-ordering on X. Let x be the least element of X. We extend this ordering to a partial ordering $\rhd$ on $S^+(X)$ in the following way: for $a, b \in S^+(X)$ we have $a \lhd b$ if and only if $m(a) = m(b)$ and a has a lesser number of inversions. It is clear that this is a semigroup ordering, i.e. $a \rhd b, c \unrhd d$ or $a \unrhd b, c \rhd d$ imply that $ab \rhd cd$.

Lemma 13.3 ([383])
Let $u \in Z(x)$. Then $u = \widetilde{u} + u^*$ where

$$u^* = \sum_{j=1}^{t} \alpha_j u_j,$$

$\alpha_j \in K$, $u_j \in S^+(X)$, $m(u_j) = m(u)$, $\widetilde{u} \rhd u_j$ for $1 \leq j \leq t$.

PROOF Since $x < y$, u is a regular monomial. By Lemma 6.6, $u = \widetilde{u} + u^*$ where

$$u^* = \sum_{j=1}^{t} \alpha_j u_j,$$

$\alpha_j \in K$, $u_j \in S^+(X)$, $m(u_j) = m(u)$, and $\widetilde{u} > u_j$ for $1 \leq j \leq t$. But, for our u and u_j, the inequalities $\widetilde{u} > u_j$ and $\widetilde{u} \rhd u_j$ are equivalent. ■

Using the induction on the length one can easily prove the following lemma.

Lemma 13.4 ([383])
Let x be the least element in X and w be a regular monomial, $w \neq x$. Then w is a monomial in the elements of $Z(x)$. Moreover, if $x \in X_-$ then $w = 2^{-k}u$, where $k \in \mathbb{Z}_+$, u is a monomial in the elements from $W(x)$.

Corollary 13.5 ([383])
Let x be the least element in a well-ordered set X, w an s-regular monomial, $w \neq x$, $\ell_X(w) = t \neq 0$. Then:

(i) *if either $x \in X_+$ or $w \neq [x,x]$ with $x \in X_-$, then w is a monomial in the elements of the set $Z = Z(x)$, $\ell_Z(w) = \ell_X(w) - t < \ell_X(w)$;*

(ii) *if $x \in X_-$ then $w = 2^{-k}u$, where $k \in \mathbb{Z}_+$ and u is a monomial in the elements of the set $W = W(x)$, $\ell_W(w) < \ell_X(w)$.*

Observe that if in Claim 2 of Corollary 13.4 one has $m(w_1) = m(w_2)$, then nevertheless it is possible to have $\ell_W(w_1) \neq \ell_W(w_2)$. Indeed, let $y_1 > y_2 > x$, $w_1 = \big[[y_1,x],[y_2,x]\big]$ and $w_2 = \big[\big[[y_1,x],x\big],y_2\big]$. Then w_1, w_2 are regular monomials, $m(w_1) = m(w_2)$, but $\ell_{W(x)}(w_1) = 2 \neq 3 = \ell_{W(x)}(w_2)$.

Lemma 13.6 ([383])
Consider a value of the word f on the elements of the sets Z and W: $u = f(z_i)$, $v = f(w_i)$, $z_i \in Z$, $w_i \in W$. Then the leading terms of u and v (relative to the ordering $\rhd$) have the following form: $u^\circ = f(\widetilde{z_i})$, $v^\circ = 2^l f(\widetilde{w_i})$, where $l \in \mathbb{Z}_+$.

PROOF Since $y > x$, $yx \rhd xy$. We remark that $[x,x] = 2xx$, $[y,x] = yx - \varepsilon(y,x)\,xy$, where $yx \rhd xy$ since $y > x$. Substituting z_i and w_i, respectively, into f and recalling Lemma 13.3 and the semigroup property of $\rhd$, we complete the proof. ∎

Theorem 13.7 ([383])
Let $L(X)$ be a free color Lie superalgebra, $x \in X$, $y \in X_-$. Then $Z = Z(x)$ and $W = W(y)$ are independent subsets of $L(X)$.

PROOF The existence of a nontrivial associative relation among elements of Z (of W, respectively) gives us a contradiction with the statements of Lemmas 13.6, 5.2. ∎

Corollary 13.8
Let $x \in X$, $L(X)$ be the free color Lie superalgebra. If $x \in X_+$, then $L(X)$ is the semidirect product of the subalgebras $L(\{x\})$ and $L(Z(x))$ with the action of $L(\{x\})$ on $L(Z(x))$ given by inner derivations. If $x \in X_-$, then $L(X)$ is the semidirect product of the subalgebras $L(\{x\})$ and $L(W(x))$ with the same adjoint action.

Subsets $ZP(x)$ and $W(z)$

In all what follows in this subsection let K be a field, $\operatorname{char} K = p > 2$.

Definition 13.9
Let $x \in X_+$. We use the notation

$$ZP(x) = \{y,\ x^p,\ y(\operatorname{Ad} x)^n \mid y \in X \setminus \{x\},\ n = 1, \ldots, p-1\} \subseteq L^p(X).$$

By analogy with the case of $L(X)$ we have the following properties of the subsets $ZP(x)$ and $W(x)$ of $L^p(X)$.

Lemma 13.10 ([386])
Let $x \in X_+$ be the least element in the totally ordered set X and $u \in ZP(x)$. Then

$$u = \widetilde{u} + \sum_i \alpha_i u_i,$$

where u_i are associative words, $m(u_i) = m(u)$, $0 \neq \alpha_j \in K$, $\widetilde{u} \triangleright u_i$ for all i.

Let $z \in X_-$ be the least element of the totally ordered set X and $u \in W(z)$. Then

$$u = \alpha\widetilde{u} + \sum_i \alpha_i u_i,$$

where u_i are associative words, $m(u_i) = m(u)$, $0 \neq \alpha_i \in K$, $\alpha = 1$ or $\alpha = 2$ and $\widetilde{u} > u_i$ for all i.

Lemma 13.11 ([386])
The subsets $ZP(x)$ and $W(z)$ are independent subsets of $L^p(X)$.

Lemma 13.12 ([386])
Let x be the least element of the totally ordered set X, w a ps-regular monomial on X such that $w \neq x$ and $\ell_X(w) \neq 0$. Then:

(i) *if $x \in X_+$ then $w = u$ in $L^p(X)$, where u is a monomial in elements of the set $ZP(x)$ and $\ell_{ZP(x)}(w) < \ell_X(w)$;*

(ii) *if $x \in X_-$, then $w = 2^{-k}u$ in $L^p(X)$, where $k \in \mathbb{Z}_+$, u is monomial in elements of the set $W(x)$ and $\ell_{W(x)}(w) < \ell_X(w)$.*

We remark that if, in Claim 1 of Lemma 13.12, we assume that $m(w_1) = m(w_2)$, then it is possible that $\ell_{ZP(x)}(w_1) \neq \ell_{ZP(x)}(w_2)$. Indeed, let $y_1 > y_2 > x$, $w_1 = \left[y_1(\operatorname{Ad} x)^{p-1}, [y_2, x]\right]$ and $w_2 = [y_1(\operatorname{Ad} x)^p, y_2]$. Then w_1, w_2 are regular monomials and $m(w_1) = m(w_2)$, but $\ell_{ZP(x)}(w_1) = 2 \neq 3 = \ell_{ZP(x)}(w_2)$.

Corollary 13.13
Let $x \in X$, $L(X)$ be the free color Lie p-superalgebra. If $x \in X_+$, then

$L^p(X)$ is the semidirect product of the subalgebras $L^p(\{x\})$ and $L^p(ZP(x))$ with the action of $L^p(\{x\})$ on $L^p(ZP(X))$ given by inner derivations. If $x \in X_-$, then $L^p(X)$ is the semidirect product of the subalgebras $L^p(\{x\})$ and $L^p(W(x))$ with the same adjoint action.

Reduced subsets

Definition 13.14
A subset $M = \{a_i\}$ of G-homogeneous elements in $L(X)$ is called *reduced* if for any i the leading part $\widehat{a_i}$ of the element a_i does not belong to the subalgebra generated by the set $\{\widehat{a_j} \mid j \neq i\}$. By analogy we define *$\varphi$-reduced subsets.*

Definition 13.15
Let $S = \{s_\alpha \mid \alpha \in I\}$ be a G-homogeneous subset of $L(X)$ (of $L^p(X)$). We say that a mapping $\omega: S \to L(X)$ ($\omega: S \to L^p(X)$, respectively) is *an elementary transformation of S* if all $\omega(s_\alpha)$ are G-homogeneous elements, $d(\omega(s_\alpha)) = d(s_\alpha)$ for all $\alpha \in I$, and either ω is a non-singular linear transformation applied to S, or there exist $\beta \in I$ such that $\omega(s_\alpha) = s_\alpha$ for all $\alpha \in I \setminus \{\beta\}$, and $\omega(s_\beta) = s_\beta + f_\omega$, where f_ω is an G-homogeneous element of the subalgebra in $L(X)$ (in $L^p(X)$, respectively) generated by the subset $\{s_\alpha \mid \alpha \in S \setminus \{\beta\}\}$, $d(f_\omega) = d(s_\beta)$.

Such elementary transformation is called *triangular* if $\ell(f_\omega) \leq \ell(s_\beta)$, and *$\varphi$-triangular* if $\varphi(f_\omega) \leq \varphi(s_\beta)$.

It is clear that elementary transformations of the set X of free generators induce automorphisms of $L(X)$ (of $L^p(X)$, respectively). Such automorphisms are called elementary.

Lemma 13.16 ([383], [386])
Let $H\{h_1, \ldots, h_q\}$ be a reduced subset of ℓ-homogeneous elements of $L(X)$ (of $L^p(X)$, respectively), $Y = \{y_1, \ldots, y_q\}$, $d(y_i) = d(h_i)$ with $i = 1, \ldots q$. Assume that there exists nonzero G-homogeneous element f of $L(Y)$ (of $L^p(Y)$, respectively) such that $f(h_1, \ldots, h_q) = 0$. Let also ω be an elementary transformation of H, $\omega(h_i) = h_i$ for $i \neq 1$, $\omega(h_1) = h_1 + s(h_2, \ldots, h_q)$, $s \in L(Y)$, $\ell(s(h_2, \ldots, h_q)) = \ell(h_1)$. Then $\omega(H) = \{\omega(h_1),\ h_2, \ldots, h_q\}$ is a reduced set and there exists a nonzero G-homogeneous element $g \in L(Y)$ ($g \in L^p(Y)$) such that $g(\omega(h_1), h_2, \ldots, h_q) = 0$.

PROOF The assertion of the lemma follows immediately from the fact that H is reduced and consists of ℓ-homogeneous elements. ■

Lemma 13.17 ([383], [386])
Let M be as in Lemma 13.16, $M' = \{b_1, \ldots, b_q\} \subseteq M$. Suppose that there exists $f \in L(Y)$ ($f \in L^p(X)$, respectively), $f \neq 0$ and at the same time $f(b_1, \ldots, b_q) = 0$. Then there exists a finite reduced multihomogeneous subset $M_1 = \{c_1, \ldots, c_q\}$ of $L(X)$ (of $L^p(X)$, respectively) such that

(i) $\ell_X(c_i) = \ell_X(b_i)$, $1 \leq i \leq q$;
(ii) $g(c_1, \ldots, c_q) = 0$ *for some nonzero G-homogeneous* $g \in L(Y)$ ($g \in L^p(X)$, *respectively*).

PROOF When we apply elementary transformations to M' it is sufficient to follow only the leading parts of elements relatively the multidegree, and to use Lemma 13.16. ■

Note that the assertions of Lemmas 13.16, 13.17 are valid also for φ-reduced subsets.

14 Main theorem

In this section we prove the main theorem of this chapter: any reduced subset of the free color Lie (p-)superalgebra is an independent subset (Theorems 14.1, 14.2, Nielsen property). As a corollary we prove that any subalgebra of the free color Lie (p-)superalgebra is free (Theorems 14.5, 14.9, Schreier property). We prove also that the automorphism group of these free algebras of finite rank is generated by elementary automorphism (Theorem 14.21).

Nielsen property

Theorem 14.1 ([380], [381], [383])
Let K be a field, $\operatorname{char} K \neq 2$. *Then any reduced subset (φ-reduced subset) of the free color Lie superalgebra $L(X)$ is independent.*

PROOF We consider only the case when $\varphi = \ell$. One can give the proof in the general case along the same lines.

Let M be a reduced subset in $L(X)$. Assume, on the contrary, that for some finite subset $M' = \{b_1, \ldots, b_q\}$ of M there exists a nontrivial relation $0 \neq f(y_1, \ldots, y_q) \in L(Y)$, $d(y_i) = d(b_i)$, $f(b_1, \ldots, b_q) = 0$. By Lemma 13.17 we may assume that M' is multihomogeneous. Let $l_k = \ell_X(b_k) > 1$ for some $b_k \in M'$. Since M' is multihomogeneous, there exists $x \in X$ such that $\ell_X(b_k) \neq 0$, $\beta x \neq b_m$, $1 \leq m \leq q$, $0 \neq \beta \in K$. Changing the ordering of X, if necessary, we may assume that x is the least element of X. Consider now the s-regular presentation of b_i for all i. Since b_k is ℓ-homogeneous and $l_k > 1$, an element βx with $0 \neq \beta \in K$ can not encounter among the summands of the s-regular presentation of b_k. On the other hand, as b_m, $b_m \neq b_k$, is multihomogeneous and $\beta x \neq b_m$, $0 \neq \beta \in K$, the elements βx can not encounter in the s-regular presentation of b_m, $m \neq k$. By Corollary 13.5, if $x \in X_+$ then all elements b_i are polynomials in the elements of $Z = Z(x)$, $\ell_Z(b_k) = \ell_X(b_k) - \ell_X(b_k) < l_k$ and $\ell_Z(b_m) \leq \ell_X(b_m) = l_m$ for $b_m \neq b_k$, $1 \leq m \leq q$. By Corollary 13.5, the elements b_i, $1 \leq i \leq q$, are also

ℓ-homogeneous relative to Z. Since $f(b_1, \ldots, b_q) = 0$, by Lemma 13.17 one can assume that M' is multihomogeneous relative to Z.

If now $x \in X_-$ then, by Corollary 13.5, all elements b_i are polynomials in the set $W = W(x)$, $\ell_W(b_k) < l_k$ and $\ell_W(b_m) \leq l_m$ for $b_m \neq b_k$, $1 \leq m \leq q$. Applying elementary transformations to the set $\{b_1, \ldots, b_q\}$ we come to a reduced set $C = \{c_1, \ldots, c_q\}$ (relative to W) with a nontrivial relation. Next we consider the following triangular elementary transformations (relative to W): if, for example, $\widehat{b_1} = -g(\widehat{b_2}, \ldots, \widehat{b_q})$, $\widehat{b_i}$ is the leading part of b_i in the alphabet W, then we apply the triangular elementary transformation ω, where $\omega(b_i) = b_i$ for $i \neq 1$ and $\omega(b_1) = b_1 + g(b_2, \ldots, b_q)$. Such transformation are also a triangular elementary transformation relative to X. By Lemma 13.16 $\{\omega(b_1), b_2, \ldots, b_q\}$ is X-reduced and satisfies to a nontrivial relation. After a finite number of such transformations we come to a set C of desired form which is X-reduced and W-reduced at the same time. By Lemma 13.17 assume that C is W-multihomogeneous.

Thus, after a finite number of such steps we come to a reduced subset $\{b_i \mid 1 \leq i \leq q\}$ such that $l_i = 1$ for $1 \leq i \leq q$, i.e. $b_i = \alpha_i x_i$, $0 \neq \alpha_i \in K$, $x_i \in X$ for $1 \leq i \leq q$. This is in contradiction with the existence of a nontrivial relation between the elements b_i, $1 \leq i \leq q$. ■

Theorem 14.2 ([384], [386])
Let K be a field, $\operatorname{char} K = p > 2$. Then any reduced subset (φ-reduced subset) of the free color Lie p-superalgebra $L^p(X)$ is independent.

PROOF We present the proof for reduced subsets (the proof for φ-reduced subsets is analogous).

By the way of contradiction assume that M is a reduced subset of $L^p(X)$ such that there exists a nontrivial relation between elements of M. By Lemma 13.17 we may assume that there is a nontrivial relation among multihomogeneous elements of a reduced subset $M_2 = \{b_1, \ldots, b_k\}$ and, moreover, $\ell_X(b_j) = \ell_X(a_j)$ for $1 \leq j \leq k$. Since M_2 is reduced and multihomogeneous, there exists an element $x \in X$ such that $b_i \neq x$ for all $i = 1, \ldots, k$ and $\ell_x(b_{i_0}) \neq 0$ for some i_0, $1 \leq i_0 \leq k$.

One can assume that x is the least element in X. By Lemma 13.10 all b_i can be written as Lie polynomials in the free generators of $ZP(x)$ in the case $x \in X_+$, and of $W(x)$ in the case $x \in X_-$,

$$\ell_{ZP(x)(W(x))}(b_i) \leq \ell_X(b_i)$$

for all i and

$$\ell_{ZP(x)(W(x))}(b_{i_0}) < \ell_X(b_{i_0}).$$

It follows from Lemma 13.16 that the set $M_2 = \{b_1, \ldots, b_k\}$ can be reduced by triangular elementary transformations to a reduced set $C = \{b'_1, \ldots, b'_k\}$ in the alphabet $ZP(x)$ (in $W(x)$, respectively) with a nontrivial relation,

and

$$\ell_X(b_i') = \ell_X(b_i), \qquad \ell_{ZP(x)(W(x))}(b_i') \le \ell_{ZP(x)(W(x))}(b_i).$$

By Lemma 13.17 we may assume that C is a multihomogeneous set. Repeating this procedure we see that there exists a nontrivial relation among the elements of the length one from a reduced multihomogeneous subset of a free color Lie p-superalgebra. This contradiction completes the proof. ∎

Lemma 14.3
Let M be a φ-reduced subset of $L(X)$ (of $L^p(X)$, respectively). Then the leading part of any element, which is a (p-)Lie polynomial in the elements of M, is a (p-)Lie polynomial in leading parts of elements in M.

PROOF Let h be a polynomial on elements $b_1, \dots, b_l$ of a reduced subset. We consider the free color Lie (p-)superalgebra $L(Z)$ ($L^p(X)$), where $Z = \{z_1, \dots, z_l\}$, $d(z_i) = d(b_i)$, $\varphi(z_i) = \varphi(b_i)$, $1 \le i \le l$. By Theorems 14.1, 14.2 the subset $Z = \{b_1, \dots, b_l\}$ is an independent subset. Hence we have the unique presentation of h in the form $h = F(b_1, \dots, b_l)$, where $F(z_1, \dots, z_l) \in L(Z)$ ($F(z_1, \dots, z_l) \in L(Z)$). Let $\widehat{F}(z_1, \dots, z_l)$ be the leading part of $F(z_1, \dots, z_l)$ relatively to φ. Since $\{\widehat{b_1}, \dots, \widehat{b_l}\}$ is a reduced subset, by Theorems 14.1, 14.2 the subset $\{\widehat{b_1}, \dots, \widehat{b_l}\}$ is an independent subset. Thus

$$\widehat{h} = \widehat{F}(\widehat{b_1}, \dots, \widehat{b_l}).$$

∎

Schreier property

Lemma 14.4
Let B be a G-homogeneous subalgebra of $L(X)$. Then there exists a reduced subset M in B such that M generates B.

PROOF We use A. G. Kurosh's method to construct the set M:

$$M = \bigcup_{j=0}^{\infty} M_j,$$

where M_j consists of G-homogeneous elements with length j, if non-empty. We set $M_0 = \emptyset$. Then, by induction, we suppose that $M_0, \dots, M_i$ have been defined. Let B_i be the subalgebra of B, generated by

$$\bigcup_{j=0}^{i} M_j$$

and $S_i = \{b \in B_i \mid \ell(b) \leq i+1\}$. Consider the G-homogeneous subsets of $\{b \in B \mid \ell(b) = i+1\}$ which are linearly independent modulo the subspace S_i, and let M_{i+1} be a maximal such subset. It is clear that M generates the subalgebra B.

Let

$$f \in \bigcup_{j=0}^{i+1} M_j.$$

Suppose that the leading part $\widehat{f}$ of f belongs to the subalgebra, generated by the set $\{\widehat{h} \mid h \in M,\, h \neq f\}$. It is obvious that $\widehat{f}$, in fact, belongs to the subalgebra, generated by the set

$$\left\{ \widehat{h} \;\middle|\; h \in \bigcup_{j=0}^{i+1} M_j,\ h \neq f \right\}.$$

Hence

$$\widehat{f} = \alpha_1 \widehat{f_1} + \cdots + \alpha_k \widehat{f_k} + \omega(\widehat{f_{k+1}}, \ldots, \widehat{f_s}),$$

where $f_1, \ldots, f_k \in M_{i+1}$ and

$$f_{k+1}, \ldots, f_s \in \bigcup_{j \leq i} M_j, \qquad \ell(w(\widehat{f_{k+1}}, \ldots, \widehat{f_s})) < i+1.$$

Considering relations among elements we derive

$$f = \alpha_1 f_1 + \cdots + \alpha_k f_k + \omega(f_{k+1}, \ldots, f_s) + g,$$

where g is an element with $\ell(g) \leq i$. Hence $c \in B$ and therefore $c \in S_i$. Thus we have a contradiction with the definition of M_{i+1}, and M is reduced. ∎

Theorem 14.5 ([380], [381], [383])
Let K be a field, $\operatorname{char} K \neq 2$, and let $L(X)$ be the free color Lie superalgebra, B a G-homogeneous subalgebra of $L(X)$. Then B is a free color Lie superalgebra.

PROOF By Lemma 14.4 we may choose a reduced generating set M of B. By Theorem 14.1, M is independent. Therefore $B = L(M)$ is the free color Lie superalgebra with the set M of free generators. ∎

Remark 14.6
It is possible to prove Theorem 14.5 using Theorem 12.4 and the fact that $A(X)$ is the universal enveloping algebra of $L(X)$. For Lie algebras the idea of such proof was suggested by P. M. Cohn in [127]; for Lie superalgebras see also A. I. Shtern's article [527]. For convenience of the reader we expose here P. M. Cohn's proof (it works also for color Lie superalgebras).

Let B be a subalgebra of $L(X)$ and $E = \{e_i\}$ a K-linear basis of B. We consider a complement $F = \{f_j\}$ of E with respect to a basis of $L(X)$. Using Theorem 8.20 we see that the monomials

$$f_I e_J = f_{i_1} \cdots f_{i_r} \cdot e_{j_1} \cdots e_{j_s}$$

with $I = (i_1, \ldots, i_r)$, $J = (j_1, \ldots, j_s)$, $i_1 \leq \cdots \leq i_r$, $j_1 \leq \cdots \leq j_s$, form a K-linear basis of $A(X)$, and the system E generates $A(X)$ as a left B-module. Moreover,

$$\begin{aligned} \ell(f_I \cdot e_J) &= \sum_{t=1}^{r} \ell(f_{i_t}) + \sum_{q=1}^{s} \ell(e_{j_q}); \\ \ell\Big(\sum \alpha_{IJ} f_I e_J\Big) &= \max_{I,J}\{\ell(f_I) + \ell(e_J)\} \end{aligned}$$

where $0 \neq \alpha_{IJ} \in K$.

Hence the system E is left $U(B)$-independent. Thus by Theorem 12.4 $U(B)$ is a free associative algebra, and it is possible to choose a set of free generators Y of $U(B)$ such that $Y \subseteq B$. It is clear that $B = L(Y)$ is the free Lie algebra with the set Y of free generators.

Definition 14.7
The G-homogeneous subalgebra B of a color Lie superalgebra L over K is called isolated if $\{a \in L \mid (K \cdot a) \cap B \neq 0\}$.

Theorem 14.8 ([394])
Let K be a commutative domain without 2-torsion over which all projective modules are free, L the free color Lie superalgebra over K, B an isolated subring of L which is a direct summand of the free K-module L. Then B is a free color Lie superalgebra over K.

PROOF We say that a G-homogeneous subset M of nonzero elements of L is reduced if for each $a \in M$ no $\alpha a'$ belongs to the subalgebra, generated by $\{b' \mid b \in M, b \neq a\}$ where a' and b' are the leading parts of a and b and $0 \neq \alpha \in K$. Using the fact that B is an isolated subalgebra of L and applying A. G. Kurosh's method (Lemma 14.4) we get a reduced G-homogeneous generating set M for the subalgebra B. Now we embed K in its field of quotients $Q(K)$. Let $L(M)$ be color Lie superalgebra generated by M in the free color Lie superalgebra $L_Q = L \otimes Q(K)$ over $Q(K)$. Since M is reduced in L, we see that M is reduced in L_Q. Applying Theorem 14.1 we complete the proof. ■

Theorem 14.9 ([384], [386])
Let K be a field, $\operatorname{char} K = p > 2$, $L^p(X)$ the free color Lie p-superalgebra, B a G-homogeneous subalgebra of $L^p(X)$. Then B is a free color Lie p-superalgebra.

PROOF One can prove along the same line as in the proof of Lemma 14.4 that there exists a reduced generating set M of B. By Theorem 14.9, M is independent. Therefore $B = L(M)$ is the free color Lie superalgebra with the set M of free generators. ■

Remark 14.10
In [286] G. V. Kryazhovskih made an attempt to develop Schreier technique in the case of subalgebras of free Lie algebras (unfortunately, there are some inaccuracies in the exposition). It would be interesting to develop in full generality Schreier type technique for subalgebras of free color Lie (p-)superalgebras.

Remark 14.11
Subalgebras of free solvable (polynilpotent) Lie algebras are not necessary free. Free subalgebras were studied in [90], [91], [92], [174], [175], [236], [518], [519], [520]. For automorphisms of such algebras see [31], [100], [159], [161], [457]. It would be also interesting to consider subalgebras and automorphisms of free solvable and free polynilpotent color Lie superalgebras.

Remark 14.12
O. G. Kharlampovich in [269] considered Lyndon property in free Lie algebras and in free solvable Lie algebras.

Recall that a variety of Lie algebras has *Lyndon property* if for any free algebra L of this variety of the rank n and an ideal J of L generated by k elements, $k < n$, there exists a subalgebra B of L such that B is a free Lie algebra of this variety of the rank $n - k$ and $B \cap J = 0$.

Theorem 14.13 (O. G. Kharlampovich [269])
Let $|X| = n < \infty$, $L(X)$ be the free Lie algebra over a field, $k < n$, $r_1, \ldots, r_k \in L(X)$, J the ideal of $L(X)$ generated by $r_1, \ldots, r_k$. Then there exists a subset Y of X such that

$$L(Y) \cap (J + L(X)^n) = L(Y)^n.$$

It follows from this theorem that the variety of all Lie algebras and the variety of all solvable Lie algebras of degree m have Lyndon property.

We mention also the following result.

Theorem 14.14 (L. M. Shneerson [521])
Let F be an associative (Lie) algebra over a field given by $n + k$ generators and k defining relations without constant terms. Then if F has a generating set S, $|S| = n$, then F is the free associative (Lie) algebra with the free generating set S.

Remark 14.15
A restricted Lie algebra is called *almost free* if it contains a free subalgebra which is an ideal of finite codimension. An element of a restricted Lie algebra is called algebraic if it generates a finite dimensional subalgebra.

Theorem 14.16 (G. P. Kukin [296])
Every almost free Lie p-algebra without nonzero algebraic elements is free.

Remark 14.17
$\mathbb{Z}$-graded free Lie superalgebras appears in homology and homotopy theory. It is possible to prove that $\mathbb{Z}$-graded subalgebras of free $\mathbb{Z}$-graded Lie superalgebras are free using methods of this theory, see [16], [17, Chapter 5] by L. L. Avramov and [330, Proposition A.1.10] by J. M. Lemaire.

Remark 14.18
V. A. Artamonov proves in [11] that projective metabelian color Lie superalgebras are free.

Problem 14.19 ([156], Problem 2.62)
Does the problem of conjugacy of subalgebras of free Lie algebras relative to the automorphism group have an algorithmic solution, or not?

Remark 14.20
Using the theorem on freeness of subalgebras A. Baudisch in [44] studied the elementary theory of a nontrivial free Lie algebra and proved that this theory is unstable and has the strict order property (in [42], [43] he studied the elementary theory of free nilpotent Lie algebra and proved that it is ω-stable).

Generators of the automorphism group

The following theorems give superanalogs of P. M. Cohn theorem on generators of the automorphism group of a free Lie algebra of finite rank.

Theorem 14.21 ([387])
The set of elementary automorphisms generates the automorphism group of a free color Lie (p-)superalgebra of finite rank.

PROOF Let $|X| < \infty$ and $L(X) = L(Y)$. By the invariance of the rank we have $|X| = |Y|$. We transform $Y = \{y_1, \ldots, y_n\}$ by elementary transformations to some X-reduced set D in the following way: if, for example, $\widehat{y_n} = g(\widehat{y_1}, \ldots, \widehat{y_{n-1}})$, then $\omega(y_i) = y_i$ and $\omega(y_n) = y_n - g(y_1, \ldots, y_{n-1})$. Since Y is an independent subset, $\omega(y_n) \neq 0$. Thus, by finitely many steps, we come to a set $D = \{d_1, \ldots, d_n\}$. We have

$$x_i = g_i(d_1, \ldots, d_n), \qquad i = 1, \ldots, n, \quad g_i \in L(T). \tag{3.1}$$

Since D is reduced, by Lemma 14.3 the leading part of $g_i(d_1, \ldots, d_n)$ is a Lie polynomial in leading parts $\widehat{d_i}$, $1 \leq i \leq n$. Therefore, it follows from (3.1) that d_i is a linear combinations of generators $x_i \in X$. By a non-degenerate linear transformation we can transform D into X. ∎

Remark 14.22
To choose generators of the automorphism group it is possible to consider only the following "elementary" automorphisms: permutations of x_i together with the automorphisms

$$x_1 \to \lambda x_1, \quad x_i \to x_i \qquad \text{with } 0 \neq \lambda \in K,\ i \neq 1,$$

and

$$x_1 \to x_1 + \omega(x_2, \ldots, x_n), \quad x_i \to x_i \qquad \text{with } i \neq 1.$$

Moreover, sometimes it is possible to use only elementary automorphisms of special type.

Theorem 14.23 (I. V. Chirkov [120])
Let K be an algebraically closed field, $4 \leq |X| < \infty$, $L(X)$ the free Lie algebra over K. Then the automorphism group of $L(X)$ is generated by the linear automorphisms and the automorphism τ defined by

$$\tau(x) = x + [y, z], \qquad \tau(t) = t$$

with fixed $x, y, z \in X$ and any $t \in X \setminus \{x\}$.

In connection with Theorem 14.21 it is interesting to consider generators of the automorphism group of the free associative algebra.

Problem 14.24 (P. M. Cohn)
Is it true that the automorphism group of the free associative algebra of finite rank is generated by the elementary automorphisms?

As for the free associative algebra of rank 2 see [140], [352].

15 Finitely generated subalgebras

In this section we study properties of finitely generated subalgebras of free color Lie (p-)superalgebras. In particular, we show that free color Lie (p-)superalgebras of finite rank have Hopf property, obtain analogs of O. Schreier's formulas, prove that the intersection of two subalgebras of finite rank in $L^p(X)$ is again a subalgebra of finite rank. Theorem 15.14 shows that a subalgebra of finite rank of $L(X)$ with nontrivial odd component is completely defined by it.

Reduced finite subsets

Theorem 15.1 (Hopf Property)
Let $X = \{x_1, \ldots, x_n\}$, and let $f_1, \ldots, f_n$ be G-homogeneous elements of $L(X)$ (of $L^p(X)$). If the subset $\{f_1, \ldots, f_n\}$ generates $L(X)$ ($L^p(X)$, respectively), then it generates the algebra freely.

PROOF Using elementary transformations we can reduce the subset $\{f_1, \ldots, f_n\}$ to a reduced subset $\{f'_1, \ldots, f'_m\}$. Since $f_1, \ldots, f_n$ generate $L(X)$ ($L^p(X)$), it is possible to realize elements $x_i, 1 \leq i \leq n$, as polynomials on elements $f'_1, \ldots, f'_m$. By Lemma 14.3 we get that $m = n$ and there exists a linear elementary transformation of the subset $\{f'_1, \ldots, f'_n\}$ such that $f''_i = x_i$, $1 \leq i \leq n$. Bearing in mind that elementary transformations are invertible we complete the proof. ■

Theorem 15.2
Let $X = \{x_1, \ldots, x_n\}$, and let ψ be an endomorphism of $L(X)$ (of $L^p(X)$), $\{\psi(x_1), \ldots, \psi(x_n)\}$ a reduced subset. If there exists i such that $\ell(\psi(x_i)) > 1$, then ψ is not an automorphism of $L(X)$ (of $L^p(X)$, respectively).

PROOF By the way of contradiction assume that $\{\psi(x_1), \ldots, \psi(x_n)\}$ is a set of free generators of $L(X)$ (of $L^p(X)$). Then by Theorem 14.21 there exists a sequence of triangular elementary transformations which reduce the subset $\psi(X)$ to the set X. But $\psi(X)$ is a reduced subset, $\ell(\psi(x_i)) > 1$, and using Lemma 14.3 we get a contradiction. ■

Remark 15.3
Any finitely generated subalgebra B of $L(X)$ (of $L^p(X)$ has a free generating set $\{f_1, \ldots, f_m\}$ with the minimal number $\varphi(f_1) + \cdots + \varphi(f_m)$. Indeed, it is sufficient using triangular elementary transformations to reduce any finite set of generators of B to the reduced subset $\{f_1, \ldots, f_m\}$.

Schreier formulas

Theorem 15.4 ([383])
Let K be a field, $\operatorname{char} K \neq 2$, $L = L(X) = L_+ \oplus L_-$ the free color Lie superalgebra, $\operatorname{rank}(L) = |X| = N < \infty$, and let M be a G-homogeneous subalgebra, $M = M_+ \oplus M_-$, $M_+ = L_+$, $\dim L_-/M_- = s < \infty$. Then $\operatorname{rank}(M) = 2^s(N-1)+1$.

PROOF We use the induction on s. The basis of induction: $s = 0$. In this case $M = L$, $\operatorname{rank}(M) = N$. Suppose now that our assertion takes place for all $s < n$ and let now $s = n$. We choose a reduced subset H of generators of M. Consider the set of leading parts $\widehat{H} = \{\widehat{h} \mid h \in H\}$. By Theorem 14.1 the set $\widehat{H}$ is independent and $|\widehat{H}| = \operatorname{rank}(L(\widehat{H})) = \operatorname{rank}(M)$.

It is clear that $\dim L_+/M_+ = \dim L_+/L_+(\widehat{H}) = 0$, i.e. $L_+(\widehat{H}) = L_+$. Thus it is sufficient to prove the formula for $\operatorname{rank}(L(\widehat{H})) = |\widehat{H}|$. We may assume that $L(\widehat{H}) \subseteq L(W(z))$ for some $z \in X_-$ Since $|W(z)| = 2(N-1)+1$, by the induction hypothesis, we obtain the wanted result. ■

Corollary 15.5
Let K be a field, $\operatorname{char} K \neq 2$, $L = L_+ \oplus L_-$ the finitely generated color Lie

superalgebra, and let $M = M_+ \oplus M_-$ *be a* G*-homogeneous subalgebra of* L *such that* $M_+ = L_+$ *and* $\dim L_-/M_- < \infty$. *Then* M *is a finitely generated color Lie superalgebra.*

PROOF It is sufficient to represent L as a factor-algebra of some free color Lie superalgebra $L(X)$ of finite rank and to consider the full inverse image of M in $L(X)$. ■

Using the same arguments for free Lie p-superalgebras we get the following result.

Theorem 15.6 ([386])
Let K *be a field,* $\operatorname{char} K = p > 2$, $L = L^p(X) = L^p_+ \oplus L^p_-$ *the free color Lie* p*-superalgebra,* $\operatorname{rank}(L) = |X| = N < \infty$. *Suppose that* $H = H_+ \oplus H_-$ *is a* G*-homogeneous subalgebra of* L, $\dim L^p_+/H_+ = t < \infty$ *and* $\dim L^p_-/H_- = s < \infty$. *Then* $\operatorname{rank}(H) = 2^s p^t(N-1)+1$.

Intersection of subalgebras of finite rank

Lemma 15.7
Let A, B *be* G*-homogeneous subalgebras of* $L(X)$ *(of* $L^p(X)$*, respectively). Suppose that* B *is a free factor of some* G*-homogeneous subalgebra* B' *of* $L(X)$ *(of* $L^p(X)$*, respectively). Then the subalgebra* $A \cap B$ *is a free factor of the subalgebra* $A \cap B'$.

PROOF We consider the case when $A, B \subseteq L(X)$ (the proof in the case when $A, B \subseteq L^p(X)$ is analogous).

Let M, N, R be reduced subsets of $L(X)$ such that $M \cup N$ is a reduced subset, $B = L(M)$, $B' = L(M \cup N)$ and $A \cap B = L(R)$. Then $A \cap B' = L(R) \oplus H$ where H is a linear subspace of elements $h \in A \cap B'$ such that in its s-regular presentation in the generators from $M \cup N$ we have each s-regular monomial containing some element of N. It is obvious that H is a subalgebra. Let $H = L(S)$ where S is an independent subset. By triangular elementary transformation, we transform $R \cup S$ to a reduced subset T. By our definition of the subsets R and S and by Theorem 14.1 we have T reduced, $T = R \cup S'$ and $A \cap B' = L(R \cup S')$. ■

Lemma 15.8
Let H *be a* G*-homogeneous subalgebra of finite rank in* $L(X)$ *(in* $L^p(X)$*). Then* H *is a free factor of some* G*-homogeneous subalgebra of finite codimension in* $L(X)$ *(in* $L^p(X)$*, respectively).*

PROOF As above, we consider only the case $H \subseteq L(X)$.

Let $M = \{m_1, \ldots, m_l\}$ be a reduced set of free generators of H. It is sufficient to find a G-homogeneous set R of ℓ-homogeneous elements such that $\widehat{M} \cup R$ is a reduced set and $\dim L(X)/L(\widehat{M} \cup R) < \infty$ where $\widehat{M} = \{\widehat{m_1}, \ldots, \widehat{m_l}\}$ is the set of leading parts. Let $k = \max\{\ell_X(m_1), \ldots, \ell_X(m_l)\}$ and let B be the subalgebra of $L(X)$ which is the linear span of all s-regular monomials w with $\ell(w) > k$, R_1 a reduced ℓ-homogeneous set of generators for B. We consider the ℓ-homogeneous set $\widehat{M} \cup R_1$ and transform it by triangular elementary transformations to the desired form. Since $\ell(y) > k$ for any $y \in B$, our transformations have the following form: $\widehat{m_i} \to \widehat{m_i}$, $1 \leq i \leq l$, $r_j \to r'_j$, $\ell(r'_j) = \ell(r_j)$, $d(r'_j) = d(r_j)$ for $r'_j \neq 0$ and either $\widehat{r'_j} = r'_j$ or $r'_j = 0$. We transform analogously the sets $R'_1 = \{r'_j \mid r_j \in R_1\}$, R''_1 and so on. Using such transformations, we obtain a reduced ℓ-homogeneous set $\widehat{M} \cup R$ generating the some subalgebra as $\widehat{M} \cup R_1$. Since this subalgebra contains the subalgebra B of finite codimension in $L(X)$, it is also of finite codimension in $L(X)$. ■

Theorem 15.9 ([386])
Let K be a field, $p = \operatorname{char} K > 2$, $L^p(X)$ the free color Lie p-superalgebra, and let A and B be some G-homogeneous finitely generated subalgebras of $L^p(X)$. Then the subalgebra $A \cap B$ is also finitely generated.

PROOF We may assume that $|X| < \infty$. By Lemma 15.8 we embed A and B as free factors in some subalgebras A' and B' of finite codimension in $L^p(X)$.

By Lemma 15.7, $A \cap B$ is a free factor of $A' \cap B'$ which is a subalgebra of finitely codimension in $L^p(X)$. By Theorem 15.6 we get $\operatorname{rank}(A' \cap B') < \infty$ and therefore $\operatorname{rank}(A \cap B) < \infty$. ■

As a corollary of Theorem 15.9 we obtain the analogous result for free color Lie superalgebras.

Theorem 15.10 ([386])
Let K be a field, $\operatorname{char} K = p > 2$, $L(X)$ the free color Lie superalgebra, and let A and B be some G-homogeneous subalgebras of $L(X)$ of finite rank. Then their intersection $A \cap B$ is a subalgebra of finite rank.

Problem 15.11
Does Theorem 15.10 take place in the case when the field K has zero characteristic?

Problem 15.12
Give an upper bound for the rank of the intersection of two finitely generated subalgebras in terms of ranks of these subalgebras.

For the case of free groups see [103], [198], [232], [442], [443], [444], [446], [501].

Note that in [293] it was made an attempt to prove that the intersection of two subalgebras of finite rank in the free Lie algebra over an arbitrary field is a free Lie algebra of finite rank, but unfortunately there are some serious erroneous statements in the proof. However using some lemmas this article and results of this chapter one can prove the following theorem.

Theorem 15.13
Let $\operatorname{char} K = 0$, $X = \{x_1, \dots, x_n\}$, $N \in \mathbb{N}$, *B a subalgebra of finite rank of the free color Lie superalgebra* $L(X)$,

$$Z_N = \{x_i,\ (\operatorname{ad} x_n)^k(x_i) \mid i \neq n,\ k \leq N\}.$$

Then $B \cap L(Z_N)$ *is a free color Lie superalgebra of finite rank.*

The odd component of a subalgebra of finite rank

Theorem 15.14 ([387])
Let K be a field, $\operatorname{char} K \neq 2$, $|X_-| \geq 1$, *and let A, B be G-homogeneous subalgebras of finite rank of the free color Lie superalgebra* $L(X)$, $A = A_+ \oplus A_-$, $B = B_+ \oplus B_-$. *If* $A_- = B_- \neq 0$, *then* $A = B$.

PROOF We assume to the contrary that $A_+ \neq B_+$. Consider the subalgebra $M = M_+ \oplus M_-$ of $L(X)$, generated by $A \cup B$. It is clear that $\operatorname{rank}(M) < \infty$. We show that $M_- = A_-$. Let $A = L(H)$ and $B = L(P)$, $|H| < \infty$, $|P| < \infty$. Taking into account our assumptions, we see that $H_- \neq \emptyset$, $P_- \neq \emptyset$. The subalgebra M is generated by $H \cup P$. Any element of M_- is a linear combination of regular monomials w_i in the elements of $H \cup P$ (possibly, there are many such presentations).

We show by induction on $l = \ell_{H \cup P}(w)$ that all $w_i \in A_-$. The basis of induction: $l = 1$. In this case $w \in H_- \cup P_- \subseteq A_-$. Let now $l = n > 1$. Then $w = [w_{11}, w_{12}]$, where $d(w) \in G_-$. We may assume that $d(w_{11}) \in G_-$, $d(w_{12}) \in G_+$. By the induction hypothesis we see that $w_{11} \in A_-$. Under the induction hypothesis we show again by induction on $\ell_{H \cup P}(w_{12})$ for even w_{12} that $[w_{11}, w_{12}] \in A_-$. The starting point as follows: $\ell_{H \cap P}(w_{12}) = 1$; it is evident that $[w_{11}, w_{12}] \in A_-$. Suppose that $[w_{11}, w'_{12}] \in A_-$ for all even monomials w'_{12} such that $\ell(w'_{12}) < \ell(w_{12})$. Let now $w_{12} = [w''_{12}, w'''_{12}]$. Consider possible cases.

Case 1. The monomials w''_{12}, w'''_{12} are odd. Then by the assumption of the first induction we have $w \in A_-$ since $w_{11}, w''_{12}, w'''_{12} \in A_-$.

Case 2. The monomials w''_{12}, w'''_{12} are even. Then

$$[w_{11}, [w''_{12}, w'''_{12}]] = \alpha_1 [[w_{11}, w''_{12}], w'''_{12}] + \alpha_2 [[w_{11}, w'''_{12}], w''_{12}],$$

where $0 \neq \alpha_1, \alpha_2 \in K$, and applying twice the second induction hypothesis we have $w = [w_{11}, w_{12}] \in A_-$.

Thus, $M_- = A_-$ and we have come to the situation when $|X| < \infty$, $|X_-| \geq 1$, and $L(X) = L_+ \oplus L_-$ is a free color Lie superalgebra, $M =$

$M_+ \oplus M_-$ is a G-homogeneous subalgebra in L such that $M_- = L_-$, $M_+ \neq L_+$ and $\operatorname{rank}(M) < \infty$. It gives a contradiction with the statement of Lemma 16.3. ■

Remark 15.15
In Theorem 15.14 the assumptions $\operatorname{rank}(A) < \infty$ and $\operatorname{rank}(B) < \infty$ are essential. Indeed, let $X_+ = \{x\}$, $X_- = \{y\}$, $A = L(X)$, $B = L(Z_X(x))$. Since $x \in A_+$ and $x \notin B_+$, we see that $A_+ \neq B_+$; but at the same time $A_- = B_-$.

The assumption $A_- \neq 0$ is also essential. Indeed, we may take any two distinct subalgebras of finite rank in L_+.

In general, from $A_+ = B_+$ it does not follow that $A = B$. Indeed, let $X_+ = \{x\}$, $X_- = \{y\}$, $A = L(X)$, $B = L(W_X(y))$, where $W_X(y) = \{x, [x,y], [y,y]\}$. Then $\operatorname{rank}(A) < \infty$, $\operatorname{rank}(B) < \infty$ and $A_+ = B_+$. Since $y \in A_-$ and $y \notin B_-$, we have at the same time $A_- \neq B_-$.

The statement of Theorem 15.14 does not take place for $L^p(X)$. Indeed, let $X_+ = \{x\}$, $X_- = \{y\}$, $A = L^p(X)$ and $B = L^p(ZP(x))$. Then $\operatorname{rank}(A) < \infty$, $\operatorname{rank}(B) < \infty$, $A_- = B_- \neq 0$, but $A_+ \neq B_+$ because of $x \in A_+$ and $x \notin B_+$.

16 Subalgebras of infinite rank

In this section we study subalgebras of infinite rank. In particular, we prove that the nontrivial even component of the free color Lie superalgebra of the rank greater than two has infinite rank. Fixed points of $L(X)$ under action of finite groups of linear automorphisms are under consideration.

We start with an example when we can give the explicit form of generators of a subalgebra.

Example 16.1 (Generators of the derived algebra)
Let X be a G-graded set, $|X| \geq 2$. Suppose that X is totally ordered in such a way that from $x \in X_+$, $y \in X_-$ it follows that $y > x$. The derived algebra $L^2(X) = [L(X), L(X)]$ is a free color Lie superalgebra, and, as a free set of generators, we can choose left-normed monomials $[x_1 x_2 \cdots x_{s-1} x_s]$ where $x_i \in X$, $s \geq 2$, $x_1 > x_2 \leq x_3 \leq \cdots \leq x_s$ and $x_i \neq x_{i+1}$ for $x_i \in X_-$, and also if $x_1, \ldots, x_s \in X_-$ then $x_1 \geq x_2 < x_3 < \cdots < x_s$. □

The even components

Theorem 16.2 ([387])
Let $|X_-| \geq 1$, $L(X) = L_+ \oplus L_-$ be a free color Lie superalgebra, $M = M_+ \oplus M_-$ a G-homogeneous subalgebra in $L(X)$ such that $M_- = L_-$ and $\dim L_+/M_+ = s < \infty$. Then $\operatorname{rank}(M) = \infty$, $M = L(Y)$ and $|Y_-| = \infty$.

PROOF We take a reduced subset H of generators of M. Consider the set of leading parts $\widehat{H} = \{\widehat{h} \mid h \in H\}$. By Theorem 14.1 the subset $\widehat{H}$ is independent and $\mathrm{rank}\,(M) = |\widehat{H}|$. Hence $\dim L_+/M_+ = \dim L_+/L_+(\widehat{H})$ and $\dim L_-/M_- = \dim L_-/L_-(\widehat{H}) = 0$, i.e. $L_-(\widehat{H}) = L_-$.

Thus it is sufficient to prove our assertion for $L(\widehat{H})$. We use the induction on s. The basis of induction: $s = 1$. We may assume that $L(\widehat{H}) \subseteq L(Z_X(z))$ for some $z \in X$. Since $L(\widehat{H})$ is an ℓ-homogeneous subalgebra of codimension one in $L(X)$, $L(\widehat{H}) = L(Z_X(z))$, $|Z_X(z)| = \infty$ and $|(Z_X(z))_-| = \infty$. Suppose now that $\dim L_+/L_+(\widehat{H}) = s > 1$. As above, we may assume that $L(\widehat{H}) \subseteq L(Z_X(z))$ for some $z \in X_+$, and $L_-(\widehat{H}) = L_-(Z_X(z)) = L_-$, $|(Z_X(z))_-| = \infty > 1$, $\dim L_+(Z_X(z))/L_+(\widehat{H}) = s - 1$. Applying the induction hypothesis we complete the proof. ∎

Lemma 16.3 ([387])
Let $|X_-| \geq 1$ and let $M = M_+ \oplus M_-$ be a G-homogeneous subalgebra of $L = L(X) = L_+ \oplus L_-$ such that $M_+ \neq L_+$ and $M_- = L_-$. Then $\mathrm{rank}\,(M) = \infty$.

PROOF By the way of contradiction assume that $\mathrm{rank}\,(M) < \infty$. Then by Lemma 15.8 we may include M as a free factor in a subalgebra M' of finite codimension in $L(X)$. It is clear that $(M')_- = L_-$. By Theorem 16.2, $M' = L(Y)$ and $|Y_-| = \infty$. But $|Y_-| = \dim M'_-/[M', M']_- = \dim L_-/[M', M']_- = \infty$. Since $M \subseteq M'$, $[M, M] \subseteq [M', M']$. As $M_- = M'_- = L_-$, we have

$$\dim M_-/[M, M]_- = \dim L_-/[M, M]_- \geq \dim L_-/[M', M']_- = \infty.$$

Thus $\mathrm{rank}\,(M) = \infty$ and we have come to a contradiction. ∎

Lemma 16.4
Any nonzero ideal $H = H_+ \oplus H_-$ of the free color Lie superalgebra $L(X) = L_+ \oplus L_-$ such that $H \neq 0$, $|X| > 1$ and $H_+ \neq L_+$, is a free color Lie superalgebra of infinite rank.

PROOF Since $H_+ \neq L_+$, there exists a G-homogeneous element $a \in L_+ \setminus H_+$. As H is an ideal of $L(X)$ and $a \in L_+$, $E = H \oplus Ka$ is a subalgebra in $L(X)$ and $[E, E] \subseteq H$. Hence we may choose a free generating set Y for the subalgebra E such that $a \in Y$ and $Y \setminus \{a\} \subseteq H$. But the set $Z_Y(a)$ is a free generating set for H and $|Z_Y(a)| = \infty$. Thus $\mathrm{rank}\,(H) = \infty$. ∎

Remark that the condition $H_+ \neq L_+$ in Lemma 16.4 is essential. Indeed, let $X_+ = \{x\}$, $X_- = \{y\}$, $L(X) = L_+ \oplus L_-$, $H = L(W_X(y)) = L(\{x, [x, y], [y, y]\}) = H_+ \oplus H_-$. Then H is a proper ideal of $L(X)$ with $\mathrm{rank}\,(H) = 3 < \infty$ but $H_+ = L_+$.

Using the results of this section one can prove that the (nontrivial) even component of the free color Lie superalgebra has infinite rank.

Theorem 16.5 ([387])
Let K be a field, char $K \neq 2$, $|X| > 1$, $|X_-| \geq 1$, and let $L(X) = L_+ \oplus L_-$ be a free color Lie superalgebra. Then L_+ is a free color Lie algebra of infinite rank. In particular, if $L(X)$ is a free Lie superalgebra, then L_+ is a free Lie algebra of infinite rank.

Note that in the case when $L(X)$ is the free Lie superalgebra Theorem 16.5 was proved also by A. S. Shtern in [527]. In [30], [387] one can find the explicit subset of free generators of L_+.

Fixed points of finite groups of linear automorphisms

In this section we prove that the set of all fixed points of the free color Lie superalgebra under the action of a finite nontrivial group of linear automorphisms is a free color Lie superalgebra of infinite rank.

Let K be a commutative ring with the unity element and L be a free Lie algebra of finite rank greater than one over K.

Theorem 16.6 (R. Bryant [99])
Let K be a principal ideal domain, L a free Lie algebra of finite rank greater than one over K, and H a finite nontrivial group of linear automorphisms of L. Then the set of fixed points L^H is a free Lie algebra of infinite rank.

Note that the proof in [99] is based on the generalized Witt formula like in Theorem 8.25 and on some facts on tensor products of representations of finite groups. In [627] R. Zerck proved this theorem in the case of relation modules. Recently, A. I. Belov in [53] proved that fixed points of the free Lie algebra under the action of finite cyclic subgroup of the automorphism group form a subalgebra of infinite rank; V. Drensky in [161] proved this theorem for any finite subgroup of the automorphism group of the free Lie algebra. As for fixed points in associative case see also monographs by S. Montgomery [432] and V. K. Kharchenko [268].

Let now X be a G-graded set, $L(X)$ the free color Lie K-superalgebra, V the free K-module spanned by the set X_g, $g \in G$, $\mathrm{GL}(V_g)$ the group of all automorphisms of the K-module V_g,

$$F = \prod_{g \in G} \mathrm{GL}(V_g)$$

the group of linear automorphisms of $L(X)$, H a finite subgroup in F. Then $L(X)$ is a KH-module, moreover, any element of H acts as an automorphism of the algebra $L(X)$. Denote by L^H the subalgebra of all fixed points in $L(X)$ relatively the action of the group H.

Theorem 16.7 ([395])
Let K be a principal ideal domain without 2-torsion, $2 < |X| < \infty$, and

H a finite and nontrivial subgroup of the group of linear automorphisms of $L(X)$. Then L^H is a free color Lie superalgebra of infinite rank.

PROOF It follows from Theorem 14.8 that L^H is a free color Lie superalgebra. One can suppose that K is a field. In the case $G = G_+$ the proof is similar to the proof of Theorem 16.6. In the general case, if $x \notin L^H$ for some $x \in X_+$ and $|X| \geq 2$, then we immediately get the assertion of the theorem. The case $X_+ = \{x\}$, $x \notin L^H$, may be considered using the fact that in this situation we have $L^H \subseteq L(Z(x))$.

Now if $X_+ \subseteq L^H$, $X = \{x_1, \ldots, x_k\}$, $k \geq 2$ and the integer r, $1 \leq r < k$, is such that the K-submodule generated by elements $x_1, \ldots, x_r$ is the set of all odd elements of the length 1 in L^H, then we consider the following independent sets:

$$Z_1 = W_X(x_1), \quad Z_2 = W_{Z_1}(x_2), \quad \ldots, \quad Z_r = W_{Z_{r-1}}(x_r).$$

It is clear that $L^H \subseteq L(Z_r)$. If there exists $z \in (Z_r)_+$, then we come to the above considered cases. Otherwise L^H contains the elements $[x_i, x_j]$ with $1 < i < r$, $1 < j < k$, $i < j$. From the fact that the elements x_j, $r + 1 \leq j \leq k$, are fixed it follows that the elements x_i , $1 \leq i \leq r$, are also fixed, and we have come to a contradiction with the definition of the number r.

If there is no such number r, then we consider the elements $[x_i, x_j]$ with $1 \leq i \leq j \leq k$. Suppose that $f \in H$, $f \neq e$, and that these elements belong to L^H. Then $f(x_i) = \pm x_i$. If $f(x_i) = x_i$ for all i, then $f = e$. Let, for example, $f(x_1) = -x_1$. Since $[x_1, x_2] \in L^H$, $f(x_2) = -x_2$. By analogy we see that $f(x_i) = -x_i$ where $1 \leq i \leq k$, and therefore $|H| = 2$, L^H is exactly the even component L_+ of $L(X)$. Bearing in mind that $|X| \geq 2$ and $|X| \geq 1$ one can use Theorem 16.5. Thus that the rank of L^H is infinite. ■

17 Algorithms and examples

In this section we give the following series of algorithms: an algorithm to reduce given subset of generators of a subalgebra to a reduced subset of generators (Algorithm 17.1); an algorithm to solve the problem of entry for a finitely generated subalgebra of free color Lie (p-)superalgebra (Algorithm 17.4); an algorithm to construct a finite generating set of the intersection of two subalgebras of finite rank in the case of prime characteristic (Algorithm 17.5), and an algorithm to decide whether an endomorphism of a free color Lie (p-)superalgebra of finite rank is an automorphism, or not (Algorithm 17.6). We present also some examples.

Algorithms

Algorithm 17.1
Let $\{a_1, \ldots, a_k\}$ be a subset of G-homogeneous elements of $L(X)$ (of $L^p(X)$, respectively). Now the problem under consideration is to construct a φ-reduced subset of elements such that it generates the same subalgebra.

We may assume that $\varphi(a_1) \leq \cdots \leq \varphi(a_k)$ and that $a_1, \ldots, a_k$ are linearly independent (otherwise we use linear transformations). We consider the first level: $\varphi(a_1) = \cdots = \varphi(a_r)$. Up to linear transformations we may assume that the subset $\{a_1, \ldots, a_r\}$ is reduced. For each element a_i, $r \leq i \leq r'$, of the second level we consider its leading word $\overline{a_i}$. By Lemma 14.3 it is sufficient to consider only the possibility of factorization of $\overline{a_i}$ of the form

$$\overline{a_i} = \overline{b_1} \cdots \overline{b_t}$$

where $b_1, \ldots, b_t \in \{a_1, \ldots, a_r\}$.

If we have a such factorization, then by Lemma 7.4 there exists an arrangement of brackets $[\overline{a_i}]$ such that

$$[\overline{a_i}] = \left[\overline{b_1} \cdots \overline{b_t}\right] = \left[\left[\overline{b_1}\right] \cdots \left[\overline{b_t}\right]\right]$$

with

$$[\overline{a_i}]^\circ = \alpha \overline{b_1} \cdots \overline{b_t},$$

where $0 \neq \alpha \in K$ and $\left[\overline{b_1}\right], \ldots, \left[\overline{b_t}\right]$ are s-regular (ps-regular) monomials.

Hence we may use the reduction

$$a_i' = a_i - \frac{1}{\alpha}\left[\left[b_1\right] \cdots \left[b_t\right]\right].$$

Then we apply these procedures on the second level (together with possible linear transformations). At the next step we take a look on $\varphi(a_i')$ and start again our procedure with the level $\varphi(a_i')$ in the case when $\varphi(a_i') < \varphi(a_i)$. Examining such reductions we finish the consideration of this level.

Then using previous levels we deal with next levels. On each step we cancel possible zero elements. After finite number of steps we get a desirable reduced set. Note that this set is a reduced set not only relatively to φ, but also relatively to φ' (firstly by φ, and then lexicographically).

Algorithm 17.2
Let $A = \{a_1, \ldots, a_k\}$ be a subset of G-homogeneous elements of $L(X)$ (of $L^p(X)$). The problem is to decide whether the subset A is independent, or not.

To solve this problem we use Algorithm 17.1. We reduce the subset A to a reduced subset $\{b_1, \ldots, b_r\}$. The subset A is independent if and only if $k = r$.

Remark 17.3
U. U. Umirbaev proved in [571] that the problem to define whether a system of elements of the free associative algebra is independent, or not, is algorithmically insolvable.

Algorithm 17.4
Let L be either the free color Lie superalgebra $L(X)$, or the free color Lie p-superalgebra $L^p(X)$. Let also $\{a_1, \ldots, a_k\}$ be a subset of G-homogeneous elements of L, A the subalgebra of L generated by $\{a_1, \ldots, a_k\}$, and $b \in L$. The problem is to decide whether $b \in A$, or not.

First of all using Algorithm 17.1 we construct a reduced set of generators $\{b_1, \ldots, b_r\}$ of A. By Lemma 14.3 it is sufficient to consider a factorization of $\bar{b}$ in the form $\bar{b} = \overline{b_{i_1}} \cdots \overline{b_{i_s}}$, $b_{i_j} \in \{b_1, \ldots, b_r\}$. Using Lemma 7.4 we find an arrangements of brackets

$$\left[\bar{b}\right] = \left[\left[\overline{b_{i_1}}\right] \cdots \left[\overline{b_{i_s}}\right]\right]$$

such that $\left[\overline{b_{i_1}}\right], \ldots, \left[\overline{b_{i_s}}\right]$ are s-regular (ps-regular) monomials and

$$\left[\bar{b}\right]^{\circ} = \alpha \bar{b},$$

$0 \neq \alpha \in K$. Then we use the reduction

$$b' = b - \frac{1}{\alpha}\left[\left[\overline{b_{i_1}}\right] \cdots \left[\overline{b_{i_s}}\right]\right].$$

After a finite number of such reductions we come either to zero element (and then $b \in A$), or to the element which can not be reduced using $b_1, \ldots, b_r$ (and then $b \notin A$).

Algorithm 17.5
Let $\operatorname{char} K = p > 2$, and let L be the free color Lie p-superalgebra, $a_1, \ldots, a_r, b_1, \ldots, b_s \in L$, A the subalgebra of L generated by $a_1, \ldots, a_r$, and B the subalgebra of L generated by $b_1, \ldots, b_s$. The problem is to construct a free generating set of $A \cap B$.

In fact, any such algorithm gives the proof of Theorem 15.9.

Algorithm 17.6
Let $|X| = n < \infty$, $L = L(X)$ or $L = L^p(X)$, and let $a_1, \ldots, a_n$ be G-homogeneous elements of L, $d(a_i) = d(x_i)$, $i = 1, \ldots, n$, ψ the endomorphism of L given by $\psi(x_i) = a_i$, $i = 1, \ldots, n$. The problem is to decide whether ψ is an automorphism of L, or not.

Using Algorithm 17.1 we construct the reduced set $B = \{b_1, \ldots, b_m\}$ such that it generates the subalgebra generated by the set $\{a_1, \ldots, a_m\}$. Multiplying if it's necessary on nonzero coefficients from K one can assume that $b_i^{\circ} \in S(X)$. Then it is easy to see that ψ is an automorphism of L if and only if $m = n$ and $B = X$.

Examples

Example 17.7
Let $X = \{x, y, z\}$, $x > y > z$, $L(X)$ be the free Lie algebra,

$$a_1 = [x, y] + z; \qquad a_2 = [yz^2] + [y, z];$$
$$a_3 = \big[x, \big[y, [yz^2]\big]\big] + x; \qquad a_4 = \big[[x, [yz^2]], y\big] + \big[[x, y], [y, z]\big].$$

The problem is to construct a reduced set of generators of the subalgebra A generated by $\{a_1, a_2, a_3, a_4\}$.

One can easily see that in this case there is only one possible factorization: $xy^2z^2 = xy \cdot yz^2$, and we set

$$\begin{aligned} a_3' &= a_3 - [a_1, a_2] \\ &= \big[x, \big[y, [yz^2]\big]\big] + x - \big[[x, y] + z, [yz^2] + [y, z]\big] \\ &= -\big[[x, [yz^2]], y\big] + [yz^3] - \big[[x, y], [y, z]\big] + \big[[y, z], z\big] + x. \end{aligned}$$

Now we can use the linear transformation

$$a_3'' = a_3' + a_4 = [yz^3] + [yz^2] + x.$$

Hence the desirable reduced set is $\{a_1, a_2, a_3'', a_4\}$ (it is possible also to consider the element $a_3''' = a_3'' - a_2 = [yz^3] - [y, z] + x$). □

Example 17.8
Let $X = \{x, y, z\}$, $x > y > z$, $G = \mathbb{Z}_2$, $d(x) = d(y) = 0$, $d(z) = 1$, and let $L(X)$ be the free Lie superalgebra,

$$a = [x, z, y] + 2[y, z]; \qquad b = [x, z] + 3x; \qquad c = [x, y] + y.$$

The problem is to decide whether the element

$$\begin{aligned} h \ = \ & \big[[x, z], [x, z, y]\big] + 3\big[x, [x, z, y]\big] + 2\big[[x, z], [y, z]\big] + 6\big[x, [y, z]\big] \\ & - 6\big[[x, y], [x, z]\big] + 18\big[x, [x, y]\big] - 12[y, z] + x - 18y \end{aligned}$$

belongs to the subalgebra of $L(X)$ generated by elements a, b, c.

To solve this problem we use Algorithm 17.4. First of all we see that the subset $\{a, b, c\}$ is a reduced subset. The leading associative word in h is $xzxzy$. It has the following factorization into products of leading words of a, b, and c: $xz \cdot xy$. Hence we consider the element

$$h_1 = h - [b, a] = -6\big[[x, y], [x, z]\big] + 18\big[x, [x, y]\big] - 12[y, z] + x - 18y,$$

and then

$$\begin{aligned} h_2 &= h_1 + 6[c, b] = -6\big[[x, z], y\big] - 18[x, y] - 12[y, z] + x - 18y, \\ h_3 &= h_2 + 6a = -18[x, y] + x - 18y, \\ h_4 &= h_3 + 18c = x. \end{aligned}$$

Thus, h does not belong to the subalgebra of $L(X)$ generated by the subset $\{a, b, c\}$. □

Example 17.9
Let $X = \{x, y, z\}$, $x > y > z$, $\operatorname{char} K = p > 2$, and let $L^p(X)$ be the free Lie p-algebra,

$$a_1 = (\operatorname{ad} x)^p(z) + [x, y] + [x, y] + [y, z] - x - z;$$
$$a_2 = x^{[p]} + y; \qquad a_3 = x - z.$$

The problem is to decide whether an endomorphism ψ of $L^p(X)$ given by $\psi(x_i) = a_i$, $i = 1, 2, 3$, is an automorphism, or not.

We use Algorithm 17.1 to construct a reduced set of generators of the algebra generated by $\{a_1, a_2, a_3\}$. According to this algorithm we need firstly to consider the element $a_2' = a_2 - a_3^{[p]}$. But now, for the simplicity, we consider at the first step the element

$$\begin{aligned} a_1' &= a_1 + [a_2, a_3] \\ &= \left[x^{[p]}, z\right] + [x, y] + [y, z] - x - z - \left[x^{[p]}, z\right] - [x, y] - [y, z] \\ &= -x - z. \end{aligned}$$

Then we use the linear transformations

$$\begin{aligned} a_1'' &= \frac{1}{2}(-a_1' + a_3) = x; \\ a_3' &= -\frac{1}{2}(a_1' a_3) = z. \end{aligned}$$

Finally, we set

$$a_2' = a_2 - (a_1'')^{[p]} = y.$$

Thus, ψ is an automorphism of $L^p(X)$. □

Comments

The theorem on subalgebras of free Lie algebras over fields was proved by A. I. Shirshov in [511] (1953), and on subrings of free Lie rings and of subalgebras of free restricted Lie algebras by E. Witt [606] (1956).

Remark that many methods of investigation of subalgebras of free Lie algebras are rather closed to group methods which are due to J. Nielsen and O. Schreier (see, for example [218], [303], [342], [351], [447], [448], [494]).

Some of this approaches were extended by A. G. Kurosh and his school for free algebras of some varieties in universal algebra. For example, some of these papers are concentrated on subalgebras of free nonassociative algebras, see [150], [151], [205], [301], [302], [628], [629].

The theorem on the intersection of two subalgebras of finite rank in free Lie (p-)algebras over a field of prime characteristic is due to G. P. Kukin [289], he also obtained the Schreier formulas for free Lie p-algebras.

The theorem on subalgebras of free (color) Lie superalgebras and p-superalgebras were obtained by A. A. Mikhalev in [380], [381], [383], [384], [385], [386] (properties of this subalgebras were studied also in [387], [390], [394], [395], [400]). A. I. Shtern in [527] also proved the theorem on subalgebras of free Lie superalgebras. Remark that the theorem on $\mathbb{Z}$-graded subalgebras of free Lie superalgebras could be find in the book of J. M. Lemaire [330].

Chapter 4

Canonical Bases

This chapter is devoted to canonical bases of Lie superalgebras. In Section 18 we prove Diamond Lemma for associative rings (Theorem 18.1) and give examples of canonical bases of associative algebras.

Section 19 is devoted to universal enveloping algebras. Here we prove Poincaré-Birkhoff-Witt theorems (Theorems 19.1, 19.2, 19.9) and show that free color Lie (p-)superalgebras are finitely separable (Theorem 19.7). We consider the structure of graded Hopf algebras on universal enveloping algebras and give Friedrichs criteria (Theorems 19.21, 19.22). Super shuffle algebras are also under considerations. We discuss also Baker-Campbell-Hausdorff formula.

In Section 20 we prove Composition Lemma (Theorem 20.5) for color Lie (p-)superalgebras. It enables us to consider canonical bases in Lie superalgebras. Theorems 20.11, 20.12 give us some connections between canonical bases of Lie superalgebras and Canonical bases of their universal enveloping algebras. We give in addition examples and discuss some open problems.

In Section 21 we construct linear bases of free products with the amalgamated subalgebra (Theorems 21.2, 21.6).

In Section 22 we present algorithms to construct Gröbner basis in the homogeneous case, to solve the equality problem employing Gröbner bases, to construct a Gröbner basis of a left ideal of free associative algebras, to decide whether a system of homogeneous elements of the free associative algebra is left dependent, or not. Some illustrative concrete examples are presented.

18 Canonical bases in associative algebras

In this section we outline Diamond Lemma (known also as Junction Lemma, Composition Lemma) for associative rings (Theorem 18.1), whereby the canonical bases for associative algebras given by generators and defining relations are constructed. We list some examples of canonical bases in associative algebras. We treat also Gröbner bases for left ideals of the free associative algebra (Proposition 18.3).

Diamond Lemma

Diamond Lemma could be found in one form or another from place to place at different times (see, for example, [63], [80], [228], [316], [445], [516]). Our exposition is based on the paper of G. Bergman [63].

Let K be a commutative and associative ring with 1, X a set, $S(X)$ the free semigroup with 1 on X, $A(X)$ the free associative K-algebra on X.

We consider a set S of pairs of the form $\sigma = (W_\sigma, f_\sigma)$, where $W_\sigma \in S(X)$ and $f_\sigma \in A(X)$. For any $\sigma \in S$ and $A, B \in S(X)$ let $R_{A\sigma B}$ denote the endomorphism of the K-module $A(X)$ that acts trivially on all elements of $S(X)$ other than $AW_\sigma B$ and at the same time $R_{A\sigma B}(AW_\sigma B) = Af_\sigma B$.

We call this set S *a reduction system* and the K-endomorphisms

$$R_{A\sigma B}: A(X) \to A(X)$$

reductions. We say that an element $a \in A(X)$ is *irreducible* if a does not contain any monomials of the form $AW_\sigma B$, $\sigma \in S$. The K-submodule of all irreducible elements of $A(X)$ will be denoted by $A(X)_{\text{irr}}$.

Suppose that for $a \in A(X)$ we have a finite sequence $R_1, \dots, R_n$ of reductions such that $R_n \cdots R_1(a) \in A(X)_{\text{irr}}$. The element $R_n \cdots R_1(a)$ will be called *a normal form* of a. The set of elements $a \in A(X)$ with unique normal form $R_S(a)$ will be denoted $A(X)_{\text{red}}$.

Consider *an overlap ambiguity* (σ, τ, A, B, C) with $\sigma, \tau \in S$ and $A, B, C \in S(X) \backslash \{1\}$ such that $W_\sigma = AB$ and $W_\tau = BC$. We say that the overlap ambiguity (σ, τ, A, B, C) is *resolvable* if there exist compositions of reductions R and R', such that $R(f_\sigma C) = R'(Af_\tau)$.

Now we look at *an inclusion ambiguity* (σ, τ, A, B, C) with $\sigma, \tau \in S$ and $A, B, C \in S(X)$ such that $W_\sigma = B$, $W_\tau = ABC$. We shall say the inclusion ambiguity is *resolvable* if $Af_\sigma C$ and f_τ can be reduced to a common element.

We say that a partial ordering $\leq$ on $S(X)$ is *a semigroup partial ordering* if $B < B'$ implies $ABC < AB'C$ for all $A, B, B', C \in S(X)$ and that it is compatible with S if for all $\sigma \in S$ the element f_σ is a linear combination of words V such that $V < W_\sigma$.

Let I be the two-sided ideal of $A(X)$ generated by the elements $W_\sigma - f_\sigma$, $\sigma \in S$. It is clear that as a K-module I is spanned by the elements $A(W_\sigma - f_\sigma)B$, where $A, B \in S(X)$.

Theorem 18.1 (Diamond Lemma [63], [80])
Let S be a reduction system for a free associative algebra $A(X)$ and $\leq$ a semigroup partial ordering on $S(X)$ compatible with S and satisfying

the descending chain condition (DCC). Then the following conditions are equivalent:

(i) *all ambiguities of S are resolvable;*

(ii) $A(X) = A(X)_{\mathrm{red}}$*;*

(iii) *a set of representatives in $A(X)$ for the elements of the algebra $R = A(X)/I$ defined in terms of generators X and relations $W_\sigma = f_\sigma$ $(\sigma \in S)$ is given by the K-submodule $A(X)_{\mathrm{irr}}$ (i.e., $A(X) = A(X)_{\mathrm{irr}} \oplus I$).*

When these conditions hold, R may be identified with the K-module $A(X)_{\mathrm{irr}}$, made into K-algebra by the multiplication $a \times b = R_S(ab)$.

PROOF Since we consider the S-compatible partial ordering with DCC every element of $A(X)$ can be reduced to a normal form under a finite sequence of reductions.

Assuming (ii) the mapping R_S will be a projection of $A(X)$ onto $A(X)_{\mathrm{irr}}$, its kernel being contained in I because every reduction replaces an element a by an element $a + b$ with $b \in I$; its kernel is contained in I since

$$R_S(A(W_\sigma - f_\sigma)B) = R_S(AW_\sigma B) - R_S(Af_\sigma B) = 0$$

for all A, B, σ (all elements of $A(X)$ have unique normal form), just proving (iii).

Now assume (iii) satisfied and suppose that $a \in A(X)$ can be reduced to distinct $b, b' \in A(X)_{\mathrm{irr}}$. Then $b - b' \in A(X)_{\mathrm{irr}} \cap I = 0$. Thus (ii) and (iii) are equivalent.

It is clear that (ii) $\Rightarrow$ (i). Now we want to show that (i) $\Rightarrow$ (ii). At the beginning observe that $A(X)_{\mathrm{red}}$ is a K-submodule of $A(X)$. Let $f, g \in A(X)_{\mathrm{red}}$, $\alpha, \beta \in K$ and R be any finite composition of reductions. Then we can find finite compositions of reductions R' and R'' such that $(R'R)(f) = R_S(f)$ and $(R''R'R)(g) = R_S(g)$.

Therefore

$$\begin{aligned}(R''R'R)(\alpha f + \beta g) &= \alpha R'' R_S(f) + \beta R_S(g)\\ &= \alpha R_S(f) + \beta R_S(g) \in A(X)_{\mathrm{irr}}.\end{aligned}$$

Thus the element $\alpha f + \beta g$ has the unique normal form. Moreover we have proved that the mapping $R_S\colon A(X)_{\mathrm{red}} \to A(X)_{\mathrm{irr}}$ is K-linear.

It remains to prove $S(X) \subset A(X)_{\mathrm{red}}$. If w is a minimal element in $S(X)$ then it is obvious that $w \in A(X)_{\mathrm{red}}$. Suppose now that w is not a minimal element in $S(X)$ and that for any $v \in S(X)$ with $v < w$ we have $v \in A(X)_{\mathrm{red}}$. We prove that $w \in A(X)_{\mathrm{red}}$. For any two reductions $R_{U\sigma V'}$ and $R_{U'\tau V}$ each acting nontrivially on w, from our inductive hypothesis we derive that $R_{U\sigma V'}(w), R_{U'\tau V}(w) \in A(X)_{\mathrm{red}}$.

We would like to show that

$$R_S(R_{U\sigma V'}(w)) = R_S(R_{U'\tau V}(w)).$$

Without loss of generality we may assume that $\ell(U) < \ell(U')$.

Consider three possible cases.

Case 1. We have an overlap ambiguity $w = UABCV$, $W_\sigma = AB$, $W_\tau = BC$. By (i) this ambiguity is resolvable, i.e. the elements $f_\sigma C$ and Af_τ can be reduced each to one and the same element f. Therefore elements $R_{U\sigma V'}(w) = Uf_\sigma CV$ and $R_{U'\sigma V}(w) = UAf_\tau V$ can be reduced to one element $g = UfV$. Thus

$$R_S(Uf_\sigma CV) = R_S(g) = R_S(UAf_\tau V).$$

Case 2. We have an inclusion ambiguity. By analogy with the first case, the present case can be handled with the solvability of inclusion ambiguities.

Case 3. Subwords W_σ and W_τ are disjoint in w (i.e., $w = AW_\sigma BW_\tau C$). Then the elements $R_{U\sigma V'}(w) = Af_\sigma BW_\tau C$ and $R_{U'\tau V}(w) = AW_\sigma Bf_\tau C$ can be reduced to one element $f = Af_\sigma Bf_\tau C$, and therefore the normal form of each of these elements is equal to $R_S(f)$. ∎

Remark 18.2

For any ideal I of the algebra $A(X)$ there exists a system of reductions S satisfying the conditions of Theorem 18.1 (I is generated by elements $W_\sigma - f_\sigma$, $\sigma \in S$). Such system of generators of I is called *a stable system of relations.* Notice that a finitely generated ideal may have no a finite stable system of relations. For example if $X = \{x, y\}$, $x > y$ and I is generated by $x^2 - yx$, then its stable system of relations is

$$\{W_\sigma - f_\sigma \mid \sigma \in S\},$$

where

$$S = \{(xy^{n+1}x, y^{n+2}x) \mid n = -1, 0, 1, 2, \dots\}.$$

Further we remark that for any ideal I there exists a stable system of relations S such that all elements $W_\sigma - f_\sigma$ are not reducible with the help of reductions $W_\tau \to f_\tau$, where $\tau \in S$, $\tau \neq \sigma$ (a stable system of relations of this sort is determined by I up to a coefficient with the leading words, and it is called *the Gröbner basis of the ideal* I).

Triangle rule

In many cases in order to construct a stable set of generators of an ideal of $A(X)$ it is not necessary to examine all ambiguities. There are some rules to separate the ambiguities which are automatically resolvable from others. An example of such rule is the so-called triangle rule. Let $S \subseteq A(X)$, $\overline{s_1} \neq \overline{s_2}$ for $s_1, s_2 \in S$, $s_1 \neq s_2$, and let we have an ambiguity ω between elements a and b from S. Suppose also that all possible ambiguities ω' between the elements of S with $\ell(\omega') \leq \ell(\omega)$ are resolvable, and that there exist $e, f \in S$ such that their overlap ambiguities ω'' where $\overline{\omega''}$ is a subword of $\overline{\omega}$, and ω'' is resolvable. Then ω is also resolvable.

Let $u, v \in S(X)$ and u a subword of v. Denote by $\deg_u v$ the number of all occurences of u in v. If $U \subseteq S(X)$ then we set

$$\deg_U v = \sum_{u \in U} \deg_u v$$

(if u is not a subword in v then $\deg_u v = 0$). Now we formulate the triangle rule in the following form. In order to construct Gröbner basis in the way when we apply all possible reductions after each step (as in Algorithm 22.1) it is possible to resolve only such ambiguities $e_1e_2e_3$ that $\deg_{\overline{S}}(e_1e_2e_3) = 2$, where $\overline{S}$ is the set of all leading words of elements from S. There are some generalizations of this rule (see, for example [316], [561]). Note that in homogeneous case D. J. Anick [8] suggested another rules to separate ambiguities using special resolutions.

Gröbner bases of left ideals

Let $A(X)$ be the free associative algebra over a field K, $S \subseteq A(X)$, $\mathrm{Id}_{\mathrm{left}}(S)$ the left ideal of $A(X)$ generated by S.

For $b \in S(X)$, $b = b'\overline{a}$, $b' \in S(X)$, we consider the reduction

$$R_{b,a}\colon A(X) \to A(X)$$

given by

$$R_{b,a}(b) = b - \frac{1}{\alpha}a,$$

where $a^\circ = \alpha\overline{a}$, $\alpha \in K$,

$$R_{b,a}(c) = c$$

for a word $c \neq b$. It is clear that

$$\overline{R_{b,a}(b)} < \overline{b}.$$

We say that a subset S of $A(X)$ is *left-stable* if for each $a \in S$ the coefficient at $\overline{a}$ is equal to 1 and it is impossible to reduce $\overline{a}$ using $b \in S \backslash \{a\}$.

The following proposition is quite clear.

Proposition 18.3
Let K be a field, S a left-stable subset of $A(X)$, $a \in \mathrm{Id}_{\mathrm{left}}(S)$. Then $\overline{a}$ contains as an ending $\overline{b}$ for some $b \in S$. Moreover, the set of all words $c \in S(X)$, $c \neq u\overline{b}$, $b \in S$, $u \in S(X)$, gives us a linear basis of the K-factorspace $A(X)/\mathrm{Id}_{\mathrm{left}}(S)$.

If for any word b from the presentation of a ($b \neq \overline{a}$) of a left-stable subset $S \subseteq A(X)$ it is impossible to reduce b using $c \in S \setminus \{a\}$, then we say that this system is *the Gröbner basis of the left ideal* $\mathrm{Id}_{\mathrm{left}}(S)$ of $A(X)$. It is clear that for any left ideal J of $A(X)$ there exists the uniquely determined Gröbner basis. To construct the Gröbner basis of a finitely generated left ideal of $A(X)$ effectively, see Algorithm 22.3.

Note that any finitely generated left ideal of $A(X)$ has the finite Gröbner basis (this is not the case for two-sided ideals, see for example Remark 18.2). P. M. Cohn's theorem that the product of free ideal rings is again a free ideal ring shows that $A(X)$ is a fir (see [126], [130]). Using the concept of Gröbner bases it is possible to prove directly that any left ideal of $A(X)$ is a free $A(X)$-module. Indeed, consider the Gröbner basis S of a left ideal I of $A(X)$. We prove that S is a free basis of I as a left $A(X)$-module. At first, $u\overline{a} \neq v\overline{b}$ with $a, b \in S$, $a \neq b$, $u, v \in A(X)$. Indeed, if $\ell(u) \geq \ell(v)$, $u\overline{a} = v\overline{b}$, then $\overline{b} = w\overline{a}$, $w \in S(X)$, and we get a contradiction. Hence, it follows from

$$\sum_i b_i a_i = 0,$$

where $b_i \in A(X)$, $a_i \in S$, $a_i \neq a_j$ with $i \neq j$, that $b_i = 0$ for all i. Thus, S is a free basis of I.

Lemma 18.4
Let K be a field, $a, a_1, \ldots, a_l$ G-homogeneous elements of the free G-graded associative K-algebra $A(X)$, J the left ideal of $A(X)$ generated by the set $\{a_1, \ldots, a_l\}$. Suppose that $a \notin J$. Then there exists a G-graded finitely generated left ideal J' of finite codimension in $A(X)$ such that $a \notin J'$.

PROOF Let $W = \{w_1, \ldots, w_k\}$ be the left Gröbner basis of the left ideal J. We may suppose that $\overline{a} \neq t\overline{w_i}$, $w_i \in W$, $t \in S(X)$ (otherwise we look at the element $a' = a - \alpha t w_i$, where $\alpha \in K$ is such that $\overline{a'} < \overline{a}$ and $a' \notin J$). Let $N = \max(\ell(a), \ell(w_1), \ldots, \ell(w_k))$. Examine the set $\{w_{k+1}, \ldots, w_l\}$ of all words from $S(X)$ such that $\ell(w_s) = N + 1$ for all $k + 1 \leq s \leq l$ and $\overline{w_{k+i}} \neq s\overline{w_j}$ for all $s \in S(X)$, $1 \leq i \leq l - k$ and $1 \leq j \leq k$. The set $\{w_1, \ldots, w_l\}$ is left-reduced and for the left ideal J' of $A(X)$ generated by this set we have $J \subseteq J'$ and $\dim A(X)/J' < \infty$. It follows from $\ell(a) \leq N$ that $a \notin J'$. ■

Examples of canonical bases in associative algebras

Embeddings into simple associative algebras

In [80] L. A. Bokut' used Diamond Lemma as a starting point for his investigation of embeddings of associative algebras into simple associative algebras. Let A be an associative algebra with 1 over a field K. Then it is possible to embed A into a simple associative algebra A' over K with the same unit element. Indeed, let 1 and $\{e_i \mid i \in I\}$ form a K-linear basis of A and

$$e_i \bullet e_j = \alpha_{ij} + \sum_k \alpha_{ij}^k e_k$$

the multiplication rule in A. Let also

$$\begin{aligned} M &= \{a \in A \mid a \notin K \cdot 1\}, \\ X &= \{e_i \mid i \in I\} \cup \{x_a, y_a \mid a \in A\}. \end{aligned}$$

We consider the free associative K-algebra $A(X)$ and the ideal J of $A(X)$ generated by the set S of all elements

$$\begin{aligned} e_i e_j - e_i \bullet e_j, &\qquad i, j \in I; \\ x_a a y_a - 1, &\qquad a \in M. \end{aligned}$$

The set S is stable, therefore by Diamond Lemma the algebra A is a subalgebra of $A_1 = A(X)/J$. By analogy, for A_1 we construct the algebra A_2, and so on. The algebra

$$A' = \bigcup_{i \geq 1} A_i$$

is a simple associative algebra with 1, and A is a subalgebra of A'.

In a similar manner one can prove P. M. Cohn's theorem (see [123]) to the effect that any associative algebra A with 1 and without zero divisors is embedable into a simple associative algebra A' with the same 1 and without zero divisors. Indeed, as above let

$$X = \{e_i \mid i \in I\} \cup \{x_a, y_a, z_a \mid a \in A\},$$

and let J be the ideal of $A(X)$ generated by the elements

$$\begin{aligned} e_i e_j - e_i \bullet e_j, &\qquad i, j \in I; \\ x_a a y_a + z_a a v_a - 1, &\qquad a \in A. \end{aligned}$$

We construct the algebras $A_1 = A(X)/J$, A_2, and so on. The algebra

$$A' = \bigcup_{i \geq 1} A_i$$

is a simple associative algebra without zero divisors, and A is a subalgebra of A' (for the details see [81]).

In [80] L. A. Bokut' using Diamond Lemma proves also that it is possible to embed any associative algebra into a simple associative algebra which is a sum of three nilpotent subalgebras of degree 3 (or a sum of four subalgebras with zero multiplication). Remark that O. H. Kegel in [263] proves that an associative algebra which is a sum of two nilpotent subalgebras, is nilpotent.

In [29] Yu. A. Bahturin and A. Giambruno show that if an associative ring R is a sum of two commutative subrings, then R satisfy the identity $[x_1, y_1][x_2, y_2] = 0$. K. I. Beidar and A. V. Mikhalev prove in [49] that if an associative ring R is a sum of two subrings (as additive subgroups) which satisfy the identity $[x_1, y_1] \cdots [x_n, y_n] = 0$, then R satisfies a polynomial identity.

Universal coaction

Let K be a field, $\operatorname{char} K \neq 2$, $P = (p_{ij})$, $Q = (q_{ij})$ matrices of parameters,

$$\begin{gathered} p_{ij}p_{ji} = q_{ij}q_{ji} = p_{ii} = q_{ii} = 1, \\ X = \{x_1, \ldots, x_n\}, \qquad Y = \{y_1, \ldots, y_n\}, \\ A = A(X)/\mathrm{Id}(\{x_i x_j - q_{ij}^{-1} x_j x_i \mid 1 \leq i, j \leq n\}), \\ B = A(X)/\mathrm{Id}(\{y_i y_j + p_{ij} y_j y_i,\ y_i^2 \mid 1 \leq i, j \leq n\}), \end{gathered}$$

where $\mathrm{Id}(R)$ denotes the ideal generated by the set R. *The universal coaction algebra* $M(A, B)$ can be given as the factor algebra $A(Z)/J$ where $Z = \{z_s^t \mid 1 \leq s, t \leq n\}$, $J = \mathrm{Id}(S)$, and S is the set of the following elements (with $i < j$ and $k < l$):

$$\begin{gathered} z_i^k z_j^k - q_{ij} z_j^k z_i^k; \\ z_i^k z_i^l - p_{kl}^{-1} z_i^l z_i^k; \\ z_i^k z_j^l - q_{ij}^{-1} q_{kl} z_j^l z_i^k + q_{kl} z_i^l z_j^k - q_{ij}^{-1} z_j^k z_i^l; \\ z_i^k z_j^l - p_{ij} p_{kl}^{-1} z_j^l z_i^k + p_{kl}^{-1} z_i^l z_j^k - p_{ij} z_j^k z_i^l. \end{gathered}$$

Suppose that $z_i^k < z_j^l$ if either $i > j$, or $i = j$ and $k > l$. In the case when $p_{ij} q_{ij} \neq -1$ for all i, j, these defining relations can be written in the following form (where we denote $z_i^k = a$, $z_i^l = b$, $z_j^k = c$, $z_j^l = d$, $i < j$ and $k < l$; $p_h = p_{kl}$, $q_h = q_{kl}$, $p_v = p_{ij}$, $q_v = q_{ij}$):

$$\begin{gathered} ab = p_h^{-1} ba; \qquad cd = p_v^{-1} dc; \\ ac = q_v^{-1} ca; \qquad bd = q_h^{-1} db; \\ bc = \frac{q_v + p_v^{-1}}{q_h^{-1} + p_h} cb + \frac{p_v^{-1} p_h - q_v q_h^{-1}}{q_h^{-1} + p_h} da; \\ ad = p_h q_h^{-1} \frac{q_v + p_v^{-1}}{q_h^{-1} + p_h} da + \frac{q_h^{-1} p_v^{-1} - p_h q_v}{q_h^{-1} + p_h} cb. \end{gathered}$$

If $p_{ij} q_{ij} = c^{\operatorname{sign}(i-j)}$ for some $c \in K^*$, $c \neq -1$, then S is a stable set, and the algebra $M(A, B)$ is a deformation of the ring of functions on the matrix semigroup $\mathrm{Mat}(n, k)$. For the proofs and details see [145], [146], [360].

Bases of the free metabelian associative algebra

Diamond Lemma is also useful in order to construct bases of relatively free associative algebras. Let $\operatorname{char} K = 0$, $X = \{x_1, \dots, x_n\}$, and let A_n be *the free metabelian associative algebra* over K, i.e. A_n is the free algebra of the variety given by the identity $\big[[u, v], w\big] = 0$.

Theorem 18.5 (L. A. Bokut' and L. G. Makar-Limanov [88]) *Let $x_1 = a$, $x_2 = b$, $x_3 = c$. The algebra A_3 can be defined as a factor algebra of the free associative algebra $A(a, b, c)$ by the ideal generated by the following stable set of defining relations:*

$$
\begin{gathered}
a^2b - 2aba + ba^2; \qquad a^2c - 2aca + ca^2; \qquad b^2c - 2bcb + cb^2; \\
ab^2 - 2bab + b^2a; \qquad ac^2 - 2cac + c^2a; \qquad bc^2 - 2cbc + c^2b; \\
(ab)^2 - (ba)^2; \qquad (ac)^2 - (ca)^2; \qquad (bc)^2 - (cb)^2; \\
b^2aba - b^3a^2; \qquad c^2aca - c^3a^2; \qquad c^2bcb - c^3b^2; \\
baba^2 - b^2a^3; \qquad caca^2 - c^2a^3; \qquad cbcb^2 - c^2b^3; \\
abc - bac - cab + cba; \qquad acb - bac - cab + bca; \\
bcab - bcba - cbab + cb^2a; \qquad bcac - cbca + c^2ba - cbac; \\
acab - baca + bca^2 - 2caba + cba^2; \\
abac - 2baca + bca^2 - caba + cba^2; \\
bcb^nab - bcb^{n+1}a - cb^{n+1}ab + cb^{n+2}a, \quad n \geq 1; \\
bcbac + (cb)^2a - cb^2ac - c^2b^2a; \qquad (cb)^2aba - c^2b^3a^2; \\
bcb^nac - cb^{n+1}ac, \quad n \geq 2; \qquad c^2b^naca - c^3b^na^2, \quad n \geq 2; \\
c^2baca - c^2bca^2; \qquad cbaca^2 - cbca^3; \\
cb^2aca^2 - 2(cb)^2a^3 + c^2b^2a^3; \qquad cb^naca^2 - c^2b^na^3, \quad n \geq 3.
\end{gathered}
$$

In the general case one can find in [88] a very interesting case when a subset S of $A(X)$ is not stable, but joining a new generators one gets a stable set. To be more precise, consider the additional generators y_{ij}, $1 \leq i < j \leq n$, and the defining relations $y_{ij} = x_ix_j - x_jx_i$. Then in generators $X \cup \{y_{ij}\}$ the algebra A_n is given by the set S' of the defining relations of the form

$$
\begin{gathered}
x_ix_j - x_jx_i + y_{ji}, \quad i > j; \qquad y_{ij}x_k - x_ky_{ij}, \quad i > j; \\
y_{ij}y_{kl}, \quad i < j,\ k < l,\ |\{i, j, k, l\}| \leq 3; \\
y_{ij}y_{kl} - y_{kl}y_{ij}, \quad i < j,\ k < l,\ i > k,\ |\{i, j, k, l\}| = 4; \\
y_{ij}y_{kl} + y_{ik}y_{jl}', \quad i < j,\ k < l,\ j > k,\ |\{i, j, k, l\}| = 4,
\end{gathered}
$$

where $y_{ij}' = y_{ij}$ for $i < j$, and $y_{ij}' = -y_{ji}$ for $i > j$.

Theorem 18.6 (L. A. Bokut' and L. G. Makar-Limanov [88]) *The subset S of $A(X \cup \{y_{ij}\})$ is stable. Moreover, the monomials*

$$x_1^{m_1} \cdots x_n^{m_n} y_{i_1 j_1} y_{i_2 j_2} \cdots y_{i_s j_s}$$

with $m_i \geq 0$, $1 \leq i_1 < j_1 < i_2 < j_2 < \cdots < i_s < j_s \leq n$, $s \geq 0$, form a linear basis of A_n.

The proof was given in [88] using Diamond Lemma and V. N. Latyshev's theorem to the effect that the ideal of the identity $[[u,v],w] = 0$ is generated as an ideal of $A(X)$ by the elements of the form

$$[[x,y],z]; \qquad [x,y] \cdot [z,t] + [x,z] \cdot [y,t]$$

with $x, y, z, t \in X$ (see [313]).

Recognizable properties of associative algebras

V. N. Latyshev in [191], [314], [315], [316], [317] used Diamond Lemma and Gröbner bases in order to prove commutativity of strongly partial ordered PI-algebras, to construct an algorithm to solve the membership problem in Lie nilpotent associative algebras, and to find various recognizable properties of associative algebras (i.e., such properties of associative algebras given by Gröbner bases that there exist algorithms to recognize them).

Another examples of stable systems of relations

Example 18.7 (Weil algebra)
Generators: $x_1, \ldots, x_n, y_1, \ldots, y_n$.
Defining relations: $[x_i, y_j] = [x_i, x_j] = 0$, $i, j = 1, \ldots, n$. □

Example 18.8 (Grassmann algebra)
Generators: $x_1, \ldots, x_n$.
Defining relations: $x_i^2 = 0$; $x_i x_j = -x_j x_i$, $i, j = 1, \ldots, n$. □

Example 18.9 (Clifford algebra, char$K \neq 2$)
Let $f(x_1, \ldots, x_n) = \sum a_{ij} x_i x_j$ be a quadratic form, $a_{ij} = a_{ji}$.
Generators: $x_1, \ldots, x_n$.
Defining relations: $x_i x_j + x_j x_i - 2a_{ij} = 0$. □

Example 18.10 (Steenrod algebra, mod p, $p > 2$, see [8])
Generators: x_i, $i = 1, 2, \ldots$; y.
Defining relations:

$$\begin{aligned} y^2 &= 0; \\ x_k x_l &= \sum_{i=0}^{[k/l]} (-1)^{k+i} C_{l-pi}^{(p-1)(n-i)-1} x_{k+l-i} \quad \text{with} \quad k < pl; \end{aligned}$$

$$\begin{aligned} x_k y x_l &= \sum_{i=0}^{[k/l]} (-1)^{k+i} C_{l-pi}^{(p-1)(n-i)-1} y x_{k+l-i} x_i \\ &+ \sum_{i=0}^{[(k-1)/l]} (-1)^{k+i} C_{l-pi-1}^{(p-1)(n-i)-1} x_{k+l-i} y x_i \quad \text{with} \quad k \leq pl. \end{aligned}$$

□

19 Universal enveloping algebras

In this section we obtain Poincaré-Birkhoff-Witt theorems for universal enveloping algebras of Lie superalgebras (Theorems 19.1, 19.2, 19.9). We prove that free color Lie (p-)superalgebras are finitely separable (Theorem 19.7). We discuss some properties of universal enveloping algebras. In particular, we consider universal enveloping algebras as graded Hopf algebras. Theorems 19.21, 19.22 give us super Friedrichs criteria. We consider Baker-Campbell-Hausdorff problem for Lie superalgebras in connection with shuffle algebras and s-regular words.

PBW-theorems for Lie superalgebras

It was proved in Corollary 8.3 that $\pi: L(X) \to A(X)$ is an embedding and $\pi(L(X)) = [X]$, $A(X) = U(L(X))$. Suppose now that K has no 2-torsion, L is a color Lie superalgebra which is a free K-module with a G-homogeneous basis $X = X_+ \cup X_-$, $A(X)$ is the free G-graded associative algebra, I is the two-sided ideal in $A(X)$, generated by the elements of the form $ab - \varepsilon(a,b)\, ba - [a,b]$ for all homogeneous $a, b \in L$. Consider the canonical mapping $\delta: L \to A(X)/I = U(L)$. It is clear that $U(L)$ is the universal enveloping algebra of L.

Let $\leq$ be a total ordering of X. Since multiplication $[\,,]$ is K-linear, elements $yx - \varepsilon(y,x)\, xy - [y,x]$ with $x < y \in X$ and $yy - \frac{1}{2}[y,y]$ with $y \in X_-$ generated the ideal I.

Theorem 19.1
Let K have no 2-torsion. The universal enveloping algebra $U(L)$ constructed above is a free K-module with a G-homogeneous basis consisting of 1 and all monomials of the form $\delta(x_1)\cdots\delta(x_n)$ where $n \in \mathbb{N}$, $x_i \in X$, $x_i \leq x_{i+1}$, $x_i \neq x_{i+1}$ with $x_i \in X_-$ for all $i = 1, \ldots, n-1$ (in particular, δ is an embedding).

PROOF Let S be the reduction system in $A(X)$ consisting of the pairs $\sigma_{yx} = (yx, \varepsilon(y,x)\, xy + [y,x])$ for all $y > x$ in X and $\sigma_{yy} = (yy, \frac{1}{2}[y,y])$ for all $y \in X_-$. It is clear that the ideal I is generated by all elements $W_\sigma - f_\sigma$, $\sigma \in S$. Let us define the misordering index of an element $x_1, \ldots, x_n \in S(X)$

as the number of pairs (i,j) such that $i<j$ but $x_i>x_j$. Let us introduce a partially order on $S(X)$ by setting $A<B$ if A is of smaller length than B, or if $m(A)=m(B)$ (i.e., A and B have the same multidegrees), but A has a lesser misordering index than B. It is clear that this ordering is a semigroup one, it is S-compatible and satisfies DCC.

Consider possible ambiguities.

Case 1. $(\sigma_{zy},\sigma_{yx},z,y,x)$ with $x,y,z\in X$, $z>y>x$. In order to resolve such an ambiguity relative to $\le$ we investigate the element

$$R_{\sigma_{zy}x}(zyx)-R_{z\sigma_{yx}}(zyx)=(\varepsilon(z,y)\,yzx+[z,y]x)-(\varepsilon(y,x)\,zxy+z[y,x])$$

more closely. In order to reduce further this element we apply reductions $R_{y\sigma_{zx}}$, $R_{\sigma_{zx}y}$, $R_{\sigma_{yx}z}$, $R_{x\sigma_{zy}}$.

Thus, we have

$$\begin{aligned}&\varepsilon(z,y)(\varepsilon(z,x)(\varepsilon(y,x)\,zyx+[y,x]z)+y[z,x])\\&\qquad+[z,y]x\,\varepsilon(y,x)(\varepsilon(z,x)(\varepsilon(z,y)\,xyz+x[z,y])+[z,x]y)-z[y,x]\\&=\quad\big[[z,y],x\big]-\big[z,[y,x]\big]+\varepsilon(z,y)\,\big[y,[z,x]\big]\equiv 0\pmod I.\end{aligned}$$

Case 2. $(\sigma_{zy},\sigma_{yy},z,y,y)$ with $z\in X$, $y\in X_-$, $z>y$.

Consider the element

$$R_{\sigma_{zy}y}(zyy)-R_{z\sigma_{yy}}(zyy)=\varepsilon(z,y)\,yzy+[z,y]y\tfrac{1}{2}z[y,y].$$

We apply now the reductions $R_{y\sigma_{zy}}$ and $R_{\sigma_{yy}z}$ to this element. Then we have $\big[[z,y],y\big]-\frac{1}{2}\big[z,[y,y]\big]\equiv 0\pmod I$.

Case 3. $(\sigma_{yy},\sigma_{yz},y,y,z)$ with $z\in X$, $y\in X_-$, $y>z$.

We consider this situation by analogy with the second case. It is obvious that the ambiguity $(\sigma_{yy},\sigma_{yy},y,y,y)$ is resolvable. Remark that in our system of reductions S there are no inclusion ambiguities.

The claim of our theorem now follows from Theorem 18.1 and from the fact that the irreducible words have the form, described in the statement of our theorem. ■

Let K be a field, $\operatorname{char}K=p>2$, L a color Lie p-superalgebra with a G-homogeneous linear basis X, $A(X)$ a free G-graded associative K-algebra, I the two-sided ideal of $A(X)$ generated by elements $ab-\varepsilon(a,b)\,ba-[a,b]$ (where a,b are homogeneous, $a,b\in L$) and $a^p-a^{[p]}$ ($a\in L_g$, $g\in G_+$).

Let $\delta\colon L\to A(X)/I$ be the canonical mapping. It is clear that $A(X)/I=u(L)$ is the restricted universal enveloping algebra for L, $u(L)\cong U(L)/J$ where $U(L)$ is the universal enveloping algebra for L, constructed above, and J is the two-sided ideal of $U(L)$ generate by elements $a^p-a^{[p]}$ ($a\in L_g$, $g\in G_+$).

Let $\le$ be a total order on X. Since $[\,,\,]$ is a K-linear multiplication we see that elements $yx-\varepsilon(y,x)\,xy-[y,x]$ where $x,y\in X$, $y>x$ and $x^p-x^{[p]}$ where $x\in X_g$, $g\in G_+$, generate the ideal I.

Theorem 19.2
Let K be a field, $\operatorname{char} K > 2$. *The monomials of the form*

$$\delta(x_1)^{\lambda_1} \cdots \delta(x_n)^{\lambda^n}$$

where $x_i \in X$, $x_1 < \cdots < x_n$, $0 \leq \lambda_i \leq p-1$ *for* $x_i \in X_+$, $\lambda_i = 0, 1$ *for* $x_i \in X_-$ *give us a linear basis of* $u(L)$.

PROOF Let S be the reduction system in $A(X)$ consisting of the pairs $\sigma_{yx} = (yx, \varepsilon(y,x)\,xy + [y,x])$ for all $y, x \in X$, $x < y$, $\sigma_{yy} = (yy, \frac{1}{2}[y,y])$ for all $y \in X_-$, and $\sigma_{x^p} = (x^p, x^{[p]})$ for all $x \in X_+$.

Using the proof of Theorem 19.1 and some properties of the operation $[p]$ one can easily see that all possible ambiguities are resolvable.

Thus, using now Theorem 18.1 and the fact that irreducible words have the form described in the statement of our theorem, we complete the proof. ∎

Separability

Theorem 19.3 ([396])
Let K be a field, L a color Lie superalgebra, B a subalgebra of L. Then B is diffential separable, i.e. if J is the left ideal of $U(L)$ generated by B, then $L \cap J = B$.

In the case when $\operatorname{char} K = p > 2$ *let M be a color Lie p-superalgebra, D a subalgebra of M. Then B is differential separable (relative to $u(M)$).*

PROOF Let $E = \{e_1, e_2, \ldots\}$ be a G-homogeneous ordered K-linear basis of the subalgebra B. We extend the subset E up to a G-homogeneous ordered K-linear basis $E \cup F$ of L such that $F = \{f_1, f_2, \ldots\}$ and $f_i < e_j$ for all i, j. By Theorem 19.1 the set of all monomials

$$f_{i_1} \cdots f_{i_m} \cdot e_{i_1} \cdots e_{i_n}$$

where

$$f_{i_1} \leq \cdots \leq f_{i_m}, \qquad e_{i_1} \leq \cdots \leq e_{i_n},$$

$i_r \neq i_{r+1}$ with $d(f_{i_r}) \in G_-$, $j_s \neq j_{s+1}$ with $d(e_{j_s}) \in G_-$, give us a K-linear basis of $A(X)$. Since B is a subalgebra of L, these monomials with $n \geq 1$ form a K-linear basis of J. Therefore $L \cap J = B$.

The same arguments work for the color Lie p-superalgebra M (we need only use Theorem 19.2). ∎

Remark 19.4
It is possible to prove Theorems 14.1 and 14.2 with the use of Theorem 19.3. We present a sketch of the proof. By the way of contradiction assume that we have a reduced system $\{h_1, \ldots, h_k\}$ of elements of $L(X)$ (of $L^p(X)$)

such that it is not an independent system. Then we get the associative presentation of the relation of the form

$$\sum_{i=1}^{k} a_i h_i = 0,$$

$a_i \in A(X)$. One can assume that

$$\sum_{i=1}^{k} \widehat{a_i}\widehat{h_i} = 0.$$

Using Theorem 12.2 we get

$$\widehat{h_i} = \sum_{j \neq i} b_j \widehat{h_j},$$

$b_j \in A(X)$. It follows from Theorem 19.3 that $\widehat{h_i}$ belongs to the subalgebra generated by the set $\{\widehat{h_j} \mid j \neq i\}$, and we have come to a contradiction.

Theorem 19.5
Let L be a color Lie (p-)superalgebra, U its (restricted) universal enveloping algebra, $I_L = L \cdot U$. Then L is the free color Lie (p-)superalgebra with the free G-homogeneous generating set $\{f_1, \ldots, f_k\}$ if and only if I_L is the free U-module with the basis $\{f_1, \ldots, f_k\} \subseteq L$.

PROOF Let I_L be the free U-module with the basis $\{f_1, \ldots, f_k\} \subseteq L$. Denote by H the subalgebra of L generated by $\{f_1, \ldots, f_k\}$. Let $I_H = H \cdot U$. By Theorem 19.3, $L \cap I_H = H$. Since $f_1, \ldots, f_k \in I_H$, $I_H = I_L$. Hence $L \cap I_H = L$, and $L = H$. Thus L is generated by the elements $f_1, \ldots, f_k$.

Let now $Z = \{z_1, \ldots, z_k\}$, $d(z_i) = d(f_i)$, and let $F(z_1, \ldots, z_k)$ be a nonzero G-homogeneous element of minimal length of the free G-graded associative algebra $A(Z)$ such that $F(f_1, \ldots, f_k) = 0$. Since $f_i \in L$, F has no constant term. Let

$$F = z_1 F_1 + \cdots + z_k F_k.$$

As $F \neq 0$, $F_i \neq 0$ for some i. It follows from $\ell_Z(F_i) < \ell_Z(F)$ that $F_i(f_1, \ldots, f_k) \neq 0$. Therefore we have a nontrivial relation

$$F(f_1, \ldots, f_k) = f_1 F_1(f_1, \ldots, f_k) + \cdots + f_k F_k(f_1, \ldots, f_k) = 0.$$

It gives a contradiction with the fact that the subset $\{f_1, \ldots, f_k\}$ is a basis of the free U-module I_L.

At the other hand, it is clear that X is a basis of the free $A(X)$-module $L(X) \cdot A(X)$ $(L^p(X) \cdot A(X)$, respectively). ∎

Remark 19.6

Now it is possible to prove Theorems 15.4 and 15.6 using Theorem 19.5. For example, if $|X| = n < \infty$, B is a subalgebra of $L(X)$, $B_+ = L(X)_+$, $\dim L(X)/B = k < \infty$, then, by Theorem 19.1, $\dim A(X)/I_B = 2^k$. By Theorem 12.3, $\operatorname{rank}(I_B) = 2^k(n-1)+1$. Using Theorem 19.5 we see that $\operatorname{rank}(B) = \operatorname{rank}(I_B)$.

Theorem 19.7

Let $L(X)$ be a free color Lie superalgebra over a field K (in the case $\operatorname{char} K = p > 2$ we consider also a free color Lie p-superalgebra $L^p(X)$). Then the algebra $L = L(X)$ ($L = L^p(X)$, respectively) is finitely separable, i.e. if a is a G-homogeneous element of L, B is a G-homogeneous finitely generated subalgebra of L, $a \notin B$, then there exists a finite dimensional color Lie (p-)superalgebra H and an epimorphism $\varphi: L \to H$ such that $\varphi(a) \notin \varphi(B)$.

PROOF It is sufficient to prove the statement in the case when $|X| < \infty$. At the beginning we consider the free color Lie superalgebra $L = L(X)$. The free G-graded associative algebra $A(X)$ is the universal enveloping algebra of L. Consider the left ideal J of $A(X)$ generated by the subalgebra B. Since B is a finetely generated subalgebra, J is a finitely generated left ideal of $A(X)$. Since $a \in L$ and $a \notin B$, by Theorem 19.3 $a \notin J$. By Lemma 18.4 there exists a left-G-graded ideal J' of finite codimension in $A(X)$ such that $a \notin J'$.

We now first look more closely at the finite dimensional left $U(L)$-module $M = U(L)/J'$, where $U(L) = A(X)$. Let $\operatorname{Ann}(M)$ be the annihilator of the module M and $R = U(L)/\operatorname{Ann}(M)$. Then R is a finite dimensional G-graded associative algebra, M is a faithful R-module and $\operatorname{Ann}(M) \subseteq J'$ since $1 \in U(L)$. Let $\xi: U(L) \to R$ be the canonical epimorphism. It is obvious that $H = [\xi(L)]$ is a finite dimensional color Lie superalgebra. Assuming now that $\xi(a) \in \xi(B)$, one obtains for $\{u_1, \ldots, u_r\}$ is a set of generators of the subalgebra B, that $\xi(a)$ belongs to the left ideal of R generated by $\xi(u_1), \ldots, \xi(u_r)$. Therefore there exist elements $v_1, \ldots, v_r \in U(L)$ such that

$$\xi(a) = \sum_{i=1}^{r} \xi(v_i)\,\xi(u_i),$$

in other words

$$a - \sum_{i=1}^{r} \xi(v_i)\,\xi(u_i) \in \operatorname{Ann}(M) \subseteq J',$$

and thus $a \in J'$. This contradiction completes the proof of our theorem in the case $L = L(X)$.

In the case $\operatorname{char} K = p > 2$ and $L = L^p(X)$ the proof is underway in the same way using the fact that the free G-graded associative algebra $A(X)$ is the restricted universal enveloping algebra of L. ∎

Remark 19.8

S. A. Agalakov in [3] proved that free associative algebras which rank is more than one contain subalgebras which are not finite separable. He also constructed in [4] an example of Lie algebra over a field of zero characteristic given by one defining relation such that it is not residually finite: if K is a field, $\operatorname{char} K = 0$, $X = \{x, y\}$, $h = \big[[x,y],x\big] - [x,y] - x$, $L = L(X)/\mathrm{id}(h)$, then L is not a residually finite algebra; if $\operatorname{char} K > 0$, then L is a residually finite algebra. This is true also if $L = L'(X)/\mathrm{id}(h)$, where $L'(X)$ is the free solvable Lie algebra of degree 3, see [2]. For the varieties of Lie algebras see also [616].

U. U. Umirbaev in [9] proved that for universal enveloping algebras of free metabelian algebras the membership problem for finitely generated right ideals is unsolvable. At the same time it follows from Yu. A. Bahturin's theorem [24] that the entry problem for finitely generated subalgebras of free metabelian Lie algebras is solvable (in [567] U. U. Umirbaev proved that this problem is unsolvable for free solvable Lie algebras of degree greater than 2; the insolvability of the equality problem for these algebras was proved by G. P. Kukin in [294]).

PBW-theorems for differential Lie superalgebras

Suppose that a (restricted) differential Lie F-superalgebra D is free as a G-graded left F-module and F is free as a G-graded Φ-module. Denote by D' the algebra D as G-graded set. Let $\Phi(D')$ be the free G-graded associative algebra with 1 over Φ on D'. Consider the free product $P = F * \Phi(D')$ over Φ of G-graded associative algebras with 1. Let I be the ideal in P generated by all elements of the form

$$(\delta+\mu)' - \delta' - \mu', \qquad (c\delta)' - c\delta', \qquad [\delta,\mu]' - [\delta',\mu'], \qquad [\delta',c] - (\partial(\delta))(c)$$

for all homogeneous $\delta, \mu \in D$, $c \in F$; if D is a restricted differential Lie F-superalgebra, then we add to the generating set of I the elements

$$\left(\delta^{[p]}\right)' - (\delta')^p$$

for all homogeneous $\delta \in D_+$.

One can easily see that the algebra P/I is the (restricted) universal enveloping algebra of the (restricted) differential Lie F-superalgebra D.

Theorem 19.9 ([399])

Let $E = E_+ \cup E_-$ be a totally ordered G-graded basis of the free F-module D, $x < y$ for all $x \in E_+$ and $y \in E_-$, F a free Φ-module, $\xi\colon D \to U(D)$ the (restricted) universal enveloping algebra of the (restricted) differential Lie F-superalgebra D. Then $U(D)$ is a free F-module with the G-homogeneous basis consisting of 1 *and all monomials of the form*

$$(\xi(e_1))^{\alpha_1}(\xi(e_2))^{\alpha_2}\cdots(\xi(e_m))^{\alpha_m}(\xi(f_1))^{\beta_1}\cdots(\xi(f_n))^{\beta_n}$$

with $e_i \in E_+$, $f_j \in E_-$, $e_1 < e_2 < \cdots < e_m$, $f_1 < f_2 < \cdots < f_n$, $\alpha_i \in \mathbb{Z}_+$, $\beta_j \in \{0,1\}$ *(in the case when D is a restricted differential Lie F-superalgebra, additionally we suppose that $\alpha_i < p$).*

For differential Lie algebras PBW-theorems may be found also in [50].

Remark 19.10
In connection with differential forms pseudo Lie algebras were under consideration (see for example the article [482] by G. Rinehart). If K is a commutative ring, then a K-module L is a pseudo Lie algebra if there are a mapping $L \times L \to L$, $(u,v) \to [u,v]$, and a K-homomorphism $D: L \to \mathrm{Der}K$ such that

$$[u,u] = 0; \qquad [u+v,w] = [u,v] + [v,w];$$
$$\big[u,[v,w]\big] + \big[v,[w,u]\big] + \big[w,[u,v]\big] = 0;$$
$$[u,\lambda v] = \lambda[u,v] + (D_u\lambda)v; \qquad D_{[u,v]} = D_uD_v - D_bD_a$$

for all $u,v,w \in L$, $\lambda \in K$ with $D_u = D(u) \in \mathrm{Der}K$.

In [221] Ya. Halluf was dealing with pseudo Lie algebras and their representations, in particular, the PBW-theorem for universal enveloping algebras of such algebras has been derived.

Now we examine graded algebras of universal enveloping algebras. We look at the filtration on the universal enveloping algebra $U(L)$ of the color Lie superalgebra L defined by $U^{-1} = \{0\}$, $U^0 = K \cdot 1$ and let U^n denote the K-submodule generated by all monomials of length $\leq n$ (see Theorem 19.1). As usual the associated graded module

$$\mathrm{gr}\, U(L) = \bigoplus_{n=-1}^{\infty} U^{n+1}/U^n$$

with multiplication

$$(a + U^{n-1})(b + U^{m-1}) = ab + U^{n+m-1}$$

where $a \in U^n$, $b \in U^m$ is a G-graded associative algebra. It is clear that $\mathrm{gr}\, U(L)$ is generated by all elements $a + U^0$, $a \in L$. For the G-homogeneous elements a and b we have

$$(a + U^0)(b + U^0) = ab + U^1 = \varepsilon(a,b)\, ba + [a,b] + U^1 = \varepsilon(a,b)\, ba + U^1$$

because $[a,b] \in L \subseteq U^1$.

As in the case of Lie algebras (see for example [30], [241], [367]) we have the following properties of the graded algebras of universal enveloping algebras.

Proposition 19.11
Let K be a field, $\operatorname{char} K \neq 2$, L a color Lie superalgebra with a G-homogeneous basis E and let $U(L)$ be its universal enveloping algebra. Then the associated graded algebra $\operatorname{gr} U(L)$ is isomorphic to the free ε-commutative associative algebra $K[E]$. Therefore, the subalgebra L_+ has no zero divisors, and the Jacobson radical of $U(L_+)$ is equal to zero.

If $\operatorname{char} K = p > 2$ and L is a color Lie p-superalgebra, then the associated graded algebra $\operatorname{gr} U(L)$ is isomorphic to $K[E]/I$ where I is the ideal of $K[E]$ generated by the elements e^p, $e \in E_+$.

Remarks

Remark 19.12
Many authors considered generalizations of the universal enveloping algebras from the point of view of computer algebra. For example, in [46] the following algebras (PBW-algebras) are under consideration. Let K be a field, B a K-algebra with a filtration

$$0 \subseteq B_0 \subseteq B_1 \subseteq \cdots \subseteq B, \qquad B = \bigcap_{i \geq 0} B_i, \qquad B_i B_j \subseteq B_{i+j}$$

for all $i, j \in \mathbb{N}$,

$$\operatorname{gr} B = \bigoplus_{i \geq 0} B_i / B_{i-1} \qquad (B_{-1} = 0)$$

the associated graded algebra. If the graded K-algebras $\operatorname{gr} B$ and the polynomial algebra $K[x_1, \ldots, x_n]$ are isomorphic, then B is called a finitely generated PBW-algebra. Taking into account the fact that B is Noetherian and applying the natural degree function, P. Beckmann and J. Stückrad in [46] consider Gröbner bases for ideals of B. It would be interesting to consider the situation when

$$\operatorname{gr} B \cong K[x_1, \ldots, x_n] \otimes \Lambda(y_1, \ldots, y_n)$$

(a super-PBW algebra).

Remark 19.13
In the theory of quantum groups the algebras with quadratic relations (it means that the degree of each relation is equal to two) play an important role (see for example [163]). In [585] A. M. Vershik considered associative algebras over $\mathbb{C}$ with generators and filtrations

$$A_0 = \mathbb{C}, \quad A_1 = \mathbb{C}x_1 \oplus \cdots \oplus \mathbb{C}x_n,$$
$$A^0 = A_0, \quad A^1 = A_0 + A_1, \quad \ldots, \quad A^{k+1} = A_1 A_k, \quad \ldots,$$
$$A^0 \subseteq A^1 \subseteq \cdots \subseteq A = \bigcup_{k=0}^{\infty} A_k.$$

Such algebra was called a PBW-algebra if $\dim_{\mathbb{C}} A_n = C^n_{n+m-1}$, $n \geq 0$, and a PBW-type algebra if $\dim_{\mathbb{C}} A_n \leq C^n_{n+m-1}$, $n \geq 2$. In [585] we may find some necessary and sufficient conditions for algebra with quadratic relations to be PBW-algebra and PBW-type algebra.

Remark 19.14
In [59], [60] relations of the form $yx = q_{xy}xy + (x, y)$, where $q_{xy} \in K$ and the length of (x, y) is less than 2, are under consideration. For algebras given by such relations necessary and sufficient conditions to have standard bases were given.

For further information on algebras with quadratic relations see for example [562].

Problem 19.15
A very interesting problem is to prove that the Jacobson radical of a finitely generated Lie superalgebra is nil, i.e. consists of nilpotent elements.

Note that in the case of associative algebras there are counterexamples of K. I. Beidar [48] and V. T. Markov [364], but the algebras from these counterexamples are rather far from to be universal enveloping algebras. Note also the following result.

Theorem 19.16 (A. A. Baranov [37])
Let L be a color Lie superalgebra over a field such that $\dim(L_-) < \infty$. *Then the Jacobson radical $J(U(L))$ is nilpotent.*

Remark 19.17
We mention some articles which are devoted to enveloping algebras of Lie superalgebras. In some of them one can find results on primeness, identities, representations (see for example [25], [30], [58], [246], [366], [393], [461], [464]; for universal enveloping algebras of Lie algebras see [155]). Some of the most intriguing questions involve connections between types of growth of algebras and their algebraic properties. In [462], [463] V. M. Petrogradsky introduced some generalizations of growth of Lie algebras and their universal enveloping algebras (as for general problems on growth see [285]).

Remark 19.18
Let $\operatorname{char} K \neq 2$, and let L be a graded Lie algebra over K. We say that L is torsion free if $[x, x] \neq 0$ for any non-zero odd element x, and absolutely torsion free if $L' = L \otimes_K K'$ is torsion free (K' is the algebraic closure of K). M. Aubry and J.-M. Lemaire prove in [15] that the universal enveloping algebra $U(L)$ of an absolutely torsion free graded Lie algebra L has no zero divisors; the proof is valid also for color Lie superalgebras.

We mention also a problem posed by V. M. Petrogradsky.

Problem 19.19 ([156], Problem 3.59)
Suppose that L is a Lie p-algebra without algebraic elements (relatively to the p-mapping). Is it true that the restricted universal enveloping algebra $u(L)$ has no zero divisors? It is also interesting to consider this problem for color Lie p-superalgebras.

Remark 19.20
If L is not a free K-module then the canonical mapping $L \to U(L)$ need not to be injective, as A. I. Shirshov's example from [510] shows. Let $K = \mathbb{Z}_2[x_1, x_2, x_3]/I$ be the factor algebra of the algebra of polynomials over $\mathbb{Z}_2$ by the ideal I generated by x_1^2, x_2^2, x_3^2. Let $L(y_1, y_2, y_3)$ be the free nilpotent of index 2 Lie algebra over K, and J the ideal of L generated by the element

$$(x_1 + I)y_1 + (x_2 + I)y_2 + (x_3 + I)y_3.$$

It is impossible to embed the Lie algebra $L(y_1, y_2, y_3)/J$ into any algebra of the form $[A]$ where A is an associative K-algebra (if φ is a homomorphism from $L(y_1, y_2, y_3)$ into $[A]$, then for

$$v = (x_1 + I)(x_2 + I)y_1y_2 + (x_1 + I)(x_3 + I) + (x_2 + I)(x_3 + I)y_2y_3 \neq 0$$

we have that $\varphi(v) = 0$ in $[A(X)]$).

For Lie superalgebras the general problem of the existence of $U(L)$ with the embedding of L into $U(L)$ is open. However in [176] M. El-Agawani and A. Micali using Baer invariants present a number of sufficient conditions for the existence of such embedding (for example, if K is a commutative ring, $1/2 \in K$, L is a Lie superalgebra over K, $M = L_K$ as K-module, then if either M is flat, or M is a direct sum of cyclic modules, or M is uniquely divisible Abelian group, then there exists an embedding; if either K is absolutely flat, or K is a $\mathbb{Q}$-algebra, or K is a generalized valuation ring, then also there exists an embedding). For the case of Lie algebras see also [110], [230], [318].

Universal enveloping algebras as graded Hopf algebras

Let K be a field, $\operatorname{char} K \neq 2$, L a color Lie superalgebra, $U(L)$ its universal enveloping algebra, $U(L) \mathbin{\widehat{\otimes}} U(L)$ the color tensor product. We define *the diagonal mapping* $\delta \colon L \to U(L) \mathbin{\widehat{\otimes}} U(L)$ setting $\delta(a) = a \otimes 1 + 1 \otimes a$ for any $a \in L$.

Assuming $\delta(1) = 1 \otimes 1$, by Theorem 19.1 we derive that this mapping has the unique extension to a homomorphism $\delta \colon U(L) \to U(L) \mathbin{\widehat{\otimes}} U(L)$ of G-graded associative algebras called *the coproduct on the universal enveloping algebra $U(L)$.*

It is clear that the coproduct is coassociative, i.e. the following diagram is commutative:

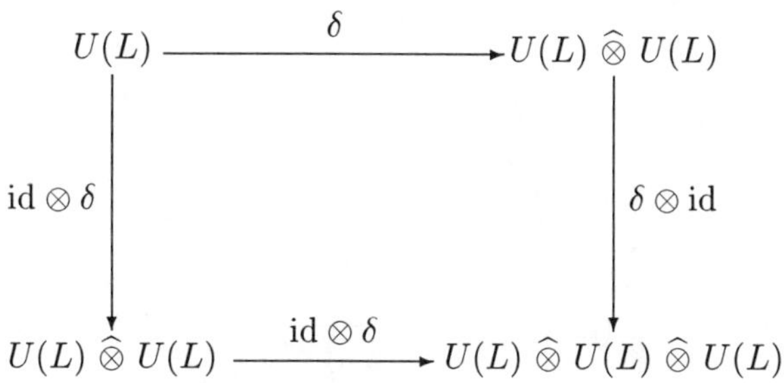

where $\mathrm{id} = \mathrm{id}_{U(L)}$.

We consider now the unique linear homomorphism $e\colon U(L) \to K$ such that $e(L) = 0$ and $e(1) = 1$. Then e is *the counit* of $U(L)$,

$$(e \otimes \mathrm{id}) \circ \delta = \mathrm{id} = (\mathrm{id} \otimes e) \circ \delta$$

(identifying canonically $K \mathbin{\widehat{\otimes}} U(L)$ and $U(L) \mathbin{\widehat{\otimes}} K$ with $U(L)$), i.e. we have the following commutative diagram:

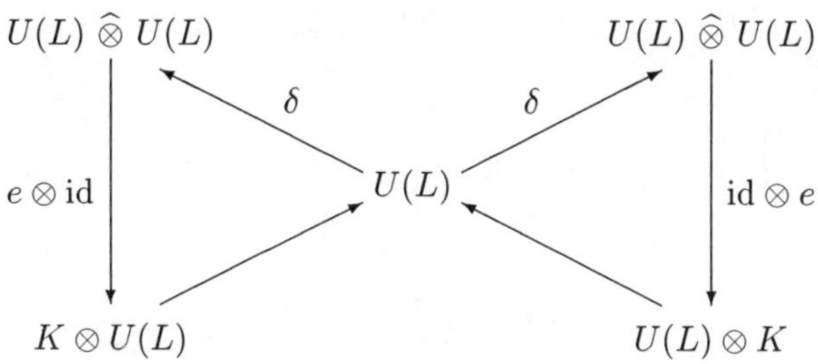

Thus, $U(L)$ is a color coalgebra with the counit. Since δ and e are homomorphisms and $\delta(1) = 1 \otimes 1$, $e(1) = 1$, we see that $U(L)$ with multiplication and comultiplication is a color bialgebra. Observe that $U(L)$ is cocommutative, i.e. for the unique automorphism τ of the G-graded associative algebra $U(L) \mathbin{\widehat{\otimes}} U(L)$ with $\tau(x \otimes y) = \varepsilon(x, y)\, y \otimes x$ with homogeneous $x, y \in U(L)$

we have $\tau \circ \delta = \delta$, i.e. the following diagram is commutative:

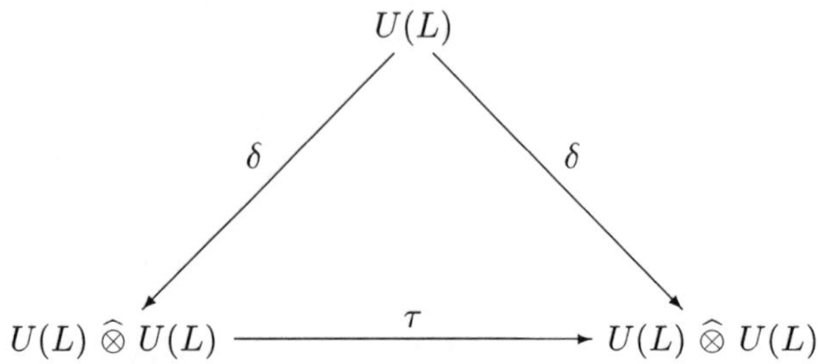

Consider now the linear mapping $\mu: U(L) \widehat{\otimes} U(L) \to U(L)$ induced by the multiplication in $U(L)$. We define the mapping $e': U(L) \to U(L)$ by setting $e'(x) = e(x) \cdot 1_U$ for all $x \in U(L)$. Let $\eta: U(L) \to U(L)$ be the linear mapping such that $\eta(1) = 1$, $\eta(x) = -x$ for all $x \in L$, $\eta(xy) = \varepsilon(x, y)\, \eta(y)\, \eta(x)$ for all homogeneous $x, y \in U(L)$. It is clear that $\eta^2 = \mathrm{id}$ and

$$\mu \circ (\eta \otimes \mathrm{id}) \circ \delta = e' = \mu \circ (\mathrm{id} \otimes \eta) \circ \delta,$$

that is the following diagram is commutative:

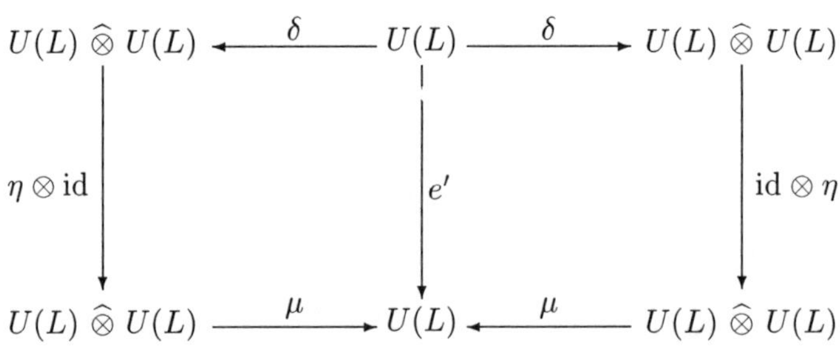

Thus the constructed linear mapping η is *an antipode* and our color bialgebra $U(L)$ with the antipode η is *a cocommutative color Hopf algebra.*

Analogously, we check that in the case of a field K, $\operatorname{char} K = p > 2$, the restricted universal enveloping algebra for a color Lie p-superalgebra L also has the structure of a cocommutative color Hopf algebra.

For any color Hopf algebra we denote the set of all its primitive elements by $\mathcal{P}(H) = \{a \in H \mid \delta(a) = a \otimes 1 + 1 \otimes a\}$.

Theorem 19.21
If K is a commutative ring which is torsion free as $\mathbb{Z}$-module, L is a color

Lie superalgebra which is a free K-module, then the subset of primitive elements $\mathcal{P}(U(L))$ of the universal enveloping algebra $U(L)$ of the color Lie superalgebra L coincides with L.

Note that Friedrichs type criterion for generalized Lie algebras were proved by R. Ree [476].

Theorem 19.22
Let K be a field, $\operatorname{char} K = p > 2$, L a color Lie superalgebra, M a color Lie p-superalgebra, $U(L)$ the universal enveloping algebra, $u(M)$ the restricted universal enveloping algebra. Then

(i) $\mathcal{P}(u(M)) = M$;

(ii) $\mathcal{P}(U(L))$ *coincides with the color Lie p-superalgebra generated by L in $U(L)$.*

Remark 19.23
In [181] D. Fischman and S. Montgomery considered graded Hopf algebras in connection with double centralizer theorem and proved a double centralizer theorem for color Lie superalgebras. In [28] Yu. A. Bahturin, D. Fischman, and S. Montgomery deal with the generalized Lie structure of associative algebras.

A. N. Koryukin in [276] examined pointed color Lie superalgebras and ε-bialgebras in connection with Hopf envelopes of color Lie superalgebras. A color Lie superalgebra L over a field K with the graded group G and the form ε is *a pointed color Lie superalgebra* if there is a mapping

$$\mathrm{In}\colon G_0 = \{g \in G \mid \varepsilon(g, G) = 1\} \to L$$

such that for all $g \in G_0$

$$\begin{array}{c} \mathrm{In}(g) \in L_g; \qquad [\mathrm{In}(g), L] = 0; \\ \mathrm{In}(g) = 0 \quad \Leftrightarrow \quad g = 1. \end{array}$$

In the case when $\operatorname{char} K = p > 0$ a color Lie p-superalgebra is called a pointed color Lie p-superalgebra if additionally

$$(\mathrm{In}(g))^{[p]} = \mathrm{In}(g^p).$$

Let H be a K-bialgebra with the comultiplication Δ and the counit e,

$$\begin{aligned} G(H) &= \{g \in H \mid \Delta(g) = g \otimes g\}, \\ L_f(H) &= \{g \in H \mid \Delta(g) = g \otimes f + f \otimes g\}, \quad f \in G(H). \end{aligned}$$

A bialgebra H is a bialgebra over G if H is a bialgebra and G is a subgroup of the semigroup $G(H)$. For $f \in G$ we put

$$\begin{aligned} L_f^{\varepsilon}(H) &= \{h \in L_f(H) \mid g^{-1}hg = h\,\varepsilon(f, g) \ \ \forall g \in G\}; \\ L^{\varepsilon}(H) &= \bigoplus_{g \in G} L_g^{\varepsilon}(H). \end{aligned}$$

A bialgebra H over G is called *an ε-bialgebra* if it is generated by all homogeneous elements from L^ε and by the elements of G.

Theorem 19.24 (A. N. Koryukin [276])
(i) *In the case when* $\operatorname{char} K = 0$ *the category of ε-bialgebras is equivalent to the category of pointed color Lie superalgebras.*
(ii) *In the case when* $\operatorname{char} K = p > 0$ *the category of ε-bialgebras is equivalent to the category of pointed color Lie p-superalgebras.*

Now we analyze connections between residual finiteness of Lie superalgebras and their enveloping algebras. Recall that a (graded) algebra is called residually finite if the intersection of all its (graded) ideals of finite codimension is equal to zero.

Theorem 19.25 ([393])
Let K be a field, $\operatorname{char} K \neq 2$. *Then:*
(i) *if L is a color Lie superalgebra over K, then L is residually finite if and only if $U(L)$ is residually finite;*
(ii) *if* $\operatorname{char} K = p > 2$, *$M$ is a color Lie p-superalgebra over K, then M is residually finite if and only if $u(M)$ is residually finite.*

One can give a proof of this theorem using PBW-theorems, Friedrichs criteria, the structure of graded Hopf algebras on $U(L)$ (on $u(M)$), and Ado-Iwasawa theorem for color Lie (p-)superalgebras. For the details for Lie algebras see [375], [376].

BCH problem for Lie superalgebras

Let K be an algebra over $\mathbb{Q}$, $L(X)$ the free Lie algebra on X over K. Consider the algebra $A(X)^\infty$ of all formal associative series of the form

$$\sum_{n=0}^{\infty} \alpha_n a_n,$$

where $\alpha_n \in K$, a_n are ℓ-homogeneous elements of $A(X)$, $\ell(a_n) = n$. Analogously we consider the Lie algebra $L(X)^\infty$ of all Lie series

$$\sum_{n=0}^{\infty} \beta_n b_n,$$

where $\beta_n \in K$, b_n are ℓ-homogeneous elements of $L(X)$, $\ell(b_n) = n$. We assume that $L(X)^\infty \subseteq A(X)^\infty$. For $z \in A^+(X)$ we define the following functions from $A^+(X)$ to $A(X)^\infty$:

$$\exp(z) = \sum_{i=0}^{\infty} \frac{z^i}{i!}$$

and

$$\log(1+z) = \sum_{i=1}^{\infty} \frac{(-1)^{i+1} z^i}{i}.$$

Then one can easily see that

$$\exp\big(\log(1+z)\big) = 1+z; \qquad \log\big(\exp(z)\big) = z.$$

If $[a,b]=0$, then

$$\begin{aligned} \exp(a+b) &= \exp(a)\exp(b); \\ \log\big((1+a)(1+b)\big) &= \log(1+a)\log(1+b). \end{aligned}$$

Theorem 19.26 (Baker-Campbell-Hausdorff Formula)
If $X=\{x,y\}$, *then* $\exp(x)\exp(y)=\exp(z)$ *for some* $z \in L(X)^{\infty}$.

PROOF We consider the complete tensor product $(A(X)\otimes A(X))^{\infty}$ and extend the diagonal homomorphism δ to the homomorphism $\delta' \colon A(X)^{\infty} \to (A(X)\otimes A(X))^{\infty}$. It's easy to see that Friedrichs criterion (Theorem 19.21) takes place for δ'. Applying δ' to $\exp(x)\exp(y)$ we get

$$\delta'\big(\exp(x)\exp(y)\big) = \exp(x)\exp(y)\otimes\exp(x)\exp(y).$$

Therefore

$$\begin{aligned} \delta'\big(\log(\exp(x)\exp(y)\big) &= \log\big((\exp(x)\exp(y)\otimes 1)(1\otimes\exp(x)\exp(y))\big) \\ &= \log\big(\exp(x)\exp(y)\big)\otimes 1 + 1\otimes\log(\exp(x)\exp(y)\big). \end{aligned}$$

It follows from Friedrichs criterion that

$$z = \log\big(\exp(x)\exp(y)\big) \in L(X)^{\infty}.$$

■

Now using Theorem 9.1 we obtain Dynkin's exact formula for z:

$$\log\big(\exp(x)\exp(y)\big) = \sum_m \sum_{\substack{p_i,q_i \\ p_i+q_i>0}} \frac{(-1)^{m-1}[x^{p_1}y^{q_1}\cdots x^{p_m}y^{q_m}]}{mp_1!q_1!\cdots p_m!q_m!\sum(p_i+q_i)}$$

where we use left normed monomials.

The first five components are as follows

$$\begin{aligned} \log\big(\exp(x)\exp(y)\big) &= x+y+\frac{1}{2}[x,y]+\frac{1}{12}([xy^2]-[xyx]) \\ &\quad -\frac{1}{24}[xy^2x]+\frac{1}{120}([xy^2x^2]-[xy^2xy]) \\ &\quad +\frac{1}{360}([xy^3x]-[xyx^2y]) \\ &\quad +\frac{1}{720}([xyx^3]-[xy^4])+\cdots \end{aligned}$$

The original Hausdorff formula is given by

$$\sum_{n=1}^{\infty} z_n(x,y)$$

with

$$\begin{aligned} z_1(x,y) &= x+y, \\ z_{n+1}(x,y) &= \frac{1}{n+1}\Bigg([x-y, z_n(x,y)] \\ &\quad + \sum_{\substack{m\geq 1 \\ 2m\leq n}} \alpha_{2m} \sum_{\substack{k_1,\dots,k_{2m}>0 \\ k_1+\cdots+k_{2m}=n}} [x+y, c_{k_{2m}}(x,y),\dots,c_{k_1}(x,y)]\Bigg), \\ \alpha_{2m} &= (-1)^{m-1}\frac{2}{(2\pi)^{2m}}\sum_{k=1}^{\infty}\frac{1}{k^{2m}} = (-1)^{m-1}\frac{B_m}{(2m)!}, \end{aligned}$$

where B_m are Bernoulli numbers.

Note that J. Katriel and A. Solomon in [261] gave a q-analog of the BCH-formula. For the function

$$E_q(x) = \sum_{n=0}^{\infty}\frac{x^n}{[n]!}$$

with

$$[n]! = [1][2]\cdots[n], \qquad [k] = \frac{q^k}{q-1},$$

they proved that $E_q(x)E_q(y) = E_q(z)$, where z is a q-series relative to commutators $[a,b] = ab - qba$.

The BCH-formula plays a very important role in Lie algebra theory (it gives the correspondence between Lie algebras and Lie groups). It would be very useful to have such correspondence between Lie superalgebras and (Lie) supergroups (i.e., to have such type presentation for elements of Lie superalgebra). Unfortunately up to now there are no general formulas. Remark only that there is the natural correspondence between free groups and free Lie rings (see [26], [45], [218], [309], [342], [351], [378]). It is interesting also to define the free supergroup with an analogous correspondence to the free Lie superalgebra. In [141] J. Czyz considered some super generalizations of BCH-formula and gave connections between Lie superalgebras and Lie supergroups in the following cases:

(i) $[L_0,L_0] = [L_0,L_1] = 0$;

(ii) $\big[[L,L_1],L_1\big] = 0$;

(iii) $\big[L_0,[L_1,L_1]\big] = 0$.

For some other aspects of BCH-formula see also [34], [35], [94], [106], [107], [109], [118], [171], [172], [188], [199], [224], [279], [280], [321], [348], [357], [373], [458], [541].

Free shuffle superalgebras

Let V and Z be G-graded sets, $V \cap Z = \emptyset$, $v_1, \dots, v_k$ pairwise distinct elements of V, $z_1 \dots, z_l$ pairwise distinct elements of Z. We say that a word $w \in S^+(V \cup Z)$ is *a shuffle word* of the words $v_1 \cdots v_k$ and $z_1 \cdots z_l$ if

$$\begin{aligned} m(w) &= v_1 + \cdots + v_k + z_1 + \cdots + z_l; \\ w\Big|_{\substack{z_1=1 \\ \cdots \\ z_l=1}} &= v_1 \cdots v_k; \\ w\Big|_{\substack{v_1=1 \\ \cdots \\ v_k=1}} &= z_1 \cdots z_l. \end{aligned}$$

The parity $\sigma(w)$ of the shuffle word w of words $v_1 \cdots v_k$ and $z_1 \cdots z_l$ is the sum of all $\varepsilon(z_i, v_j)$ such that z_i is situated before v_j in w.

Let X be a G-graded set. Consider now G-graded sets V and Z, such that $|V| = |Z| = |X|$, $d(v_i) = d(z_i) = d(x_i)$. For $x_{i_1}, \dots, x_{i_k}, x_{j_1}, \dots, x_{j_l} \in X$ we define *the shuffle product* $(x_{i_1} \cdots x_{i_k}) * (x_{j_1} \cdots x_{j_l})$ as the following linear combination of words of multidegree $x_{i_1} + \cdots + x_{i_k} + x_{j_1} + \cdots + x_{j_l}$:

$$(x_{i_1} \cdots x_{i_k}) * (x_{j_1} \cdots x_{j_l}) = \sum_w \sigma(w) w\Big|_{\substack{v_{i_s} = x_{i_s},\ s=1,\dots,k; \\ z_{j_t} = x_{j_t},\ t=1,\dots,l}}$$

where w is running through all shuffle words of the words $v_{i_1} \cdots v_{i_k}$ and $z_{j_1} \cdots z_{j_l}$. Taking $1 * u = u * 1$ for all $u \in S^+(X)$ and extending $*$ on the free G-graded associative algebra $A(X)$ on X over a commutative associative ring K with the unity element by linearity, we define *the shuffle product* $*$ on $A(X)$. Then $A(X)$ with this product is an ε-commutative and associative algebra.

For example, if $G = \mathbb{Z}_2$, $\varepsilon(f, g) = (-1)^{fg}$, $d(x) = 0$, $d(y) = 1$, then

$$xxy * xy = -xxyxy - xyxxy;$$

if $d(x) = 1$, $d(y) = 0$, then

$$xxy * xy = 2xxxyy + xxyxy + xyxxy.$$

One can define the shuffle product $*$ on $A(X)$ in the following way:

$$\begin{aligned} 1 * u &= u * 1 = u; \\ (xu) * (yv) &= x(u * (yv)) + \varepsilon(y, x)\varepsilon(y, u)\, y((xu) * v) \end{aligned}$$

for all $x, y \in X$, $u, v \in S(X)$ (with the extension $*$ on $A(X)$ by linearity).

If we consider $A(X)$ as the universal enveloping algebra of the free color Lie superalgebra $L(X)$, then $*$ is the adjoint operation to the coproduct δ on $A(X)$, i.e.

$$\langle u * v, w \rangle = \langle u \otimes v, \delta(w) \rangle$$

for all $u, v, w \in A(X)$ with the standard scalar product $\langle\, , \rangle$ on $A(X)$ ($S(X)$ forms an orthonormal basis of $A(X)$). Therefore we have the adjoint Friedrich's criterion:

Theorem 19.27
Let char $K = 0$, $w \in A^+(X)$. *Then* $w \in L(X)$ *if and only if* $\langle u * v, w\rangle = 0$ *for all* $u, v \in A^+(X)$.

In the case of prime characteristic such criterion for the free Lie algebra was given by R. Block.

Theorem 19.28 (R. Block [76])
At arbitrary characteristic, an element w *of* $A(X)$ *belongs to the free Lie algebra* $L(X)$ *if and only if* w *is annihilated by the unity element and by the divided power shuffle square of* $A^+(X)$ *(i.e., by the subalgebra of the shuffle algebra* $A(X)$ *generated by all products* $a*b$ *and divided powers* $\gamma_n a = a^n/n!$ *with* $a, b \in A^+(X)$ *and* $n > 1$).

Furthermore, one can give Friedrichs criterion in the following form.

Theorem 19.29
Let char $K = 0$, *and let* X *be a* G*-graded set. Consider a* G*-graded set* Y, *such that* $|Y| = |X|$, $d(y_i) = d(x_i)$ *for all* i, $Y \cap X = \emptyset$. *Suppose that* $x_i y_j = \varepsilon(x_i, y_j)\, y_j x_i$. *Then a polynomial* $f(x_1, \ldots, x_n)$ *from* $A(X)$ *belongs to* $L(X)$ *if and only if*

$$f(x_1 + y_1, \ldots, x_n + y_n) = f(x_1, \ldots, x_n) + f(y_1, \ldots, y_n).$$

PROOF It is sufficient to note only that

$$A(X) \mathbin{\widehat{\otimes}} A(X) \cong A(X \cup Y)/I$$

where I is the ideal generated by the set

$$\{xy - \varepsilon(x, y)\, yx \mid x \in X, y \in Y\}.$$

■

D. Wigner in [602] gave a very nice short proof of Friedrichs and Dynkin-Specht-Wever criteria in one formula. Let A(X) be the free associative algebra over a field K, $L(X) \subseteq A(X)$ the free Lie algebra, μ the multiplication and Δ the comultiplication on $A(X)$, respectively, $\gamma\colon A(X) \to L(X)$ the linear mapping defined by $\gamma(x_1 \cdots x_n) = [x_1 \cdots x_n]$, where $[x_1 \cdots x_n]$ is the right normed monomial, $x_i \in X$.

Proposition 19.30 (D. Wigner [602])
Let a *be an* ℓ*-homogeneous element of* $A(X)$, $\ell(a) = n$. *Then*

$$\mu \circ (\gamma \otimes \mathrm{Id}_A(X)) \circ \Delta(a) = na.$$

D. Radford in [468] proved that in the case of trivial grading group the free associative algebra as a shuffle algebra is the free commutative associative algebra with a set of free generators consisting of Lyndon words. In the general case we have the following theorem.

Theorem 19.31 ([412])
If K is a $\mathbb{Q}$-algebra, then $A(X)$ with the shuffle product $*$ is a free ε-commutative algebra with a set of free generators consisting of s-regular words.

For applications of shuffle algebras to free Lie algebras see, for example, [481].

20 Canonical bases in Lie superalgebras

In this section we introduce compositions of elements of the free color Lie (p-)superalgebra. We prove Composition Lemma (Theorem 20.5) for Lie (p-)superalgebras. Theorems 20.11, 20.12 describe connections between Canonical bases of Lie (p-)superalgebras and Canonical bases of their universal enveloping algebras. We give some examples of applications of Composition Lemma for Lie algebras and discuss some problems for Lie superalgebras.

Compositions

Definition 20.1
Let a, b be some elements in $L(X)$. Assume that their leading terms a_0 and b_0 in the s-regular presentation (first relative to weight φ and then lexicographically) have coefficients which are equal to 1 and $\widetilde{a_0} = e_1e_2$, $\widetilde{b_0} = e_2e_3$ where $e_1, e_2, e_3 \in S(X)$. By Lemmas 5.22, 5.24 the word $e_1e_2e_3$ is regular. Now we define the element $(a, b) \in L(X)$ which is called *the composition of intersection for elements* $a, b \in L(X)$ (relative to a subword e_2):

(i) in the case when $\widetilde{a_0}$, $\widetilde{b_0}$ are regular words, $\widetilde{a_0} \neq \widetilde{b_0}$, we set

$$(a, b) = [(a)e_3] - [e_1(b)]$$

where $[(a)e_3]$ is obtained from the monomial $\big[[a_0]e_3\big]$ with the arrangement of brackets described in Point (i) of Lemma 7.1 by replacing a_0 by a and analogously $[e_1(b)]$ is obtained from $\big[e_1[b_0]\big]$ by replacing b_0 by b;

(ii) in the case when u, $\widetilde{b_0}$ are regular words, $u \neq \widetilde{b_0}$, $d(u) \in G_-$, $\widetilde{a_0} = uu$, $e_1 = ue_1'$, $u = e_1'e_2$ and $e_1' \in S(X)$, we set

$$(a, b) = [(a)e_3] - 2[e_1(b)]$$

where $[(a)e_3]$ is the element obtained from the monomial $\big[[u, u]e_3\big]$ with the arrangement of brackets described in Point (ii) of Lemma 7.1 by replacing

$[u,u]$ by a and $[e_1(b)]$ is the element obtained from the monomial $[e_1[b_0]]$ with the arrangement of brackets described in Point (i) of Lemma 7.1 by replacing b_0 by b;

(iii) in the case when $\widetilde{a_0}$, u are regular words, $d(u) \in G_-$, $\widetilde{a_0} \neq u$, $\widetilde{b_0} = uu$, $e_3 = e_3'u$, $u = e_2e_3'$, $e_3' \in S(X)$ we set

$$(a,b) = 2[(a)e_3] - [e_1(b)]$$

where the element $[(a)e_3]$ is obtained from the monomial $[[a_0]e_3]$ with the arrangement of brackets describes in Point (i) of Lemma 7.1 by replacing a_0 with a and $[e_1(b)]$ is the element obtained from the monomial $[e_1[u,u]]$ with the arrangement of brackets described in Point (ii) of Lemma 7.1 by replacing $[u,u]$ with b;

(iv) in the case when u, v are regular words with $u \neq v$, $d(u)$, $d(v) \in G_-$, $\widetilde{a_0} = uu$, $\widetilde{b_0} = vv$, $e_1 = ue_1'$, $e_3 = e_3'v$, $u = e_1'e_2$, $v = e_2e_3'$ and $e_1', e_3' \in S(X)$, we set

$$(a,b) = [e_1(b)] - [(a)e_3]$$

where the element $[e_1(b)]$ is obtained from the monomial $[e_1[v,v]]$ by replacing of $[v,v]$ with b and the element $[(a)e_3]$ is obtained from the monomial $[[u,u]e_3]$ by replacing $[u,u]$ with a (the arrangements of brackets are described in Point (ii) of Lemma 7.1;

(v) for $a = b$, $\widetilde{a_0} = uu$, where u is a regular word and $d(u) \in G_-$ we set

$$(a,a) = [a,[u]] = [a - [u,u],[u]]$$

where $[u]$ is a regular monomial.

Definition 20.2
Let a, b be some elements of $L^p(X)$. Suppose that their leading terms a_0 and b_0 in ps-regular presentation (first relative to weight φ and then lexicographically) have coefficients which are equal to 1 and $\widetilde{a_0} = e_1e_2$, $\widetilde{b_0} = e_2e_3$ where $e_1, e_2, e_3 \in S(X)$. By Lemma 5.24 the word $e_1e_2e_3$ is regular. Now we define the element $(a,b) \in L^p(X)$ which is called *the composition of intersection for elements $a, b \in L^p(X)$* (relative to a subword e_2):

(i) in the case when $\widetilde{a_0}$ and $\widetilde{b_0}$ are either regular words or squares uu of regular words u with $d(u) \in G_-$ the composition $(a,b) \in L^p(X)$ is defined as in Points (i)–(v) of Definition 20.1;

(ii) in the case when $\widetilde{a_0}$, v are regular words, $\widetilde{a_0} \neq v$, $\widetilde{b_0} = v^{p^t}$, $e_3 = e_3'v^{p^t-1}$, $t \in \mathbb{N}$, $v = e_2e_3'$, $d(v) \in G_+$, we set

$$(a,b) = [(a)e_3] - [e_1(b)]$$

where $[(a)e_3]$ is obtained from the monomial $[[a_0]e_3]$ with the arrangement of brackets described in Point (i) of Lemma 7.1 by replacing a_0 with a and

$[e_1(b)]$ is obtained from the monomial $\left[e_1[v]^{p^t}\right]$ with the arrangement of brackets from Point (iii) of Lemma 7.1 by replacing $[v]^{p^t}$ with b;

(iii) in the case when u, $\widetilde{b_0}$ are regular words, $u \neq \widetilde{b_0}$, $\widetilde{a_0} = u^{p^t}$, $e_1 = u^{p^t-1}e_1'$, $t \in \mathbb{N}$, $u = e_1'e_2$, $d(u) \in G_+$, we set

$$(a,b) = [(a)e_3] - [e_1(b)]$$

where we consider the arrangement of brackets on $\left[[a_0]e_3\right]$ in accordance with Point (iii) and on $\left[e_1[b_0]\right]$ in accordance with Point (i) of Lemma 7.1;

(iv) in the case when u, $\widetilde{b_0}$ are regular words, $u \neq \widetilde{b_0}$, $\widetilde{a_0} = (uu)^{p^s}$, $e_1 = u^{2p^s-1}e_1'$, $u = e_1'e_2$, $d(u) \in G_-$, then we set

$$(a,b) = [(a)e_3] - 2^{p^s}[e_1(b)]$$

where we consider the arrangement of brackets on $\left[[a_0]e_3\right]$;

(v) if $\widetilde{a_0}$, v are regular words, $\widetilde{a_0} \neq v$, $\widetilde{b_0} = (vv)^{p^t}$, $e_3 = e_3'v^{2p^t-1}$, $v = e_2e_3'$, $d(v) \in G_+$, then we set

$$(a,b) = 2^{p^t}[(a)e_3] - [e_1(b)];$$

(vi) in the case when u, v are regular words, $d(u) \in G_+$, $d(v) \in G_-$, $\widetilde{a_0} = u^{p^s}$, $\widetilde{b_0} = (vv)^{p^t}$, $t \geq 0$, $e_1 = u^{p^s-1}e_1'$, $e_3 = e_3'v^{2p^t-1}$, $u = e_1'e_2$, $v = e_2e_3'$, we set

$$(a,b) = 2^{p^t}[(a)e_3] - [e_1(b)]$$

where we consider the arrangement of brackets on $\left[[a_0]e_3\right]$ in accordance with Point (iii) and on $\left[e_1[b_0]\right]$ in accordance with Point (i) of Lemma 7.1;

(vii) in the case when u, v are regular words, $u \neq v$, $d(u), d(v) \in G_+$, $\widetilde{a_0} = u^{p^s}$, $\widetilde{b_0} = v^{p^t}$, $s,t \in \mathbb{N}$, $e_1 = u^{p^s-1}e_1'$, $e_3 = e_3'v^{p^t-1}$, $u = e_1'e_2$ and $v = e_2e_3'$, we set

$$(a,b) = [(a)e_3] - [e_1(b)]$$

where both arrangements of brackets on $\left[[a_0]e_3\right]$ and on $\left[e_1[b_0]\right]$ are as in Point (iii) of Lemma 7.1;

(viii) in the case when u, v are regular words, $d(u) \in G_-$, $d(v) \in G_+$, $\widetilde{a_0} = (uu)^{p^s}$, $s \geq 0$, $\widetilde{b_0} = v^{p^t}$, $t \in \mathbb{N}$, $e_1 = u^{2p^t-1}e_1'$, $e_3 = e_3'v^{p^t-1}$, $u = e_1'e_2$, $v = e_2e_3'$, we set

$$(a,b) = [(a)e_3] - 2^{p^t}[e_1(b)]$$

where we consider the arrangement of brackets on $\left[[a_0]e_3\right]$ in accordance with Point (ii) and on $\left[e_1[b_0]\right]$ in accordance with Point (iii) of Lemma 7.1;

(ix) if u, v are regular words, $u \neq v$, $d(u), d(v) \in G_-$, $\widetilde{a_0} = (uu)^{p^s}$, $\widetilde{b_0} = (vv)^{p^t}$, $s,t \geq 0$, $e_1 = u^{2p^s-1}e_1'$, $e_3 = e_3'v^{2p^t-1}$, $e_1'e_2 = u$, $e_2e_3' = v$, then we set

$$(a,b) = [(a)e_3] - 2^{p^s-p^t}[e_1(b)]$$

in the case when $s \geq t$, and

$$(a, b) = 2^{p^s - p^t}[(a)e_3] - [e_1(b)]$$

otherwise;

(x) for $a = b$ where u is a regular word, $d(u) \in G_+$, $\widetilde{a_0} = u^{p^t}$, $t \in \mathbb{N}$, we set

$$(a, a) = [a, [u]] = [a[u]^{p^t}, [u]]$$

where $[u]$ is a regular monomial.

(xi) if v is a regular word, $d(v) \in G_-$, $\widetilde{a_0} = (vv)^{p^t}$, $a = b$, then we set

$$(a, a) = [a, [v]] = [a - [v, v]^{p^t}, [v]]$$

where $[v]$ is a regular monomial and $[v, v]$ is an s-regular monomial.

Definition 20.3
Let a, b be some elements of $L(X)$ (of $L^p(X)$, respectively). Assume that their leading terms a_0 and b_0 in the s-regular (in ps-regular) presentation (first relative to the weight φ and then lexicographically) have coefficients which are equal to 1 and $\widetilde{a_0} = e\widetilde{b_0}f$, $e, f \in S(X)$. Then we define the element $\langle a, b\rangle$ of $L(X)$ (of $L^p(X)$, respectively) which is called *the composition of inclusion* by the following way:

(i) if $a_0 = b_0$, then we set

$$\langle a, b\rangle = a - b;$$

(ii) if $\overline{a}\overline{b} \neq \overline{b}\overline{a}$, then we set

$$\langle a, b\rangle = a - \alpha[e(b)f],$$

where $[e(b)f]$ is obtained from the monomial $[e[b_0]f]$ with the arrangements of brackets described in Lemma 7.1 by replacing b_0 by b, and $\alpha \in K$ is such that

$$a^\circ = \alpha[e(b)f]^\circ \qquad (\alpha = 2^\beta, \quad \beta \in \mathbb{Z});$$

(iii) if $\overline{a} = u^{p^s}$, $\overline{b} = u^{p^t}$, u is a regular monomial, $d(u) \in G_+$, $s > t$, then we set

$$\langle a, b\rangle = a - b^{p^{s-t}};$$

(iv) if $\overline{a} = [v, v]^{p^s}$, $\overline{b} = [v, v]^{p^t}$, v is a regular monomial, $d(v) \in G_-$, $s > t$, then we set

$$\langle a, b\rangle = a - b^{p^{s-t}}.$$

Definition 20.4
A subset $S \in L(X)$ ($S \in L^p(X)$) is called *stable* if the following conditions are satisfied:

(i) the coefficients of the leading monomials in s-regular (ps-regular) presentation for all elements from S are equal to 1;

(ii) if $a, b \in S$ and the composition $c = (a, b)$ (or $c = \langle a, b \rangle$) does exist, then either $\alpha(a, b) \in S$, $0 \neq \alpha \in K$ ($\beta\langle a, b\rangle \in S$, $0 \neq \beta \in K$, respectively), or after expressing the elements associatively we have that

$$c = \sum_i \alpha_i A_i s_i B_i,$$

where $\alpha_i \in K$, $s_i \in S$, $A_i, B_i \in S(X)$, $A_i \overline{s_i} B_i \leq \overline{c}$ for all i.

Composition Lemma

Theorem 20.5 (Composition Lemma [388], [389], [401])
Let K be a field, $\operatorname{char} K \neq 2$, $L(X)$ the free color Lie superalgebra (in the case when $\operatorname{char} K = p > 0$ let $L^p(X)$ be the free color Lie p-superalgebra), S a stable subset of $L(X)$ (of $L^p(X)$), $\operatorname{id}(S)$ an ideal of $L(X)$ (of $L^p(X)$) generated by S. Then for any element $a \in \operatorname{id}(S)$ the word $\overline{a}$ contains a subword $\overline{b}$ for some $b \in S$. Moreover, s-regular (ps-regular) monomials u such that $\overline{u}$ does not contain subwords $\overline{b}$, $b \in S$, form a linear basis of the algebra $L(X)/\operatorname{id}(S)$ (of $L^p(X)/\operatorname{id}(S)$, respectively).

PROOF We prove our statement for elements of the associative ideal $\operatorname{Id}(S)$ generated by S in the free associative algebra $A(X)$. Let $a \in \operatorname{Id}(S)$. Then

$$a = \sum_i \alpha_i A_i s_i B_i$$

where $A_i, B_i \in S(X)$, $s_i \in S$ and $\alpha_i \in K$. Let $A_1 \overline{s_1} B_1$ be the highest order term among all $A_i \overline{s_i} B_i$. If $A_1 \overline{s_1} B_1 \neq A_i \overline{s_i} B_i$ for all $i \neq 1$ then our theorem is proved. If for example $\alpha_1' A_1 \overline{s_1} B_1 = -\alpha_2' A_2 \overline{s_2} B_2$, $\alpha_1', \alpha_2' \in K$, then we have the following cases.

Case 1. Subwords $\overline{s_1}$ and $\overline{s_2}$ have empty intersection in $A_1 \overline{s_1} B_1$. Then we may assume that

$$A_1 s_1 B_1 = A s_1 B \overline{s_2} C, \qquad A_2 s_2 B_2 = A \overline{s_1} B s_2 C$$

where $A, B, C \in S(X)$. Therefore

$$\begin{aligned}
\alpha_1' A_1 s_1 B_1 + \alpha_2' A_2 s_2 B_2 &= \alpha_1' A s_1 B \overline{s_2} C + \alpha_2' A \overline{s_1} B s_2 C \\
&= -\alpha_1' A s_1 B (s_2 - \overline{s_2}) C - \alpha_2' A (s_1 - \overline{s_1}) B s_2 C, \\
\overline{A \overline{s_1} B (s_2 - \overline{s_2}) C} &< A \overline{s_1} B \overline{s_2}, \\
\overline{A (s_1 - \overline{s_1}) B \overline{s_2} C} &< A \overline{s_1} B \overline{s_2}.
\end{aligned}$$

We consider now possible ambiguities. We start with intersections.

Case 2. The words $\overline{s_1}$ and $\overline{s_2}$ are different and they are not subwords of each other, $s_1 \neq s_2$, $\overline{s_1} = e_1e_2$, $\overline{s_2} = e_2e_3$. Then

$$As_1e_3B - A[(s_1)e_3]B = \sum_i \gamma_i E_i s_1 F_i,$$

with $\gamma_i \in K$, $E_i, F_i \in S(X)$, $E_i\overline{s_1}F_i < A\overline{s_1}e_3B$ for all i;

$$Ae_1s_2B - A[e_1(s_2)]B = \sum_i \delta_i T_i s_2 H_i,$$

with $\delta_i \in K$, $T_i, H_i \in S(X)$, $T_i\overline{s_2}H_i < Ae_1\overline{s_2}B$ for all i, where the arrangements of brackets as in Lemma 7.1. Thus we have

$$\begin{aligned}
\alpha_1' As_1e_3B + \alpha_2' Ae_1s_2B &= \alpha_1' A[(s_1)e_3]B + \alpha_2' A[e_1(s_2)]B \\
&\quad + \sum_i \alpha_1'\gamma_i E_i s_1 F_i + \sum_i \alpha_2'\delta_i T_i s_2 H_i \\
&= \alpha A(s_1, s_2)B \\
&\quad + \sum_i \alpha_1'\gamma_i E_i s_1 F_i + \sum_i \alpha_2'\delta_i T_i s_2 H_i \\
&= \sum_i \beta_i C_i g_i D_i,
\end{aligned}$$

$\alpha, \beta_i \in K$, $g_i \in S$, $C_i, D_i \in S(X)$, $C_i\overline{g_i}D_i < A_1\ \overline{s_1}B_1$ for all i by the stability of S and the fact $\overline{(s_1, s_2)} < e_1e_2e_3$.

Case 3. If $s_1 = s_2$, $\overline{s_1}$ is a regular word, then by Lemma 5.23 there are no intersections between $\overline{s_1}$ and $\overline{s_1}$. If $s_1^\circ = 2uu$, u is a regular word, $d(u) \in G_-$, then by Lemma 5.24 the overlaps have the form $\alpha_1'(As_1uB - Aus_1B)$ (it is clear that $\alpha_1' = -\alpha_2'$). Note that if $[u]$ is a regular monomial, then by Lemma 6.6

$$As_1uB - As_1[u]B = \sum_i \gamma_i E_i s_1 F_i$$

with $\gamma_i \in K$, $E_i, F_i \in S(X)$, $E_i\overline{s_1}F_i < AuuuB$ for all i;

$$Aus_1B - A[u]s_1B = \sum_i \delta_i T_i s_1 H_i,$$

$\delta_i \in K$, $T_i, H_i \in S(X)$, $T_i\overline{s_1}H_i < AuuuB$ for all i.

Therefore

$$\begin{aligned}
As_1uB - Aus_1B &= As_1[u]B - A[u]s_1B + \sum_i \gamma_i E_i s_1 F_i - \sum_i \delta_i T_i s_1 H_i \\
&= A[s_1, [u]]B + \sum_i \gamma_i E_i s_1 F_i - \sum_i \delta_i T_i s_1 H_i
\end{aligned}$$

since

$$\begin{aligned}[][s_1,[u]] &= s_1[u]-\varepsilon(d(s_1),d(u))\,[u]s_1 = s_1[u]-\varepsilon(2d(u),d(u))\,[u]s_1\\ &= s_1[u]-(\varepsilon(d(u),d(u)))^2[u]s_1 = s_1[u]-[u]s_1.\end{aligned}$$

Taking into account the stability of S, as above we have

$$\alpha_1'(As_1uB - Aus_1B) = \sum_i \beta_i C_i g_i D_i,$$

$\beta_i \in K$, $C_i, D_i \in S(X)$, $C_i\overline{g_i}D_i < A_1\overline{s_1}B_1$ for all i.

Now if $\overline{s_1} = u^{p^t}$, u is a regular word, $d(u) \in G_+$, then by Lemma 5.24 the overlaps have the form $e_2 = u^l$, $l \in \mathbb{N}$. At first we consider the case $e_2 = u$. Then the overlap has the form $As_1uB - Aus_1B$. As above we get

$$As_1uB - As_1[u]B = \sum_i \gamma_i E_i s_1 F_i,$$

$\gamma_i \in K$, $E_i, F_i \in S(X)$, $E_i\overline{s_1}F_i < Au^{p^t+1}B$ for all i;

$$Aus_1B - A[u]s_1B = \sum_i \delta_i T_i s_1 H_i,$$

$\delta_i \in K, T_i, H_i \in S(X)$, $T_i\overline{s_1}H_i < Au^{p^t+1}B$ for all i, with $[u]$ being a regular monomial.

Since

$$\begin{aligned}\varepsilon(d(u),d(u)) &= 1,\\ [s_1,[u]] &= s_1[u]-\varepsilon(d(s_1),d(u))\,[u]s_1\\ &= s_1[u]-\varepsilon(p^t d(u),d(u))\,[u]s_1\\ &= s_1[u]-(\varepsilon(u,u))^{p^t}[u]s_1 = s_1[u]-[u]s_1,\end{aligned}$$

we get

$$\begin{aligned}As_1uB - Aus_1B &= A[s_1,[u]]B + \sum_i \gamma_i E_i s_1 F_i - \sum_i \delta_i T_i s_1 H_i\\ &= \sum_i \beta_i C_i g_i D_i,\end{aligned}$$

$\beta_i \in K$, $C_i, D_i \in S(X)$, $C_i\overline{g_i}D_i < A_1\overline{s_1}B_1 = Au^{p^t+1}B$ for all i (because of S is stable).

Let now $e_2 = u^l$, $l > 1$. We have just shown that

$$s_1u = us_1 + \sum_i \beta_i' C_i' g_i' D_i',$$

$\beta_i' \in K$, $C_i', D_i' \in S(X)$, $C_i'\overline{g_i'}D_i' < \overline{us_1} = \overline{s_1u}$. Therefore

$$As_1u^lB - Au^ls_1B = \sum_i \beta_i C_i g_i D_i,$$

$\beta_i \in K$, $C_i, D_i \in S(X)$, $C_i\overline{g_i}D_i < A\overline{s_1}u^lB = Au^l\overline{s_1}B = Au^{p^t+l}B$.

If $s_1^\circ = 2^{p^t}v^{2p^t}$ where v is a regular word, $d(v) \in G_-$, then by Lemma 5.24 the intersections have the form $e_2 = v^m$, $m \in \mathbb{N}$. If $e_2 = v$, then as above

$$\begin{aligned} As_1vB - As_1[v]B &= \sum_i \delta_i T_i s_1 H_i; \\ [s_1,[v]] &= s_1[v] - \varepsilon(d(s_1), d(v))\,[v]s_1 \\ &= s_1[v] - \varepsilon(2p^t d(v), d(v))\,[v]s_1 \\ &= s_1[v] - (\varepsilon(d(v), d(v)))^{2p^t}\,[v]s_1, \\ s_1[v] - (-1)^{2p^t}[v]s_1 &= s_1[v] - [v]s_1. \end{aligned}$$

Let now $e_2 = v^l$, $l \geq 1$. Since

$$s_1v = vs_1 + \sum_i \beta_i' C_i' f_i' D_i', \qquad C_i'\overline{f_i'}D_i' < v^{2p^t+1},$$

we get

$$As_1v_lB - Av_ls_1B = \sum_i \beta_i C_i f_i D_i, \qquad C_i\overline{g_i}D_i < Av^{2p^t+l}B.$$

Case 4. Let $\overline{s_1} = u^{p^s}$, $\overline{s_2} = u^{p^t}$, where $s > t$, u is a regular word, $d(u) \in G_+$. By Lemma 5.24 the intersection has the form $e_2 = u^l$, $l \in \mathbb{N}$. As in Case 3,

$$As_1u^{p^t-l}B = Au^{p^t-l}s_1B + \sum_i \gamma_i E_i s_1 F_i, \qquad E_i\overline{s_1}F_i < u^{p^s+p^t-l}.$$

At the same time,

$$Au^{p^t-l}s_1B - Au^{p^s-l}s_2B = Au^{p^t-l}\langle s_1, s_2\rangle B + \sum_i \delta_i C_i f_i D_i,$$

where $f_i \in S$, $C_i\overline{f_i}D_i < u^{p^s+p^t-l}$.

Now one can analogously touch the case when $s_1^\circ = 2^{p^s}v^{p^s}$, $s_2^\circ = 2^{p^t}v^{p^t}$, $s > t$, v is a regular word, $d(v) \in G_-$.

At the end of the proof we deal with possible inclusions.

Case 5. If $s_1 \neq s_2$, $\overline{s_1} = \overline{s_2}$, then

$$As_1B - As_2B = A\langle s_1, s_2\rangle B.$$

Case 6. We have an inclusion $s_1^\circ = \alpha a s_2^\circ b$, $\alpha \in K$, $a, b \in S(X)$. Then

$$As_1B - \alpha A' a s_2 b B' = \beta A\langle s_1, s_2\rangle B + \sum_i \delta_i E_i F_i F_i,$$

$0 \neq \beta \in K$, $f_i \in S$, $E_i \overline{f_i} F_i < A\overline{s_1}B$.

Using Lemma 5.24 we see that we have considered all possible ambiguities of intersection and of inclusion between $\overline{s_1}$ and $\overline{s_2}$.

Thus in the case of cancellation of $\alpha_1 A_1 \overline{s_1} B_1$ in the sum $a = \sum \alpha_i A_i s_i B_i$ we may rewrite

$$a = \sum_i \beta_i C_i g_i D_i,$$

$\beta_i \in K$, $C_i, D_i \in S(X)$, $C_i \overline{g_i} D_i < A_1 \overline{s_1} B_1$ for all i. It is clear that our statement follows from this fact for each $a \in \mathrm{Id}(S)$. ■

For any ideal J of $L(X)$ (of $L^p(X)$) there exists the uniquely determined stable subset S of J such that for each $a \in S$ there are no compositions of inclusion of the type $\langle b, c\rangle$, where b is an s-regular (a ps-regular) monomial in s-regular (in ps-regular) presentation of a, and $c \in S \setminus \{a\}$. Such system is called *the Gröbner basis of the ideal J*. Note that it is possible to define compositions and to prove Composition Lemma for algebras over Euclidean rings (as for commutative case see [252], [597]).

Remark 20.6

In connection with Composition Lemma it is interesting to study Lie superalgebras given by one defining relation. Recall two problems for Lie algebras with one defining relation.

Problem 20.7 ([156], Problem 1.176)

Is the membership problem resolvable for Lie algebras with one defining relation?

The same problem is interesting for color Lie p-superalgebras.

Problem 20.8 ([156], Problem 1.87)

Is the isomorphism problem resolvable for Lie algebras with one defining relation?

The same question is also interesting for color Lie p-superalgebras.

Using Theorem 20.5 one can easily see that if $z \in X$, $a \in L(X)$ ($a \in L^p(X)$), $\ell_z(a) \neq 0$, a is a G-homogeneous element (we may assume that the weight φ is such that $\varphi_z(\overline{a}) \neq 0$). Then if $\overline{a}$ is a regular word (in particular, if a is odd: $d(a) \in G_-$) then the leading term of any element of $\mathrm{id}(a)$ contains the subword $\overline{a}$ and therefore the Freiheitssatz is valid and there exists an algorithm for solving the equality problem for color Lie superalgebras and for color Lie p-superalgebras with one defining relation $a = 0$ (recall that the Freiheitssatz means that $\mathrm{id}(a) \cap L(X \setminus \{z\}) = 0$).

The general problems of Freiheitssatz in Lie superalgebras, restricted Lie algebras and superalgebras are open. Note that the Freiheitssatz in group theory is due to W. Magnus [344]. In associative algebras these problems were solved in the case of the field of zero characteristic by L. G. Makar-Limanov [354] and in the case of homogeneous relation by M. Hedges in [226]. We mention also papers [152], [195], [333] devoted to the study of associative algebras with one defining relation.

We cite another example of Lie superalgebras with one defining relation. Let $a \in L(X)$, $a^\circ = 2uu$, where u is a regular word, $d(u) \in G_-$,

$$a = [u, u] + \sum_i \alpha_i v_i$$

where $\alpha_i \in K$, v_i is an s-regular monomial, $\overline{v_i} < uu$, $\overline{v_1} > \overline{v_i}$ for $i \neq 1$ (first by weight φ and then lexicographically), $\overline{v_1} \neq au$ for $\overline{v_1} > u$, $\overline{v_1} \neq ua$ for $u > \overline{v_1}$, $a \in S(X)$, $z \in X$, $\varphi_z(\overline{a}) \neq 0$. Then $\mathrm{id}(a) = \mathrm{id}(S)$, $S = \{a, (a, a)\}$ where (a, a) is the composition, S is a stable set.

Thus $\mathrm{id}(a) = \mathrm{id}(S)$, where $S = \{a, (a, a)\}$, and S is a stable subset of $L(X)$.

Therefore, for $L(X)/\mathrm{id}(a)$, we have an algorithmic solution of equality problem. It is clear that the Freiheitssatz takes place in this situation.

In connection with above mentioned results we point out the following problems.

Problem 20.9
Does the Freiheitssatz take place for color Lie (p-)superalgebras given by one defining relation?

Problem 20.10
Describe elements a of $L(X)$ (of $L^p(X)$) such that the ideal $\mathrm{id}(a)$ has the finite Gröbner basis.

If a color Lie superalgebra is given by a set of generators and defining relations, then its universal enveloping algebra is given by the same set of generators and defining relations (as an associative algebra). The same situation takes place for the restricted universal enveloping algebra of a color Lie p-superalgebra. We consider now connections between stabilities of the above mentioned set in the free color Lie (p-)superalgebra and in the free associative algebra.

Theorem 20.11 ([401])
Let $\mathrm{char}\, K \neq 2$, $S \subseteq L(X)$ (in the case when $\mathrm{char}\, K = p \geq 3$ we suppose that $S \subseteq L^p(X)$), $\mathrm{id}(S)$ the ideal of $L(X)$ (of $L^p(X)$) generated by S, $\mathrm{id}(S)$ the ideal of $A(X)$ generated by S. Then S is a stable subset in $L(X)$ (in $L^p(X)$) if and only if S is a stable subset in $A(X)$. In other words, a color

Lie (p-)superalgebra and its (restricted) universal enveloping algebra have the same canonical bases.

Now we look at PBW-theorem as one of the corollaries of Theorem 20.11.

Corollary 20.12
The monomials of the form $u_1u_2\cdots u_k$ where u_i are S-irreducible s-regular monomials, $u_1 \le u_2 \le \cdots \le u_k$, $u_i \ne u_{i+1}$ with $d(u_i) \in G_-$, form a K-linear basis of $U(L)$; in the case when $\operatorname{char} K = p > 2$, the monomials of the form $u_1^{\xi_1}u_2^{\xi_2}\cdots u_k^{\xi_k}$ where u_i are S-irreducible ps-regular monomials, $u_1 \le u_2 \le \cdots \le u_k$, $\xi_i = 1$ with $d(u_i) \in G_-$, $\xi_j < p$ with $d(u_j) \in G_+$, form a K-linear basis of $u(L)$.

Theorem 20.11 and Corollary 20.12 for Lie algebras were proved by L. A. Bokut' in [84].

Note that in general the above mentioned theorem follows from the form of the associative presentation of a regular monomial (see Lemma 6.6). One can also prove Theorem 20.11 using the general theory of H-presentations, see for example [316, Theorem 2.11, Point 2].

Applications of Composition Lemma

It has become clear that Composition Lemma is a very useful tool to solve a number of algorithmic problems in the theory of Lie algebras. In this subsection we give some examples.

Algebraically closed Lie algebras

Let K be a field, L a Lie algebra over K, $X = \{x_1, \dots, x_n\}$, $L(X)$ the free Lie algebra. A Lie algebraic equation over L is an equation of the form $v(x_1, \dots, x_n) = 0$ where v is an element of the free product $L * L(X)$, $n \notin L$. A Lie algebra L is *algebraically closed* if any Lie algebraic equation over L has a solution in L. Using Composition Lemma L. A. Bokut' in [77] (see also [82]) proved that it is possible to embed any Lie K-algebra L into an algebraically closed Lie K-algebra. In connection with this result it would be interesting to touch the same problem for Lie superalgebras.

Presentation of the free square of a Lie algebra

In [79] L. A. Bokut' proves that there exists a finitely presented Lie algebra over a field with unsolvable equality problem. The proof was based on Composition Lemma. In particular, he considered a presentation of the free square of a Lie algebra with an amalgamated subalgebra. Let L be a Lie algebra, B a subalgebra of L, $\{b_j \mid j \in J\}$ a basis of B, $\{a_i \mid i \in I\}$ a complement of $\{b_j\}$ with respect to a basis of L. Consider the free Lie algebra $L(\{a_j\} \cup \{b_j\} \cup \{d\})$ and the following ordering on the set of free generators: $a_i > b_j > d$ for all i, j; $a_i > a_j$, $b_i > b_j$ with $i > j$. Let S be the set of all elements ($n \geq 3$):

$$[a_i, a_j] - [a_i, a_j]', \quad i > j; \qquad [a_i, b_j] - [a_i, b_j]';$$
$$[b_i, b_j] - [b_i, b_j]', \quad i > j; \qquad [a_i d^n]; \qquad [b_i, d];$$
$$[a_i d^k] b_j - [[a_i, b_j]' d^k], \quad 1 \neq k \neq n-1;$$
$$\sum_{k=i+1}^{n-1} C_{n+i}^k [[a_i d^k], [a_j d^{n+i-k}]], \quad 1 \leq i \leq n-2, \quad i > j$$

(where we use left normed monomials, $[a_i, a_j]'$, $[a_i, b_j]'$, $[b_i, b_j]'$ has the same form as in L).

The factor algebra $L(\{a_j\} \cup \{b_j\} \cup \{d\})/\mathrm{id}(S)$ is isomorphic to the algebra L' given by generators $L \cup \{d\}$ and defining relations

$$[ad^n] = 0, \qquad [b, d] = 0,$$

$a \in L$, $b \in B$. The assertion of the theorem on presentation is in fact that $\overline{L}$ contains the free product of L and Le^d with the amalgamated subalgebra Be^d where e^d is an isomorphism of the form

$$e^d = 1 + d + \frac{d^2}{2!} + \cdots + \frac{d^{n-1}}{(n-1)!}.$$

Using similar ideas, in [83] L. A. Bokut' considered connections between the properties of a Lie algebra to be solvable and of its diagonal monomorphism to be inner.

Canonical bases of classical simple Lie algebras

It is well-known that semisimple Lie algebra can be presented using Cartan matrix by generators and Serre's relations (see for example [500]). L. A. Bokut' and A. A. Klein in [85] complemented this set of defining relations with respect to a stable set of defining relations in the case of algebras of type A_n. To be more precise, consider the Lie algebra of the type A_n. Let (a_{ij}), $1 \leq i, j \leq n$, be its Cartan matrix. Let also $\operatorname{char} K \neq 2, 3$, $H = \{h_1, \ldots, h_n\}$, $X = \{x_1, \ldots, x_n\}$, $Y = \{y_1, \ldots, y_n\}$, and let L be the free Lie algebra on $H \cup X \cup Y$. Consider *Serre's relations* ($1 \leq i, j \leq n$):

$$[h_i, h_j], \quad i > j; \qquad [x_i, y_i] - h_i; \qquad [x_i, y_j], \quad i \neq j;$$
$$[h_i, x_j] - a_{ij} x_j; \qquad [h_i, y_j] + a_{ji} y_j;$$
$$(\operatorname{ad} x_i)^{1-a_{ji}}(x_j), \quad i \neq j; \qquad (\operatorname{ad} y_i)^{1-a_{ji}}(y_j), \quad i \neq j.$$

As it was shown in [85], to come to a stable set of defining relations for A_n it is sufficient to add to Serre's relations the following relations:

$$\begin{aligned} &\big[[x_{i+k}, x_{i+k-1}, \ldots, x_{i-1}], x_{i+k-1}\big], && k \geq 1,\ i \geq 2;\\ &\big[[x_{i+k}, \ldots, x_i], [x_{i+k}, \ldots, x_{i-1}]\big], && n \geq i+k;\\ &\big[[y_{i+k}, y_{i+k-1}, \ldots, y_{i-1}], y_{i+k-1}\big], && k \geq 1,\ i \geq 2;\\ &\big[[y_{i+k}, \ldots, y_i], [y_{i+k}, \ldots, y_{i-1}]\big], && n \geq i+k, \end{aligned}$$

where we use left normed monomials and $h_i > x_j > y_k$.

It would be interesting to write down the canonical bases for other types of classical Lie algebras and also for Kac-Moody algebras and for Lie superalgebras (see [250], [329], [492], [528]).

Composition technique for relatively free Lie algebras

Composition Lemma for free metabelian Lie algebras was proved in [553] by V. V. Talapov. Using this lemma he proved that if two elements a and b generate the same ideal of this algebra, then $a = \lambda\varphi(b)$, where $\lambda \in K$, φ is an inner automorphism. Also he proved that any metabelian Lie algebra can be embedded into an algebraically closed metabelian Lie algebra over the same field.

See also [552], [553], [554] for other applications of Composition Lemma.

The free center-metabelian Lie algebra is the relatively free Lie algebra of the variety given by the identity $\big[[[a,b],[c,d]],e\big] = 0$. In [563] U. U. Umirbaev introduced compositions of elements in the free center-metabelian Lie algebra and proved Composition Lemma in this case. Using this result he constructed an algorithm to solve the equality problem for such algebras.

Problem 20.13

Define compositions for elements of metabelian and center-metabelian color Lie superalgebras and to prove Composition Lemma in these cases.

It would be also very interesting to consider other relatively free color Lie superalgebras and (p-)superalgebras.

21 Free products

In this section we construct bases of free products of color Lie superalgebras (Theorem 21.2) and of color Lie p-superalgebras (Theorem 21.6) with the amalgamated subalgebra. We consider also subalgebras of such free products. In all what follows in this section K is a field, $\operatorname{char} K \neq 2$.

Bases of free products

Let T be a set, $H^0, H_\alpha, \alpha \in T$, color Lie K-superalgebras (with the same group G and commutation factor ε), H^0_α be a G-homogeneous subalgebra of H_α, $\delta_\alpha\colon H^0 \to H^0_\alpha$ an isomorphism of color Lie superalgebras.

Let also $B^0 = \{f_\beta \mid \beta \in T^0\}$ be a G-homogeneous basis in H^0,

$$B^0_\alpha = \{e_{\alpha\beta} \mid e_{\alpha\beta} = \delta_\alpha(f_\beta),\ \alpha \in T,\ \beta \in T^0\},$$

and let B_α be a homogeneous basis in H_α, $B^0_\alpha \subseteq B_\alpha$ and $B_\alpha = \{e_{\alpha\beta} \mid \beta \in T_\alpha\}$ for all $\alpha \in T$. Then the multiplication in H_α for all $\alpha \in T$ can be given using structure constants $p^\tau_{\alpha\beta\gamma} \in K$:

$$[e_{\alpha\beta}, e_{\alpha\gamma}] = \sum_\tau p^\tau_{\alpha\beta\gamma} e_{\alpha\tau},$$

$\beta, \gamma \in T_\alpha$. For any we $\alpha \in T$ consider the G-graded set

$$X_\alpha = \{x_{\alpha\beta} \mid \beta \in T_\alpha,\ d(x_{\alpha\beta}) = d(e_{\alpha\beta})\}.$$

Let X be the set obtained from $\bigcup\limits_{\alpha \in T} X_\alpha$ by identifying $x_{\alpha\gamma} = x_{\beta\gamma}$ for $\gamma \in T^0$. Consider a total order on T^0 and a total order extension of T^0 to T_α for all $\alpha \in T$ such that $\beta > \gamma$ for $\beta, \gamma \in T_\alpha$, $\beta \notin T^0$, $\gamma \in T^0$. Let J be the ideal of the free color Lie superalgebra $L(X)$ generated by the following subset of G-homogeneous elements $D = \{d_{\alpha\beta\gamma}\}$,

$$d_{\alpha\beta\gamma} = [x_{\alpha\beta}, x_{\alpha\gamma}] - \sum_\tau p^\tau_{\alpha\beta\gamma} x_{\alpha\tau},$$

$\alpha \in T$, $\beta, \gamma \in T_\alpha$, where $\beta > \gamma$ if $x_{\alpha\beta} \in X_+$ and $\beta \geq \gamma$ if $x_{\alpha\beta} \in X_-$.

Now we show that the factor algebra $L(X)/J$ is isomorphic to the algebra $\prod\limits_{\alpha \in T}{}^{*}_{H_0} H_\alpha$. Indeed, it is sufficient to consider the following commutative diagram:

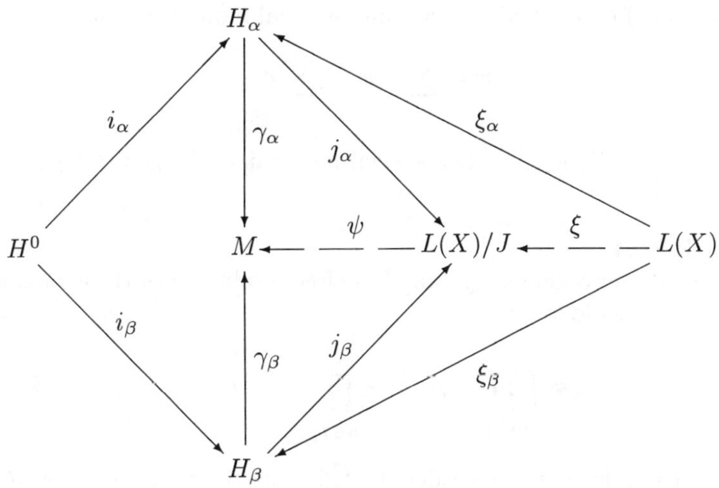

Now we order X by the following way: for $x_{\alpha\beta} \neq x_{\gamma\delta}$ we set $x_{\alpha\beta} > x_{\gamma\delta}$ if $\alpha > \gamma$ or $\alpha = \gamma$ and $\beta > \delta$.

Definition 21.1
We say that a regular monomial $w \in \Gamma(X)$ is *special* if the word $\overline{w}$ does not contain subwords $x_{\alpha\beta}x_{\alpha\gamma}$, $\beta > \gamma$, and also subwords $x_{\alpha\beta}x_{\alpha\gamma}$, where $\beta \geq \gamma$ for $x_{\alpha\beta} \in X_-$. We say that a monomial $w \in \Gamma(X)$ is *s-special* if either w is a special monomial or $w = [v, v]$ where v is a special monomial, $d(v) \in G_-$, $\ell_X(v) > 1$.

Theorem 21.2 ([382], [385])
Let K be a field, $\operatorname{char} K \neq 2$. *Then the cosets whose representatives are s-special monomials give us a linear basis of the algebra* $L(X)/J = \prod^*_{\substack{H^0\\ \alpha\in T}} H_\alpha$.

PROOF Let φ be just the length. Considering possible compositions one can easily see that $D = \{d_{\alpha\beta\gamma}\}$ is a stable subset of $L(X)$. Now the assertion of our theorem follows immediately from Theorem 20.5, Definition 21.1 and Lemma 7.1. ∎

Remark 21.3
If w is a regular monomial and w is not a special monomial then w contains as a submonomial either $[x_{\alpha\beta}, x_{\alpha\gamma}]$ with $\beta > \gamma$, or a regular monomial of

the form $[x_{\alpha\beta}, u]$ with $\overline{u} = x_{\alpha\beta}u_0$, $\gamma < \beta$ ($\gamma \leq \beta$ with $x_{\alpha\beta} \in X_-$), or $\big[[u, x_{\alpha\beta}], x_{\alpha\beta}\big]$ with $x_{\alpha\beta} \in X_-$.

If w is either a regular monomial, or $w = [v, v]$ with v a regular monomial, $\ell(v) > 1$, $d(v) \in G_-$, and w is not an s-special monomial, then

$$w = \sum_i \gamma_i w_i + \sum_j \delta_j u_j$$

where $\gamma_i, \delta_j \in K$, w_i, u_j are s-special monomials, $\ell(u_j) < \ell(w)$, $m(w_i) = m(w)$, $w_i < w$.

Remark 21.4

The universal enveloping algebra of the free product with the amalgamated subalgebra is as follows:

$$U^+\left(\prod_{\alpha\in T}{}^{*}_{H^0} H_\alpha\right) \cong \prod_{\alpha\in T}{}^{*}_{U^+(H^0)} U^+(H_\alpha).$$

Indeed, it is sufficient to consider the following commutative diagram:

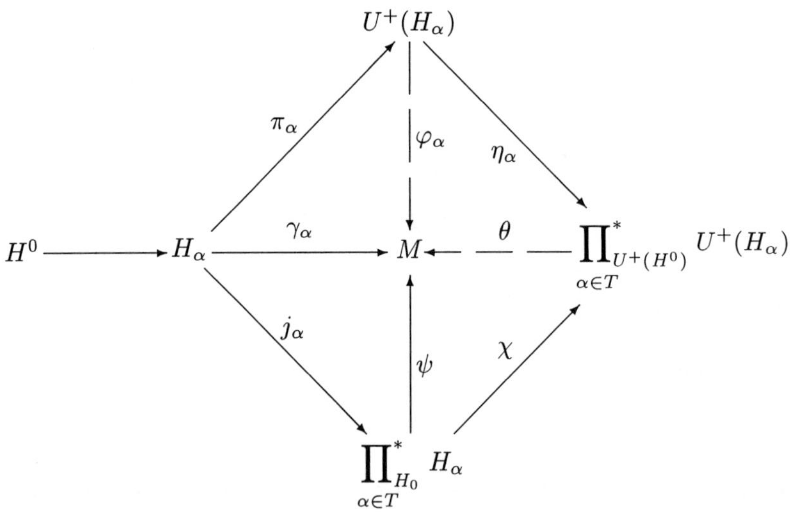

Let L and M be color Lie K-superalgebras with the same G and ε. Consider the Cartesian product $L \times M$, $(L \times M)_g = L_g \times M_g$, $g \in G$. Then the universal enveloping algebra $U(L \times M)$ is canonically isomorphic to $U(L) \mathbin{\widehat{\otimes}} U(M)$ (the canonical mapping $\xi: L \times M \to U(L) \mathbin{\widehat{\otimes}} U(M)$ is given by $\xi(l, m) = \xi_L(l) \otimes 1 + 1 \otimes \xi_M(m)$, where $\xi_L: L \to U(L)$ and $\xi_M: L \to U(M)$ are canonical embeddings).

We construct now bases of free products with the amalgamated subalgebra of color Lie p-superalgebras.

Let now H_α be color Lie p-superalgebras. Let J be an ideal of the free color Lie p-superalgebra $L^p(X)$ generated by elements $d_{\alpha\beta\gamma}$ and $h_{\alpha\beta}$ where

$$h_{\alpha\beta} = x_{\alpha\beta}^p - \sum_\tau q_{\alpha\beta}^\tau x_{\alpha\tau}$$

with $\alpha \in T$, $\beta, \tau \in T_\alpha$, $q_{\alpha\beta}^\tau \in K$, $x_{\alpha\beta} \in X_+$ and

$$e_{\alpha\beta}^p = \sum_\tau q_{\alpha\beta}^\tau e_{\alpha\tau}$$

in H_α. It is clear that the algebra $L^p(X)/J$ is the free product of color Lie p-superalgebras H_α with the amalgamated G-homogeneous subalgebra H^0.

Definition 21.5
We say that a regular monomial $w \in \Gamma(X)$ is p-special if the word $\overline{w}$ does not contain subwords of the form $x_{\alpha\beta}x_{\alpha\gamma}$, $\beta > \gamma$; $x_{\alpha\beta}x_{\alpha\gamma}$, $\beta \geq \gamma$ with $x_{\alpha\beta} \in X_-$; or $x_{\alpha\beta}^p$ with $x_{\alpha\beta} \in X_+$; or $[x_{\alpha\beta}, x_{\alpha\beta}]^p$ with $x_{\alpha\beta} \in X_-$. A monomial $w \in \Gamma(X)$ is called *ps-special* if either w is a p-special monomial or $w = [u,u]^{p^t}$, $t \geq 0$, where u is a p-special monomial, $d(u) \in G_-$, $\ell_X(u) > 1$, or $w = v^{p^t}$ where $t \in \mathbb{N}$, v is a p-special monomial, $d(v) \in G_+$, $\ell_X(v) > 1$.

Theorem 21.6 ([387], [397], [401])
Let K be a field, $\operatorname{char} K = p > 2$. Then the cosets, whose representatives are ps-special monomials, give us a linear basis of $L^p(X)/J$.

PROOF Let us consider φ as just the usual length. Using the identities of color Lie p-superalgebras it is easily to see that the subset $\{d_{\alpha\beta\gamma}\} \cup \{h_{\alpha\beta}\}$ is stable, and taking into account Composition Lemma and Definition 21.5 we complete the proof. ■

Subalgebras of free products

A. G. Kurosh in [300] proved that if $A = \prod^* A_\alpha$ is the free product of groups A_α, H a is a subgroup of G, then $H = F * \prod^* B_\beta$, where F is a free group and each B_β is conjugated in G with a subgroup of A_α for some α. Subalgebras of free products with amalgamated subalgebras of nonassociative algebras were studied by C. E. Dicidze in [150], [151].

In [301] A. G. Kurosh proved that any nonzero subalgebra H of the free nonassociative product $\prod^* E_\alpha$ is a free product of the form $F * \prod^*(H \cap E_\alpha)$ where F is a free algebra. This theorem does not take place in the case of Lie algebras, as it was shown by A. I. Shirshov in [513]. Let L_1 be the one dimensional Lie algebra with a basis x and L_2 the two dimensional Lie algebra with a basis $\{y, z\}$, $[y, z] = y$, $L = L_1 * L_2$. Consider the subalgebra L' of L generated by the elements $y, z, [y, x], [z, x]$. Then L' is isomorphic to the Lie algebra given by generators x_1, x_2, x_3, x_4 and defining relations

$$[x_4, x_1] + [x_3, x_2] - x_1 = 0; \qquad [x_4, x_2] - x_2 = 0$$

with the correspondence

$$x_1 \to [y, x]; \quad x_2 \to y; \quad x_3 \to [z, x]; \quad x_4 \to z.$$

The algebra L is not a free product as above, and L is not a subalgebra either of L_1, or of L_2 (for the details see [513]).

Such examples occur also in the case of restricted Lie algebras. Let $\operatorname{char} K = p > 2$, L_1 be the one dimensional Lie p-algebra with zero mapping $[p]$ and a basis x, L_2 the three dimensional Lie p-algebra with a basis $\{y, z, v\}$ and defining relations

$$\begin{array}{lll} [y, z] = v; & [z, v] = y; & [v, y] = z; \\ y^{[p]} = \varepsilon \cdot y; & z^{[p]} = \varepsilon \cdot z; & v^{[p]} = \varepsilon \cdot v, \end{array}$$

where $\varepsilon = -1$ with $p \equiv 3 \pmod 4$, $\varepsilon = 1$ with $p \equiv 1 \pmod 4$. Consider the subalgebra H of the free product of restricted Lie algebras $L_1 * L_2$ generated by elements y, z, v, $[y, x]$, $[z, x]$, $[v, x]$. The algebra H is not a free Lie p-algebra, H is not either a subalgebra of L_1, or of L_2, and it is impossible to present H in the form $H = H_1 * H_2$, where H_1 and H_2 are subalgebras of L_1 and L_2, respectively. For the details see [291].

The general structure theory of free products with the amalgamated subalgebra is rather complicated (see G. P. Kukin [290], [291]). Finitely generated and finitely presented subalgebras were studied by L. M. Nurieva in [455]. As for free subalgebras of free products of Lie algebras and restricted Lie algebras first of all we mention that Cartesian subalgebras of free products of Lie algebras (i.e., kernels of the canonical epimorphisms on the direct products of these algebras) are free, see [173] and [287]. If B is a subalgebra of the free product $\prod_H^* H_\alpha$ of Lie algebras H_α with the amalgamated subalgebra H such that $B \cap H_\alpha$ is a free Lie algebra for each α, and $B \cap H = 0$, then B is a free Lie algebra, see [290].

Problem 21.7
Describe subalgebras and free subalgebras of free products with the amalgamated subalgebra of color Lie superalgebras and of color Lie p-superalgebras.

Note that S. A. Agalakov and U. U. Umirbaev constructed examples of free products $L = L_1 * L_2$ of two Lie algebras L_1 and L_2 such that the entry problem in a finitely generated subalgebra is algorithmically insolvable in L, but at the same time there exists an algorithm to solve this problem in L_1 and in L_2 (it gives an answer on G. P. Kukin's problem, [156], Problem 1.86). In fact, S. A. Agalakov in [5] considered the one dimensional Lie algebra L_1 and the Lie algebra L_2 given in the variety of metabelian Lie algebras by generators a, b, a', b', d and by defining relations

$$[a', b] = [a', b'] = [d, a] = [d, b] = [d, a'] = [d, b'] = 0.$$

U. U. Umirbaev in [567] took as L_1 the one dimensional Lie algebra and as L_2 the free metabelian Lie algebra of sufficiently large rank.

Recall also the following problem.

Problem 21.8 ([156], Problem 1.85)
Is it true that the minimal number of generators of the free product of two subalgebras is equal to the sum of these numbers for multipliers?

Note that A. B. Verevkin pointed out that the answer is negative if we consider the variety of associative commutative algebras over a field. Indeed, let K be a field, $L_1 = K[x, x^{-1}]$, $L_2 = K[y, y^{-1}]$ (Laurent polynomials). Then it is clear that the associative free product $K[x, x^{-1}] * K[y, y^{-1}]$ is generated by $x, y, x^{-1}y^{-1}$.

22 Algorithms and examples

In this Section we give an algorithm to construct Gröbner bases of homogeneous ideals of color Lie (p-)superalgebras (Algorithm 22.1), an algorithm to solve the equality problem in a color Lie (p-)superalgebra given by a set of defining relations which form a Gröbner basis of the corresponding ideal (Algorithm 22.2), an algorithm to construct Gröbner basis of a left ideal of the free associative algebra (Algorithm 22.3), an algorithm to construct a Gröbner bases of a left ideal of the direct sum of k copies of the free associative algebra (Algorithm 22.4), and an algorithm to decide whether a system of homogeneous elements of the free associative algebras is left dependent, or not, and in the case of positive answer to realize this dependence (Algorithm 22.5). We add also some concrete examples.

In all what follows in this section K is a field, $\operatorname{char} K \neq 2$.

Algorithms

Algorithm 22.1
An algorithm to construct Gröbner bases for color Lie (p-)superalgebras given by φ-homogeneous set of defining relations.

Let $|X| < \infty$, $L(X)$ the free color Lie superalgebra ($L^p(X)$ the free color Lie p-superalgebra) and let R be a subset consisting of φ-homogeneous elements in $L(X)$ (in $L^p(X)$).

1) Eliminations of subwords.

If $r_1, r_2 \in R$ and there exists a composition of inclusion $\langle r_1, r_2\rangle$ then we set $r_1' = \langle r_1, r_2\rangle$. Since r_1 and r_2 are φ-homogeneous, the element r_1' is also φ-homogeneous.

2) Addition of compositions.

If $r_1, r_2 \in R$ and there exists the composition of intersection (r_1, r_2) then we join it to the set R. Remark, that r_1 and r_2 are φ-homogeneous, and therefore their composition (r_1, r_2) is also φ-homogeneous. It is clear that $\varphi((r_1, r_2)) > \varphi(r_1)$ and $\varphi((r_1, r_2)) > \varphi(r_2)$.

At the beginning we take the subset

$$\{r \in R \mid \varphi(r) = \min_{a\in R} \varphi(a)\}.$$

Then we apply all possible transformations of the first type to this subset (observe that applying a transformation of the first type we have either zero or an element with the same value of φ). Then we apply all possible transformations of the second type (all of corresponding compositions belong to higher order φ-subsets). By induction we choose the next φ-subset, with the help of the first type transformations we eliminate subwords (looking through all leading terms of all previous φ-subsets and of all leading terms of the fixed subset). By the use of the second type transformation we add all such compositions to higher order φ-subsets.

It is clear that the so constructed set S is stable and $\mathrm{id}(R) = \mathrm{id}(S)$. Since $|X| < \infty$, the number of pairwise different monomials with fixed weight φ_0 is finite. We see now that we made only a finite number of steps in our algorithm to construct φ-subsets with $\varphi \leq n \in \mathbb{N}$ because the set R is φ-homogeneous.

Now in order to construct the Gröbner basis of I it is sufficient for each $a \in S$ to reduce each s-regular (ps-regular) monomial b in the s-regular (ps-regular) presentation of a using elements $c \in S \setminus \{a\}$ (i.e., to consider all possible compositions of inclusion $\langle b, a\rangle$ and to replace b by $\langle b, a\rangle$).

Algorithm 22.2
An algorithm to solve the equality problem in a color Lie (p-)superalgebra of the form $L(X)/I$ ($L^p(X)/I'$) where the ideal I (the ideal I') is given by its Gröbner basis.

We suppose that $|X| < \infty$. Remember that to solve the equality problem means to give an algorithm to decide whether an element a of $L(X)$ (of $L^p(X)$) belongs to I (to I'), or not. Let R be the Gröbner basis of the ideal I of $L(X)$ (of the ideal I' of $L^p(X)$). It follows from Theorem 20.5 that if the word $\overline{a}$ does not contain any subword $\overline{b}, b \in R$, then $a \notin \mathrm{id}(R)$ (by stability of S, the checking procedure can be done in a finite number of steps). If $\overline{a}$ contains a subword $\overline{b}$, $b \in R$, then it is sufficient to solve the problem for the element $\langle a, b\rangle$, $\overline{\langle a, b\rangle} < \overline{a}$. Thus, after a finite number of steps we find out whether a belongs to I (to I'), or not.

Algorithm 22.3
An algorithm to construct the Gröbner basis of a left ideal of the free associative algebra given by a finite set of generators.

Let R be a finite set of generators of a left ideal I of the free associative algebra $A(X)$. If R is not a left stable subset, then we consider elements $a, b \in R$ of minimal length such that $\bar{b} = u\bar{a}$, $u \in S(X)$. We apply the reduction $R_{\bar{b},a}$ to b. After a finite number of such reductions, multiplying if it is necessary by elements from K, we come to a left-reduced set S such that $\mathrm{Id}_{\mathrm{left}}(S) = \mathrm{Id}_{\mathrm{left}}(R) = I$. To construct the Gröbner basis of I we need only to reduce all words in presentations of elements from S: if $a, b \in S$, and in the presentation of b we have a summand αu, $0 \neq \alpha \in K$, $u \in S(X)$, $u = v\bar{a}$, then we apply the reduction $R_{u,a}$ to b. After a finite number of such reductions, multiplying possibly on nonzero elements of K, we get the Gröbner basis of the left ideal I.

Note that using Algorithm 22.3 one can decide whether a left ideal I of $A^+(X)$ coincides with $A^+(X)$, or not. Indeed, $I = A^+(X)$ if and only if the Gröbner basis of I is exactly the set X.

Algorithm 22.4
We consider $A(X)^k$ as a left $A(X)$-module. Let

$$(a_{i1}, \ldots, a_{ik}) = a_{i1}e_1 + \cdots + a_{ik}e_k \in A(X)^k,$$

$i = 1, \ldots, s$. The problem is to construct a basis of the submodule I of $A(X)^k$ generated by the subset $\{(a_{i1}, \ldots, a_{ik}) \mid i = 1, \ldots, s\}$.

First of all we use the same reductions as in Algorithm 22.3 to construct the Gröbner basis of the left ideal J of $A(X)$ generated by the set $\{a_{11}, \ldots, a_{s1}\}$: if for example

$$\overline{a_{11}} = b_{21}\overline{a_{21}} + \cdots + b_{s1}\overline{a_{s1}},$$

where either $b_{i1} \in S(X)$, or $b_{i1} = 0$, $i \geq 2$, then we set

$$(a_{11}, \ldots, a_{k1})' = (a_{11}, \ldots, a_{k1}) - b_{21}(a_{21}, \ldots, a_{2k}) - \cdots - b_{s1}(a_{s1}, \ldots, a_{sk}).$$

After a finite number of such reductions we come to a subset

$$\{(c_{i1}, \ldots, c_{ik}) \mid i = 1, \ldots, s\} \tag{4.1}$$

where nonzero elements $c_{11}, \ldots, c_{s1}$ form the Gröbner basis of J.

If $c_{i1} \neq 0$ for all $i = 1, \ldots, s$, then the subset (4.1) gives us the basis of I. Otherwise we consider all $i_1, \ldots, i_r$ such that $c_{i_1 1} = \cdots = c_{i_r 1} = 0$, and the subset

$$\{(c_{i_j 1}, \ldots, c_{i_j k}) \mid j = 1, \ldots, r\}.$$

As above, we use the reductions from our construction of the Gröbner basis of the left ideal of $A(X)$ generated by $\{c_{i_j 2} \mid j = 2, \ldots, r\}$. Then we consider the third position (if it is necessary), and so on. After the inspection of the k-th position we see that the union of all above constructed elements is the wanted basis.

Algorithm 22.5
Let $a_1, \ldots, a_k$ be φ-homogeneous elements of $A(X)$. The problem is to decide whether a system $\{a_1, \ldots, a_k\}$ is left dependent, or not. In the case of positive answer it is necessary to realize this dependence in the form

$$a_i = \sum_{j \neq i} b_j a_j$$

for some i, where $b_j \in A(X)$.

We may suppose that $\varphi(a_1) \leq \cdots \leq \varphi(a_k)$ and apply Algorithm 22.3. If at some step of this algorithm we come to zero element, then we get the desirable dependence. Otherwise there are no such dependencies. Remark that at each step it is possible to reduce all words from the presentation of elements of this φ-level by elements of the previous levels and then to consider only linear dependencies.

Examples

Example 22.6
Let K be a field, $\operatorname{char} K \neq 2$, $G = \mathbb{Z}_2$, $X_0 = \{x, y\}$, $X_1 = \{z\}$, $x > y > z$, φ the usual length function, $S = \{a, b, c\}$,

$$\begin{aligned} a &= [[x,[y,z],[y,z]] + [yz^4]; \\ b &= [[y,z],[y,z]] - [y,[yz^2]]; \\ c &= [x,y] - [z,z]. \end{aligned}$$

The problem is to write down all elements of the given length of the Gröbner basis of the ideal $\operatorname{id}(S)$ of $L(X)$ (we construct them up to the length 5).

At first we consider possible inclusions. We get

$$\begin{aligned} a' &= \langle a,b\rangle = 2a - [x,b] \\ &= 2[x,[y,z],[y,z]] + 2[yz^4] - [x,[[y,z],[y,z]]] + [x,[y,[yz^2]]] \\ &= [x,[y,[yz^2]]] + 2[yz^4]; \\ a'' &= \langle a',c\rangle = a' - [c,[yz^2]] \\ &= [x,[y,[yz^2]]] + 2[yz^4] - [[x,y],[yz^2]] + [[z,z],[yz^2]] \\ &= [x,[y,[yz^2]]] - [[x,y],[yz^2]] + 3[yz^4] \\ &= -[[x,[yz^2]],y] + 3[yz^4]; \\ a''' &= \langle a'',c\rangle = 2a'' + [[c,[z,z]],y] \end{aligned}$$

$$
\begin{aligned}
&= -2\big[[x,[yz^2]],y\big] + 6[yz^4] + 2\big[[[[x,y],z],z],y\big] - \big[[[z,z],[z,z]],y\big] \\
&= 2\big[[[xz^2],y],y\big] + 6[yz^4].
\end{aligned}
$$

Now we consider possible intersections (of the length not more than 5). We have:

$$
\begin{aligned}
b' &= (b,c) = [x,b] - 2\big[[c,z],[y,z]\big] \\
&= 2\big[x,[y,z],[y,z]\big] - \big[x,[y,[yz^2]]\big] - 2\big[[[x,y],z],[y,z]\big] \\
&= -2\big[[[x,z],y],[y,z]\big] - \big[x,[y,[yz^2]]\big] \\
&= 2\big[[x,z],[y,[y,z]]\big] + 2\big[[[x,z],[y,z]],y\big] - \big[x,[y,[yz^2]]\big].
\end{aligned}
$$

There are no compositions of inclusion between elements a''', b, c, b'. At the same time all possible compositions of intersection have the length greater than 5. Hence to construct the beginning of the Gröbner basis of $\mathrm{id}(S)$ it is sufficient to reduce all s-regular monomials from s-regular presentations of a''', b, c, b'.

Since

$$
\begin{aligned}
\langle -\big[x,[y,[yz^2]]\big], c\rangle &= -\big[x,[y,[yz^2]]\big] + \big[c,[yz^2]\big] \\
&= -\big[x,[y,[yz^2]]\big] - \big[[[[x,y]-[z,z],y],z],z\big] \\
&= \big[[x,[yz^2]],y\big] + 2[yz^4], \\
\langle \big[[x,[yz^2]],y\big], c\rangle &= 2\big[[x,[yz^2]],y\big] - \big[[[x,y],[z,z]],y\big] \\
&= -2\big[[[xz^2],y],y\big], \\
\langle -2\big[[[xz^2],y],y\big], a'''\rangle &= -2\big[[[xz^2],y],y\big] + a''' = 6[yz^4],
\end{aligned}
$$

we get the following list of all elements of the Gröbner basis of $\mathrm{id}(S)$ which length is not greater than 5:

$$
[x,y]-[z,z]; \qquad \big[[y,z],[y,z]\big] - \big[y,[yz^2]\big]; \qquad \big[[[xz^2],y],y\big] + 3[yz^4];
$$
$$
\big[[x,z],[y,[y,z]]\big] + \big[[[x,z],[y,z],y\big] + 4[yz^4].
$$

□

Example 22.7
Let $X = \{x,y,z\}$, $A(X)$ the free associative algebra, φ the usual length, $S = \{a,b,c\}$,

$$
a = xz + y; \qquad b = xxz + xy + z; \qquad c = yzz + xy.
$$

The problem is to construct the Gröbner basis of the left ideal $\mathrm{Id}_{\mathrm{left}}(S)$ of $A(X)$ generated by S. We consider the following reductions:

$$
\begin{aligned}
b' &= R_a(b) = b - x\cdot a = xxz + xy + z - xxz - xy = z; \\
a' &= R_{b'}(a) = a - x\cdot b' = xz + y - xz = y; \\
c' &= R_{b'}(c) = c - yz\cdot b' = yzz + xy - yzz = xy; \\
c'' &= R_{a'}(c') = c' - x\cdot a' = 0.
\end{aligned}
$$

Thus, the Gröbner basis of $\mathrm{Id}_{\mathrm{left}}(S)$ is the set $\{y, z\}$. □

Example 22.8
Let $X = \{x, y\}$. The problem is to construct the basis of the left ideal of $A(X)^2$ generated by the elements

$$b_1 = (x, yx), \qquad b_2 = (yx, yx),$$
$$b_3 = (xyx + yx, xy^2x + yx), \qquad b_4 = (x^2yx + x, x^2y^2x).$$

We use the following reductions:

$$\begin{aligned} b_2' &= b_2 - y \cdot b_1 = (0, yx - y^2x); \\ b_3' &= b_3 - (xy + y)b_1 = (0, yx - y^2x); \\ b_4' &= b_4 - (x^2y + 1)b_1 = -(0, yx). \end{aligned}$$

Now we consider the second component:

$$\begin{aligned} b_3'' &= b_3' - b_2' = (0, 0); \\ b_2'' &= b_2' + (y + 1)b_4' = (0, 0); \\ b_1' &= b_1 + b_4' = (x, 0). \end{aligned}$$

Thus, we have constructed the following basis of the left ideal I:

$$b_1' = (x, 0), \qquad -b_4' = (0, yx).$$

□

Example 22.9
Let $X = \{x, y, z\}$, $x > y > z$, φ be the usual length function,

$$a_1 = x^2 + yx, \qquad a_2 = xy - zx,$$
$$a_3 = x^3 + z^2x, \qquad a_4 = xyx - zxy; \qquad a_5 = xyz.$$

The problem is to decide whether the subset $\{a_1, a_2, a_3, a_4, a_5\}$ is left dependent, or not, and in the case of positive answer to present the explicit dependence.

Starting with the leading words we use the following reductions:

$$\begin{aligned} a_3' &= a_3 - xa_1 = -xyx + z^2x; \\ a_4' &= a_4 + za_2 = xyx - z^2x. \end{aligned}$$

Now we see that $a_3' + a_4' = 0$. Thus we get the following dependence:

$$a_4 = xa_1 - za_2 - a_3.$$

□

Comments

The canonical forms of elements of different algebraic systems were used in the combinatorial theory to solve a big number of problems. The sources of such considerations are due to graph theory, group theory and to polynomial rings, see [101], [212], [228], [445]. Details of the theory of Gröbner bases of ideals of polynomial rings can be found in [102], [143], [192], [316]. In noncommutative case Composition Lemma (Theorem 20.5) for Lie algebras was proved by A. I. Shirshov in [516]. In associative case Diamond Lemma (Theorem 18.1) was suggested by L. A. Bokut' in [80] and by G. Bergman in [63]. Now this type of Composition Lemma is becoming to be a rather useful tool for solving algorithmic problems in Ring theory. For more general constructions of canonical bases of associative algebras from the homological point of view see [200]. In homogeneous case Canonical bases with the use of special resolution can be found in [8]. For some other aspects of canonical forms of elements in algebraic systems see for example [68], [138], [192], [208], [213], [225], [254], [316], [322], [435], [436]. Canonical bases for color free Lie (p-)superalgebras were suggested be A. A. Mikhalev in [387], [388], [389], [392], [397], [401].

We mention some papers on symbolic computation in algebras: [9], [19], [46], [68], [112], [113], [133], [134], [135], [136], [143], [144], [166], [196], [197], [206], [227], [233], [253], [275], [277], [312], [357], [428], [434], [469], [502], [544], [549], [562], [575], [631], [632], [634].

Notice that PBW-theorems for Lie superalgebras and color Lie superalgebras were pointed out by L. E. Ross in [486] and by M. Scheunert in [488]. Some aspects of PBW-theorems can be found in [10], [59], [60], [67], [110], [125], [132], [142], [176], [177], [190], [230], [318], [331], [334], [337], [338], [355], [374], [399], [453], [454], [478], [498], [534], [611].

Theorems 19.3 and 19.7 for Lie algebras was proved by U. U. Umirbaev in [565]. For the case of Lie superalgebras see [396].

As for connections with quantum groups and quantum algebras see for example [114], [147], [162], [182], [299], [359], [360], [361], [531].

Criteria for an associative element to be a Lie element may be found in [76], [122], [171], [185], [211], [341], [350], [475], [476], [512], [536], [594], [599], [602].

Relations with Hopf algebras described in [1], [12], [64], [108], [181], [215], [359], [361], [427], [433], [468], [475], [476], [548].

Shuffle algebras were introduced by R. Ree in [475], [476]. For various aspects of shuffle algebras see for example recent monograph [481] by Ch. Reutenauer (see also articles [76], [262], [371], [412], [468], [477]).

Main results of Section 20 are due to A. A. Mikhalev [385], [387], [388], [397], [401].

General algorithmic problems of Lie algebras are under consideration in [86], [87], [192], [225], [270], [572].

Bases of free products of Lie algebras with the amalgamated subalgebra were constructed by A. I. Shirshov in [513]. In the case of restricted Lie algebras such bases were suggested by G. P. Kukin in [291]. Bases of free products with the amalgamated subalgebra of color Lie (p-)superalgebras were presented by A. A. Mikhalev in [385], [387], [397].

Chapter 5

Free Differential Calculus

This chapter is devoted to the free differential calculus for free color Lie (p-)superalgebras. In Section 23 we discuss universal derivations. We prove some lemmas on properties of partial derivatives. In Section 24 we prove the Jacobian conjecture for free color Lie (p-)superalgebras. In Section 25 we collect necessary technical lemmas. In Sections 26 and 27 we prove the main theorem of this chapter on the rank of an element of the free color Lie (p-)superalgebra: in Section 26 we consider the homogeneous case relative to a functional, and in Section 27 we consider the general case. In connection with Theorem 27.9 we present also an example of a nonfree Lie algebra with cohomological dimension one (over a field of prime characteristic). Section 28 is devoted to primitive systems of elements (i.e., to such systems which are subsets of free generating sets). We present necessary and sufficient conditions for a system of elements to be a primitive system. We prove also that an endomorphism preserving primitivity of elements, is an automorphism. In Section 29 we present algorithms and examples. In Sections 25, 27 and 28 we formulate a number of open problems and discuss possible directions of the research.

In all what follows in this chapter let K be a field, $\operatorname{char} K \neq 2$, G an Abelian group. We denote $L_X = L(X)$ in the case when $\operatorname{char} K = 0$, and $L_X = L^p(X)$ in the case when $\operatorname{char} K = p > 2$. We say that an additive function $\omega\colon \mathbb{Z}_+^n \to \mathbb{Q}_+$ is *a functional* if $\omega(r) \neq 0$ for all $r \neq 0$. As in Chapter 1 where for the weight function $\varphi\colon A(X) \to \mathbb{Z}_+$ we consider φ-homogeneous elements, φ-leading parts and so on, in this chapter we use the same conceptions and notations for ω-homogeneous elements, ω-leading parts and so on.

23 Universal derivations

In this section we consider universal derivations, expose basic facts from the free differential calculus for free color Lie superalgebras and collect some elementary properties of partial derivatives.

Free differential calculus

If K is a field, $X = \{x_1, \dots, x_n\}$, $K[X]$ is the algebra of commutative polynomials, then for any endomorphism φ of $K[X]$ consider its Jacobian matrix

$$J(\varphi) = \left(\frac{\partial\varphi(x_j)}{\partial x_i}\right), \qquad 1 \le i, j \le n.$$

The well-known Jacobian conjecture proposed by O. H. Keller in [264] states that if $\operatorname{char} K = 0$, then φ is an automorphism if and only if the matrix $J(\varphi)$ is invertible over $K[X]$. The general answer is not known even in the case $n = 2$ (for more details see for example [41], [466]).

In [184] R. H. Fox introduced the free differential calculus for free groups. Let $F = F(X)$ be the free group on X, $\mathbb{Z}F$ the integral group ring of F, Δ_F the augmentation ideal of F, i.e. the kernel of the natural homomorphism $\sigma: \mathbb{Z}F \to \mathbb{Z}$,

$$\sigma\left(\sum_i (n_i f_i)\right) = \sum_i n_i,$$

$n_i \in \mathbb{Z}$, $f_i \in F$. The right Fox partial derivative on x_i is the mapping $d_i: \mathbb{Z}F \to \mathbb{Z}F$ given by

$$\begin{aligned} d_i(x_j) &= \delta_{ij};\\ d_i(uv) &= d_i(u)v + \sigma(u)\, d_i(v);\\ d_i(ku + lv) &= kd_i(u) + ld_i(v), \end{aligned}$$

$u, v \in \mathbb{Z}F$, $k, l \in \mathbb{Z}$. There is another realization of d_i as follows. The ideal Δ_F is a free right $\mathbb{Z}F$-module with the free basis $\{(x_i - 1) \mid 1 \le i \le n\}$, the mappings d_i are the projections on the corresponding free cyclic direct summands. Any element $u \in \Delta_F$ can be uniquely written in the form

$$u = \sum_{i=1}^{n} (x_i - 1)\, d_i(u).$$

J. S. Birman in [69] proved that an endomorphism φ of F is an automorphism if and only if its Jacobian matrix $(d_j(\varphi(x_i)))$ is invertible over $\mathbb{Z}F$. A system of elements of F is called primitive if it is a subset of some set of free generators of F. A $(q \times r)$-matrix A over $\mathbb{Z}F$ is called left invertible if there exists an $(r \times q)$-matrix B over $\mathbb{Z}F$ such that $BA = E_r$, where E_r is the unity matrix of the order r. U. U. Umirbaev in [568], [570] proved that a system $\{g_1, \ldots, g_k\}$ of F is primitive if and only if the matrix $\left(\frac{\partial g_j}{\partial x_i}\right)$, $1 \le i \le n$, $1 \le i \le n$, is left invertible over $\mathbb{Z}F$. For more details on free differential calculus in groups see [214].

Now we consider the universal derivations for free algebras. Let A be the free algebra of a variety $\mathcal{V}$ on $X = \{x_1, x_2, \ldots\}$, where $\mathcal{V}$ is the variety of all associative algebras or the variety of all Lie algebras. Let B be the free algebra of the same variety on $X \cup Y$, $Y = \{y_1, y_2, \ldots\}$, $|X| = |Y|$, $M(B)$ the subalgebra of $\operatorname{End}_K B$ generated by the set of left and right multiplications $\{L_b, R_b \mid b \in B\}$. *The multiplicative envelope* $U(A)$ of A is the subalgebra with 1 of $M(B)$ generated by $\{L_a, R_a \mid a \in A\}$.

Let M be a K-linear space, $l\colon A \to \operatorname{End}_K M$ a K-linear mapping. The pair (l, M) is called a left A-module if

$$l_{ab} \cdot c = l_a \cdot (l_b \cdot c)$$

in the case of associative algebras and

$$l_{ab} \cdot c = l_a \cdot (l_b \cdot c) - l_b \cdot (l_a \cdot c)$$

in the case of Lie algebras, $a, b \in A$, $c \in M$. By analogy, a pair (M, r) is called a right A-module if

$$c \cdot r_{ab} = (c \cdot r_a) \cdot r_b$$

and

$$c \cdot r_{ab} = (c \cdot r_a) \cdot r_b - (c \cdot r_b) \cdot r_a,$$

respectively. A triple (l, M, r) is called an A-bimodule relative to the variety $\mathcal{V}$ if the pairs (l, M) and (M, r) are left and right A-modules, respectively, and moreover, the algebra $A \oplus M$ with the operation

$$(a_1 + c_1) \cdot (a_2 + c_2) = a_1 a_2 + l_{a_1} \cdot c_2 + c_1 \cdot r_{a_2},$$

where $a_1, a_2 \in A$, $c_1, c_2 \in M$, is the algebra of the variety $\mathcal{V}$.

Let M be an A-bimodule relative to a variety $\mathcal{V}$. A linear mapping $\partial\colon A \to M$ is called *a derivation* if

$$\partial(ab) = \partial(a) \cdot b + a \cdot \partial(b)$$

for all $a, b \in A$. A derivation $\partial_0\colon A \to M_0$ is called *the universal derivation* of A if for any derivation $\partial\colon A \to M$ there exists the unique homomorphism of

A-bimodules $\varphi\colon M_0 \to M$ such that the following diagram is commutative:

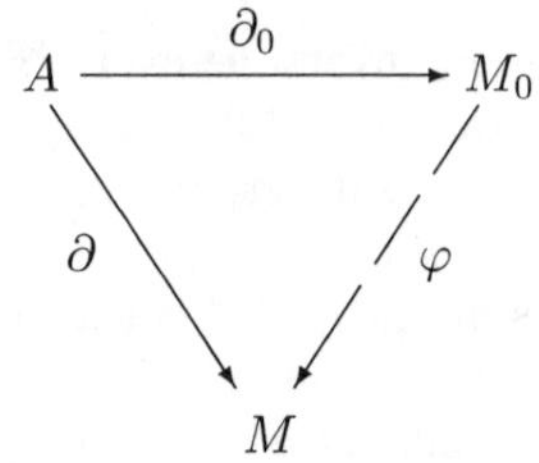

Let

$$I_A = y_1U(A) \oplus \cdots \oplus y_nU(A) \oplus \cdots$$

be the free right $U(A)$-module with the basis Y. Consider the derivation $D\colon A \to I_A$ given by $D(x_i) = y_i$, $i \geq 1$. One can easily see that D is the universal derivation of A.

Remark that in a similar way it is possible to define the universal derivation for the free group. The components of this universal derivation are Fox partial derivatives. In the case of the ring of polynomials we get the usual partial derivatives.

For the free associative algebra $A(X)$ denote by $A(X)^e$ the tensor product $A(X) \otimes_K A(X)$ with the multiplication given by $(a \otimes b)(c \otimes d) = ac \otimes db$. It is clear that $A(X)^e$ is the multiplicative envelope of $A(X)$. Let

$$I_A = \bigoplus_{i=1}^{|X|} (A(X)^e)_i$$

be the direct sum of $|X|$ copies of $A(X)^e$, $(A(X)^e)_i = A(X) \cdot dx_i \cdot A(X)$, $(a \otimes b)_i = a \cdot dx_i \cdot b$. The algebra

$$I_A = \bigoplus_{i=1}^{|X|} (A(X) \cdot dx_i \cdot A(X))_i$$

is the multiplicative envelope of $A(X)$. It is an $A(X)$-bimodule (because of the left and the right $A(X)$-actions on I_A are permutable). The universal derivation of $A(X)$ is the K-linear mapping $D\colon A(X) \to I_A$ given by $D(x_i) = 1 \cdot dx_i \cdot 1$, $D(ab) = D(a) \cdot b + a \cdot D(b)$, $D(1) = 0$. For any $a \in A(X)$ the element $D(a)$ has the unique presentation in the form $D(a) = a_1 + \cdots + a_n$, $a_i \in (A(X)^e)_i$. The component $a_i = \dfrac{\partial a}{\partial x_i}$ is called the partial derivative of a on x_i.

If $X = \{x_1, \ldots, x_n\}$ and φ is an endomorphism of $A(X)$ then φ is an automorphism if and only if the Jacobian matrix

$$\left(\frac{\partial \varphi(x_j)}{\partial x_i} \right), \qquad 1 \leq i, j \leq n,$$

is invertible over $A(X)^e$ (see [154], [493]).

Remark 23.1

In the case $n = 2$ W. Dicks in [153] gave a very simple commutator test for an endomorphism φ of $A(X)$ over a field K to be an automorphism: φ is an automorphism if and only if $[\varphi(x_1), \varphi(x_2)] = \alpha\,[x_1, x_2]$ for some nonzero $\alpha \in K$.

Definition 23.2

For any element a of the free associative algebra $A(X)$ we have the unique presentation in the form $a = \alpha \cdot 1 + x_1\, a_1 + x_2\, a_2 + \cdots$, where only finite number of elements $a_i \in A(X)$ are nonzero, $\alpha \in K$. We call the element a_i *the right Fox partial derivative of the element a by x_i*, and we use the notation $a_i = \dfrac{\partial a}{\partial x_i} = \dfrac{\partial}{\partial x_i} a$. Thus we have the operators $\dfrac{\partial}{\partial x_i}$. These operators are exactly linear mappings $\dfrac{\partial}{\partial x_i}\colon A(X) \to A(X)$ such that

$$\frac{\partial}{\partial x_i}(x_j) = \delta_{ij}$$

and

$$\frac{\partial}{\partial x_i}(uv) = \frac{\partial}{\partial x_i}(u)\, v + \sigma(u)\, \frac{\partial}{\partial x_i}(v),$$

where $\sigma\colon A(X) \to A(X)$ is the homomorphism defined by $\sigma(x_i) = 0$ for all $x_i \in X$.

Also for $a \in A(X)$ we have the unique presentation $a = \alpha \cdot 1 + a'_1\, x_1 + a'_2\, x_2 + \cdots$, where $a'_i \in A(X)$, $\alpha \in K$. We call the element a'_i *the left Fox partial derivative of the element a by x_i* and denote $a'_i = a\dfrac{\partial}{\partial x_i}$.

We show now that the restriction of operators $\dfrac{\partial}{\partial x_i}$ on $L(X)$, $x_i \in X$, are derivations of $L(X)$.

If L is a color Lie K-superalgebra, M is a G-graded K-linear space, then M is an L-bimodule if there are fixed linear mappings $l, r\colon L \to \operatorname{End}_K M$ such that

$$\begin{aligned} l_{[a,b]} \cdot c &= l_a \cdot (l_b \cdot c) - \varepsilon(a,b)\, l_b \cdot (l_a \cdot c); \\ c \cdot r_{[a,b]} &= (c \cdot r_a) \cdot r_b - \varepsilon(a,b)\, (c \cdot r_b) \cdot r_a \end{aligned}$$

for all G-homogeneous $a, b \in L$, $c \in M$, and $L \oplus M$ is a color Lie superalgebra with the operation $[\ ,\ ]$ defined by

$$[a_1 + b_1, a_2 + b_2] = [a_1, a_2] + l_{a_1} \cdot b_2 + b_1 \cdot r_{a_2}$$

for all G-homogeneous $a_1, a_2 \in A$, $b_1, b_2 \in M$, $d(a_i) = d(b_i)$, $i = 1, 2$.

Let now $L = L(X)$ and $M = A(X)$. For all $a \in L(X)$ and $b \in A(X)$ we set $l_a \cdot b = -\varepsilon(a,b)\, ba$ and $a \cdot r_b = a \cdot b$. Then $A(X)$ is an $L(X)$-bimodule. Indeed,

$$\begin{aligned} l_{[a,b]} \cdot c &= -\varepsilon([a,b],c)\, c \cdot [a,b] \\ &= -\varepsilon(a,c)\,\varepsilon(b,c)\, cab + \varepsilon(a,c)\,\varepsilon(b,c)\,\varepsilon(a,b)\, cba \\ &= l_a \cdot (-\varepsilon(b,c)\, cb) - \varepsilon(a,b)\,\varepsilon(a,c)\,\varepsilon(b,ac)\, cab \\ &= l_a \cdot (l_b \cdot c) - \varepsilon(a,b)\, l_b \cdot (l_a \cdot c), \\ c \cdot r_{[a,b]} &= cab - \varepsilon(a,b)\, cba = (c \cdot r_a) \cdot r_b - \varepsilon(a,b)\,(c \cdot r_b) \cdot r_a \end{aligned}$$

for all G-homogeneous a, b, c;

At the same time,

$$\begin{aligned} [a_1 + b_1, a_2 + b_2] &= [a_1, a_2] + l_{a_1} \cdot b_2 + b_1 \cdot r_{a_2} \\ &= -\varepsilon(a_1,a_2)\,[a_2,a_1] - \varepsilon(a_1,a_2)\, b_2 a_1 + b_1 a_2 \\ &= -\varepsilon(a_1,a_2)\,([a_2,a_1] + l_{a_2} \cdot b_1 + b_2 \cdot r_{a_1}) \\ &= -\varepsilon(a_1,a_2)\,[a_2 + b_2, a_1 + b_1], \end{aligned}$$

and we get the ε-anticommutativity.

Now we verify the Jacobi identity:

$$\begin{aligned} J_1 &= \big[[a_1 + b_1, a_2 + b_2], a_3 + b_3\big] \\ &= \big[[a_1,a_2] - \varepsilon(a_1,a_2)\, b_2 a_1 + b_1 a_2, a_3 + b_3\big] \\ &= j_1 - \varepsilon(a_1 a_2, a_3)\, b_3 (a_1 a_2 - \varepsilon(a_1,a_2)\, a_2 a_1) \\ &\quad + (-\varepsilon(a_1,a_2)\, b_2 a_1 + b_1 a_2)\, a_3; \\ J_2 &= \varepsilon(a_1,a_2)\,\big[a_2 + b_2, [a_1 + b_1, a_3 + b_3]\big] \\ &= \varepsilon(a_1,a_2)\,\big[a_2 + b_2, [a_1,a_3] - \varepsilon(a_1,a_3)\, b_3 a_1 + b_1 a_3\big] \\ &= j_2 - \varepsilon(a_1,a_2)\,\varepsilon(a_2, a_1 a_3)\,(-\varepsilon(a_1,a_3)\, b_3 a_1 + b_1 a_3) a_2 \\ &\quad + \varepsilon(a_1,a_2)\, b_2 (a_1 a_3 - \varepsilon(a_1,a_3) a_3 a_1); \\ J_3 &= \big[a_1 + b_1, [a_2 + b_2, a_3 + b_3]\big] \\ &= \big[a_1 + b_1, [a_2,a_3] - \varepsilon(a_2,a_3)\, b_3 a_2 + b_2 a_3\big] \\ &= j_3 - \varepsilon(a_1, a_2 a_3)\,(-\varepsilon(a_2,a_3)\, b_3 a_2 + b_2 a_3)\, a_1 \\ &\quad + b_1 (a_2 a_3 - \varepsilon(a_2,a_3)\, a_3 a_2), \end{aligned}$$

where $j_1 = \big[[a_1,a_2],a_3\big]$, $j_2 = \varepsilon(a_1,a_2)\,\big[a_2,[a_1,a_3]\big]$, $j_3 = \big[a_1,[a_2,a_3]\big]$.

One can easily see now that it follows from the Jacobi identity $j_3 = j_1 + j_2$ in $L(X)$ that the Jacobi identity $J_3 = J_1 + J_2$ is valid in $L \oplus M$. Therefore $L \oplus M$ is a color Lie superalgebra. In other words, $A(X)$ is an $L(X)$-bimodule. Moreover, for $x \in X$

$$\frac{\partial}{\partial x}([a,b]) = \frac{\partial a}{\partial x}\, b - \varepsilon(a,b) \frac{\partial b}{\partial x}\, a = \frac{\partial a}{\partial x} \cdot r_b + l_a \cdot \frac{\partial b}{\partial x},$$

that is $\frac{\partial}{\partial x}$ is a derivation.

It is clear that the multiplicative envelope of the free color Lie superalgebra $L(X)$ is exactly the free G-graded associative algebra $A(X)$ with the above defined actions l_a and r_a. The derivation

$$D: L(X) \to I_A = \bigoplus_{i=1}^{|X|} dx_i \cdot A(X)$$

given by

$$D(a) = \sum_{i=1}^{|X|} dx_i \cdot \frac{\partial a}{\partial x_i}$$

for all $a \in L(X)$ (only finite number of summands are nonzero) is the universal derivation of $L(X)$.

Elementary properties of derivatives

We say that an element $a \in A(X)$ *depends on* $x_i \in X$ if in a we have with nonzero coefficient a word which contains x_i.

One can easily prove the following two elementary statements.

Lemma 23.3
Let $h \in L(X)$ ($h \in L^p(X)$), $a \in A(X)$, and a does not depend on $x_k \in X$. Then

$$\frac{\partial}{\partial x_k}(h \cdot \mathrm{Ad}\,(a)) = \frac{\partial h}{\partial x_k}\, a.$$

Lemma 23.4
Let $h \in L(X)$ ($h \in L^p(X)$), $x_k \in X$, and $\frac{\partial h}{\partial x_k} = 0$. Then the element h does not depend on x_k.

We complete this section considering the left ideal of $A(X)$ generated by the right partial derivatives of given element of $L(X)$ (of $L^p(X)$).

Let $h \in L(X)$ ($h \in L^p(X)$). We consider $A(X)$ as a left $A(X)$-module relative to usual multiplication. Denote by M_h the $A(X)$-submodule of the left $A(X)$-module $A(X)$ generated by the elements $\left\{ \frac{\partial h}{\partial x_i} \middle| x_i \in X \right\}$. Since M_h is finitely generated, $\mathrm{rank}\,(M_h) < \infty$.

Lemma 23.5
Let φ be an automorphism of $L(X)$ (of $L^p(X)$). Then $M_{\varphi(h)} = \varphi(M_h)$ for all $h \in L(X)$ ($h \in L^p(X)$).

PROOF One can assume that h depends only on $x_1, \dots, x_m$, and the elements $\varphi(x_1), \dots, \varphi(x_m)$ depend only on $x_1, \dots, x_n$. Since

$$h = \sum_{i=1}^{m} x_i \frac{\partial h}{\partial x_i},$$

applying φ, we get

$$\varphi(h) = \sum_{i=1}^{m} \varphi(x_i)\,\varphi\left(\frac{\partial h}{\partial x_i}\right),$$

and

$$\begin{aligned}\frac{\partial \varphi(h)}{\partial x_j} &= \sum_{i=1}^{n} \frac{\partial \varphi(x_i)}{\partial x_j}\,\varphi\left(\frac{\partial h}{\partial x_i}\right) \\ &= \varphi\left(\sum_{i=1}^{n} \varphi^{-1}\left(\frac{\partial \varphi(x_i)}{\partial x_j}\right)\frac{\partial h}{\partial x_i}\right) \in \varphi(M_h).\end{aligned}$$

Therefore $M_{\varphi(h)} \subseteq \varphi(M_h)$ for all φ and h. In particular,

$$\varphi(M_h) = \varphi(M_{\varphi^{-1}\varphi(h)}) \subseteq \varphi\varphi^{-1}(M_{\varphi(h)}) = M_{\varphi(h)}.$$

■

Corollary 23.6
Let $h \in L(X)$ ($h \in L^p(X)$) and let φ be an automorphism of $L(X)$ (of $L^p(X)$). Then the ranks of left $A(X)$-modules M_h and $M_{\varphi(h)}$ are equal.

24 Jacobian matrices

In this section we consider the Jacobian matrix of an endomorphism of the free color Lie (p-)superalgebra of finite rank and prove that such endomorphism is an automorphism if and only if its Jacobian matrix is invertible.

In all what follows in this section let $X = \{x_1, \dots, x_n\}$, and $L = L(X)$ or $L = L^p(X)$.

Jacobians

For $a \in L$ we denote by $\partial(a) = \partial_X(a)$ the column $\left(\frac{\partial a}{\partial x_1}, \dots, \frac{\partial a}{\partial x_n}\right)^\top$. For an endomorphism φ of L by $J_k(\varphi)$, $k \le n$, we consider the $(k \times k)$-matrix

$$J_k(\varphi) = \begin{pmatrix} \frac{\partial \varphi(x_1)}{\partial x_1} & \dots & \frac{\partial \varphi(x_k)}{\partial x_1} \\ \vdots & \ddots & \vdots \\ \frac{\partial \varphi(x_1)}{\partial x_k} & \dots & \frac{\partial \varphi(x_k)}{\partial x_k} \end{pmatrix},$$

and we say that $J_k(\varphi)$ is *the k-th Jacobian matrix* of φ. If $k = n$ then $J(\varphi) = J_n(\varphi)$ is called *the Jacobian matrix* of φ.

Since $a = (x_1, \dots, x_n)\,\partial(a)$,

$$(\varphi(x_1), \dots, \varphi(x_n)) = (x_1, \dots, x_n)\,J(\varphi).$$

If $\{y_1, \ldots, y_n\}$ is another set of free generators of L, then the change rule is given by

$$\frac{\partial a}{\partial y_i} = \left(\frac{\partial x_1}{\partial y_i}, \ldots, \frac{\partial x_n}{\partial y_i}\right) \partial_X(a),$$

that is

$$\partial_Y(a) = \left(\frac{\partial x_j}{\partial y_i}\right) \partial_X(a),$$

and

$$\begin{aligned} \frac{\partial a}{\partial x_i} &= \left(\frac{\partial y_1}{\partial x_i}, \ldots, \frac{\partial y_n}{\partial x_i}\right) \partial_Y(a), \\ \partial_X(a) &= \left(\frac{\partial y_j}{\partial x_i}\right) \partial_Y(a). \end{aligned}$$

Jacobian conjecture for free Lie superalgebras

Lemma 24.1
Let $y_1, \ldots, y_n \in L$. Then the set $\{y_1, \ldots, y_n\}$ is a free generating set of L if and only if $\{y_1, \ldots, y_n\}$ is a basis of the free right $A(X)$-module $LA(X)$.

PROOF Let $\{y_1, \ldots, y_n\} \subseteq L$ be a basis of the free right $A(X)$-module $L \cdot A(X)$. Denote by H the subalgebra of L generated by $\{y_1, \ldots, y_n\}$, and $J_H = H \cdot A(X)$. By Theorem 19.3, $L \cap J_H = H$. Since $y_1, \ldots, y_n \in J_H$, we get $J_H = L \cdot A(X)$. Hence $L \cap J_H = L$, $H = L$, and L is generated by $\{y_1, \ldots, y_n\}$. By Theorem 15.1, $\{y_1, \ldots, y_n\}$ is a free generating set of L.

It is obvious that any set of free generators of L forms a basis of the free right $A(X)$-module $LA(X)$. ∎

Theorem 24.2
Let K be a field, $X = \{x_1, \ldots, x_n\}$. An endomorphism φ of the free color Lie (p-)superalgebra L is an automorphism if and only if the Jacobian matrix $J(\varphi)$ is invertible over $A(X)$. Moreover, $J(\varphi)^{-1} = \varphi(J(\varphi^{-1}))$.

PROOF Let $y_i = \varphi(x_i)$, $1 \leq i \leq n$. If φ is an automorphism of L, then for any $h \in L$

$$\begin{aligned} h &= \sum_{j=1}^{n} x_j \frac{\partial \varphi(h)}{\partial x_j}, \\ \varphi(h) &= \sum_{j=1}^{n} \varphi(x_j) \varphi\left(\frac{\partial \varphi(h)}{\partial x_j}\right), \end{aligned}$$

and

$$\begin{aligned} \frac{\partial \varphi(h)}{\partial x_i} &= \sum_{j=1}^{n} \frac{\partial \varphi(x_j)}{\partial x_i} \varphi\left(\frac{\partial h}{\partial x_j}\right), \\ \partial(\varphi(h)) &= J(\varphi)\varphi(\partial(h)). \end{aligned} \tag{5.1}$$

Hence

$$\begin{aligned}\partial(\varphi^{-1}(h)) &= J(\varphi^{-1})\,\varphi^{-1}(\partial(h));\\ \partial(\varphi^{-1}(\varphi(h))) = \partial(h) &= J(\varphi^{-1})\,\varphi^{-1}(\partial(\varphi(h))).\end{aligned} \tag{5.2}$$

It follows from (5.1) and (5.2) that

$$\partial(\varphi(h)) = J(\varphi)\,\varphi(J(\varphi^{-1}))\,\partial(\varphi(h)).$$

Since $\varphi(L) = L$ and h is an arbitrary element,

$$J(\varphi)\,\varphi(J(\varphi^{-1})) = E.$$

Let now $J(\varphi)$ be an invertible matrix over $A(X)$ and $A = (J(\varphi))^{-1}$. Then

$$(x_1, \ldots, x_n)\,J(\varphi)\,A = (x_1, \ldots, x_n).$$

Since $\{x_1, \ldots, x_n\}$ is a basis of $LA(X)$, it follows from

$$(x_1, \ldots, x_n)\,J(\varphi) = (y_1, \ldots, y_n)$$

that $\{y_1, \ldots, y_n\}$ is also a basis of $LA(X)$. By Lemma 24.1, $\{y_1, \ldots, y_n\}$ is a free generating set of $L(X)$, i.e. φ is an automorphism of L. ∎

Note that for the composition of endomorphisms φ and ψ of $L(X)$ (of $L^p(X)$) the Jacobian matrix $J(\psi \circ \varphi)$ is equal to $J(\psi)\psi(J(\varphi))$.

It is possible also to give a solution of the Jacobian problem for Lie superalgebras in the following form (the exposition of this result for Lie algebras was given by U. U. Umirbaev in [569]).

Theorem 24.3
Let $X = \{x_1, \ldots, x_n\}$, $Y = \{y_1, \ldots, y_n\}$, $d(x_i) = d(y_i)$, $X \cap Y = \emptyset$, $L = L(X)$ $(L = L^p(X))$, $L' = L(X \cup Y)$ $(L' = L^p(X \cup Y))$, and let φ be an endomorphism of L, $\varphi(x_i) = f_i$. Consider the endomorphism φ' of L' given by

$$\varphi'(x_i) = f_i{}', \quad \varphi'(y_i) = y_i, \qquad 1 \le i \le n,$$

where $f_i{}'$ are X-linear parts of the elements

$$f_i\Big|_{\substack{x_1 = x_1 + y_1 \\ \cdots \\ x_n = x_n + y_n}}, \qquad i = 1, \ldots, n.$$

Then φ' is an automorphism of L' if and only if φ is an automorphism of L.

Remark 24.4
The statement of Theorem 24.2 fails for $A(X)$ if we use Lie partial derivatives. For example, let $X = \{x_1, x_2\}$, $\varphi(x_1) = x_1 + x_2x_1x_2$, $\varphi(x_2) = x_2$. Then

$$J(\varphi) = \begin{pmatrix} 1 & 0 \\ x_1x_2 & 1 \end{pmatrix}$$

is invertible, but at the same time φ is not an automorphism. However in the case $n = 2$ V. Shpilrain in [522] using Dicks' commutator test proved that φ is an automorphism if and only if the matrix

$$\begin{pmatrix} \frac{\partial}{\partial x_2}\varphi(x_2) & -\frac{\partial}{\partial x_2}\varphi(x_1) \\ -\frac{\partial}{\partial x_1}\varphi(x_2) & \frac{\partial}{\partial x_1}\varphi(x_1) \end{pmatrix} \begin{pmatrix} \varphi(x_1)\frac{\partial}{\partial x_1} & \varphi(x_1)\frac{\partial}{\partial x_2} \\ \varphi(x_2)\frac{\partial}{\partial x_1} & \varphi(x_2)\frac{\partial}{\partial x_2} \end{pmatrix}$$

is a nonzero scalar matrix.

Let ω be a functional, φ an automorphism of $L(X)$ (of $L^p(X)$). We say that φ is *ω-homogeneous* if $\varphi(x_i)$ are ω-homogeneous elements for all $x_i \in X$ and $\omega(\varphi(x_i)) = \omega(x_i)$.

Lemma 24.5
Let $k \in \mathbb{N}$, $1 \le k \le n$, $X = \{x_1, \dots, x_n\}$, ω be a functional, and let φ be an ω-homogeneous automorphism of L such that $\varphi(x_i) = x_i$ for all $i > k$. Then the Jacobian matrix $J_k(\varphi)$ is invertible. Moreover, if $B = (b_{ij}) = J_k(\varphi)^{-1}$, then b_{ij} are ω-homogeneous elements, $i, j = 1, \dots, k$, and $\omega(b_{ij}) = \omega(x_j) - \omega(x_i)$ (if $\omega(x_j) < \omega(x_i)$, then $b_{ij} = 0$).

PROOF By Theorem 24.2 the matrix $J_n(\varphi)$ is invertible. For all $i > k$ and $j \le k$

$$\frac{\partial \varphi(x_i)}{\partial x_j} = 0$$

by our assumption. Therefore matrices $J_n(\varphi)$ and $J_k(\varphi)$ are invertible at the same time.

Consider the matrix $B = J_k(\varphi)^{-1}$. It is easy to see that there exist pairwise distinct numbers $\alpha_1, \dots, \alpha_r \in \mathbb{Q}$ and matrices $B_1, \dots, B_r$ such that $B = B_1 + \dots + B_r$ and for all $i, j = 1, \dots, k$; $s = 1, \dots, r$ the elements $b_{ij}^{(s)}$ at the position (i, j) of the matrix B_s are ω-homogeneous with $\omega(b_{ij}^{(s)}) = \omega(x_j) - \omega(x_i) + \alpha_s$. Then the element at the position (m, l) of the matrix $J_k(\varphi)\, B_s$ has the form

$$\sum_{i=1}^{k} \frac{\partial \varphi(x_i)}{\partial x_m} b_{il}^{(s)}.$$

Therefore, taking into account that

$$\begin{aligned} \omega\left(\frac{\partial \varphi(x_i)}{\partial x_m} b_{il}^{(s)}\right) &= \omega(x_i) - \omega(x_m) + \omega(x_l) - \omega(x_i) + \alpha_s \\ &= \omega(x_l) - \omega(x_m) + \alpha_s, \end{aligned}$$

we see that all elements of the matrices $J_k(\varphi)\, B_s$ are ω-homogeneous and

$$\omega\left(\sum_{i=1}^{k} \frac{\partial \varphi(x_i)}{\partial x_m} b_{il}^{(s)}\right) = \omega(x_l) - \omega(x_m) + \alpha_s. \tag{5.3}$$

On the other hand, $J_k(\varphi)\,B = J_k(\varphi)\,B_1 + \cdots + J_k(\varphi)\,B_r = E$, and therefore

$$\sum_{s=1}^{r}\left(\sum_{i=1}^{k}\frac{\partial\varphi(x_i)}{\partial x_m}b_{il}^{(s)}\right) = \delta_{ml}\cdot 1.$$

Bearing in mind that $\alpha_1,\ldots,\alpha_r$ are pairwise distinct numbers, it follows from (5.3) that

$$\sum_{i=1}^{k}\frac{\partial\varphi(x_i)}{\partial x_m}b_{il}^{(s)} = 0$$

for all m, l, s such that $m \neq l$ or $\alpha_s \neq 0$. Thus $s = 1$ and $\alpha_1 = 0$. ∎

25 Admissible elements

In this section we collect some necessary technical lemmas.

In all what follows in this section let ω be a fixed functional. We suppose that all elements in this section are both ω-homogeneous and G-homogeneous.

Main auxiliary lemma

Lemma 25.1 ([414], [418])
Let $Z = \{z_1,\ldots,z_n\}$ and Y be G-graded sets, $Z \cap Y = \emptyset$, and let u be a G-homogeneous element of $L_{Y\cup Z}$ such that $\frac{\partial}{\partial z_i}u\frac{\partial}{\partial z_j} = 0$ for all $z_i, z_j \in Z$. Then u can be written in the following form:

$$u = \sum_k [v_k, w_k] + w + \sum_i \alpha_i z_i,$$

where G-homogeneous $w, w_k \in L_Y$, $v_k \in L_{Y\cup Z}$, $\alpha_i \in K$, $z_i \in Z$.

PROOF Without loss of generality we can assume u to be m-homogeneous (relative to $Y \cup Z$). We use the induction on the length $\ell(u)$. The basis of induction is $\ell(u) = 1, 2$. It is clear that the statement of the lemma is valid in these cases.

Suppose now that $\ell(u) = m > 2$ and that our statement is valid for all $m' < m$. We may suppose also that u depends on $z = z_n$. Denote

$$z_{ik} = z_i \cdot (\operatorname{Ad} z)^k, \quad z_{i0} = z_i \in Z, \quad i \neq n, \quad k \in \mathbb{Z}_+;$$
$$y_{jl} = y_j \cdot (\operatorname{Ad} z)^l, \quad y_{j0} = y_j \in Y, \quad l \in \mathbb{Z}_+;$$
$$W = \{z_{ik} \mid z_i \in Z,\, i \neq n,\, k \in \mathbb{Z}_+\} \cup \{y_{jl} \mid y_j \in Y,\, l \in \mathbb{Z}_+\}.$$

Then W is an independent set (see Theorem 13.7 and Lemma 13.11). Moreover, W is an associative independent set.

Consider the subalgebra L_W. This subalgebra is exactly the linear span of all s-regular (ps-regular) monomials which are different from z. Since u is homogeneous, we see that $u \in L_W$, $\ell_W(u) < \ell_X(u)$. We set

$$Z' = \{z_{ik} \mid z_{ik} \neq z\} \cup \{y_{jl} \mid l \neq 0\}, \qquad Y' = Y.$$

Then $W = Y' \cup Z'$.

Now we show that $u \in L_W$ is under the conditions of the lemma (and then using the induction hypothesis we immediately prove the lemma). In the case when $\ell_W(u) = 1$ we have the following possibilities:

a) $u \in Y \cup Z$;

b) $u = z_{ik} = z_i \cdot (\mathrm{Ad}\, z)^k, \quad k \geq 1$;

c) $u = y_{jl} = y_j \cdot (\mathrm{Ad}\, z)^l, \quad l \geq 1$.

In a) we have a contradiction with $\ell_{Y \cup Z}(u) > 1$. In b) we again get a contradiction because of $\frac{\partial}{\partial z_i} u \frac{\partial}{\partial z_n} = z^{k-1} \neq 0$. In c) if $l = 1$, then we have the desirable form. Otherwise, if $\operatorname{char} K = 0$, then $\frac{\partial}{\partial z} y_{jl} \frac{\partial}{\partial z} \neq 0$, and we get a contradiction. If $\operatorname{char} K = p > 2$, then $\frac{\partial}{\partial z} y_{jl} \frac{\partial}{\partial z} = 0$ if and only if either $z \in Z_-$ and $l = 2p^k$, or $z \in Z_+$ and $l = p^k$ (because of equality to zero of the binomial coefficients). Therefore one can rewrite u either in the form $\left(\frac{1}{2}\right)^{p^k} \left[y, [z, z]^{p^k}\right]$ in the case when $z \in Z_-$, or in the form $[y, z^{p^k}]$ in the case when $z \in Z_+$. In other words, in both cases we obtain the desirable presentation of u.

Thus we may assume that $\ell_W(u) \geq 2$. Consider the unique associative presentation of u in the following form:

$$\begin{aligned} u \quad = \quad & \sum_{i,j \neq n} z_{ik}\, a_{ik}^{jl}\, z_{jl} + \sum_{i \neq n} z_{ik}\, b_{ik}^{jl}\, y_{jl} \\ & + \sum_{j \neq n} y_{ik}\, c_{ik}^{jl}\, z_{jl} + \sum y_{ik}\, d_{ik}^{jl}\, y_{jl}, \end{aligned} \tag{5.4}$$

where $a_{ik}^{jl}, b_{ik}^{jl}, c_{ik}^{jl}, d_{ik}^{jl} \in A(W)$.

Assume that there exist nonzero summands in (5.4) with $k \neq 0$, $l \neq 0$. We choose such summands with maximal (k, l), comparing firstly by $k + l$ and then by k. For each such summand we single out the leading term in its presentation:

$$\begin{aligned} z_{ik}\, a_{ik}^{jl}\, z_{jl} &= \alpha\, z^k\, z_i\, a_{ik}^{jl}\, z_j\, z^l + a; \\ z_{ik}\, b_{ik}^{jl}\, y_{jl} &= \beta\, z^k\, z_i\, b_{ik}^{jl}\, y_j\, z^l + b; \\ y_{ik}\, c_{ik}^{jl}\, z_{jl} &= \gamma\, z^k\, y_i\, c_{ik}^{jl}\, z_j\, z^l + c; \\ y_{ik}\, d_{ik}^{jl}\, y_{jl} &= \delta\, z^k\, y_i\, d_{ik}^{jl}\, y_j\, z^l + d, \end{aligned}$$

where $0 \neq \alpha, \beta, \gamma, \delta \in K$, and a, b, c, d are linear combinations of associative words from $S(Y \cup Z)$ with less than $k + l$ letters z counting together in left and right sides.

It is clear that every such leading term can not be cancelled in (5.4). Therefore $\frac{\partial}{\partial z} u \frac{\partial}{\partial z} \neq 0$ in $A(Y \cup Z)$, and we have come to a contradiction. Thus all summands in (5.4) with $k \neq 0$ and $l \neq 0$ are equal to zero.

Assume now that we have summands in (5.4) with $k \neq 0$, $l = 0$, and $a_{ik}^{j0} \neq 0$ or $c_{ik}^{j0} \neq 0$. Consider such summands with maximal k. As above, we single out the leading terms:

$$
\begin{aligned}
z_{ik}\, a_{ik}^{j0}\, z_j &= \alpha\, z^k\, z_i\, a_{ik}^{j0}\, z_j + a; \\
y_{ik}\, c_{ik}^{j0}\, z_j &= \gamma\, z^k\, y_i\, c_{ik}^{j0}\, z_j + c,
\end{aligned}
$$

where $0 \neq \alpha, \gamma \in K$, and a, c are linear combinations of associative words with less than k times of letter z in the left side.

One can easily see that under the assumption $a_{ik}^{jl} = b_{ik}^{jl} = c_{ik}^{jl} = d_{ik}^{jl} = 0$ with $k \neq 0$ and $l \neq 0$, every such leading term can not be cancelled in (5.4). Hence, $\frac{\partial}{\partial z} u \frac{\partial}{\partial z_j} \neq 0$, and we have come to a contradiction again.

Suppose now that there are summands in (5.4) with $l \neq 0$, $k = 0$, and $a_{i0}^{jl} \neq 0$ or $b_{i0}^{jl} \neq 0$. As above, we consider in such summands with maximal l the leading terms:

$$
\begin{aligned}
z_i\, a_{i0}^{jl}\, z_{jl} &= z_i\, a_{i0}^{jl}\, z_j\, z^l + a; \\
z_i\, b_{i0}^{jl}\, y_{jl} &= z_i\, b_{i0}^{jl}\, y_j\, z^l + b,
\end{aligned}
$$

where a, b are linear combinations of associative words with less than l times of letter z from the right side. As above, we see that in this case $\frac{\partial}{\partial z_i} u \frac{\partial}{\partial z} \neq 0$, and again we have come to a contradiction.

Finally, taking into account that $a_{ik}^{jl} = 0$ for all (k, l) with $k^2 + l^2 \neq 0$, $b_{ik}^{jl} = 0$ for all (k, l) with $l \neq 0$, $c_{ik}^{jl} = 0$ for all (k, l) with $k \neq 0$, $d_{ik}^{jl} = 0$ for all (k, l) with $k \neq 0$ or $l \neq 0$, we consider the following summands: $z_i\, a_{i0}^{j0}\, z_j$ with $z_i, z_j \in Z$, $z_i, z_j \neq z$. If $a_{i0}^{j0} \neq 0$, $i, j \neq n$, then $\frac{\partial}{\partial z_i} u \frac{\partial}{\partial z_j} \neq 0$, which gives a contradiction.

Now one can rewrite (5.4) in the following form:

$$
\begin{aligned}
u &= \sum_{i \neq n} z_{ik}\, b_{ik}^{j0}\, y_j + \sum_{j \neq n} y_i\, c_{i0}^{jl}\, z_{jl} \\
&\quad + \sum y_i\, d_{i0}^{jl}\, y_{jl} + \sum_{k \neq 0} y_{ik}\, d_{ik}^{j0}\, y_j.
\end{aligned}
$$

Thus in $A(W)$, $W = Y' \cup Z'$, we have $\frac{\partial}{\partial v} u \frac{\partial}{\partial w} = 0$ for all $v, w \in Z'$. Recalling that $\ell_W(u) < \ell_{Y \cup Z}(u)$ and using the induction hypothesis, we complete the proof of the lemma. ■

Corollary 25.2
Let h be a G-homogeneous element of L_X, $\frac{\partial}{\partial x_n} h \frac{\partial}{\partial x_n} = 0$. Then there exist G-homogeneous elements $h_1, \dots, h_{n-1} \in L_X$ and $\alpha_1, \dots, \alpha_n \in K$ such that

$$h = \sum_{i=1}^{n} \alpha_i x_i + \sum_{i=1}^{n-1} [h_i, x_i].$$

In connection with Lemma 25.1 we formulate the following problem.

Problem 25.3
What is the form of elements h of $L(X)$ (of $L^p(X)$) such that
(i) $\frac{\partial}{\partial x} h = h \frac{\partial}{\partial x}$ *for some $x \in X$;*
(ii) $\frac{\partial h}{\partial x} = h \frac{\partial}{\partial x}$ *for all $x \in X$.*
For example, for $h = [x, y, y, x, x] - 2\big[x, [x, y], [x, y]\big]$ we have that

$$\frac{\partial}{\partial x} h = h \frac{\partial}{\partial x}.$$

Admissible sets

Definition 25.4
We say that the set of ω-homogeneous elements $a_1, \dots, a_k$ of $A(X)$ such that they do not depend on $x_n \in X$, is *admissible* if from the equality

$$\frac{\partial h}{\partial x_n} = \sum_{i=1}^{k} m_i a_i,$$

where $h \in L_X$, $m_1, \dots, m_k \in A(X)$, it follows that there exist $h_1, \dots, h_k \in L_X$ such that

$$\frac{\partial h}{\partial x_n} = \sum_{i=1}^{k} \frac{\partial h_i}{\partial x_n} a_i.$$

The following statement is immediately evident from this.

Lemma 25.5 ([414], [418])
(i) *Zero element is an admissible set.*
If $\{a_1, \dots, a_k\}$ is an admissible set, then
(ii) *the set $\{a_1, \dots, a_k, 0\}$ is admissible;*
(iii) *the set $\{a_{\sigma(1)}, \dots, a_{\sigma(k)}\}$ is admissible for any element σ of the symmetric group S_k;*
(iv) *if $0 \neq \alpha \in K$, then the set $\{a_1, \dots, a_{k-1}, \alpha a_k\}$ is admissible;*
(v) *if $b \in A(X)$, b does not depend on x_n, and $a_k + b a_{k-1}$ is both G-homogeneous and ω-homogeneous, then the set $\{a_1, \dots, a_{k-1}, a_k + b a_{k-1}\}$ is admissible.*

Lemma 25.6
Let $h \in L_X$, $m_1, \ldots, m_k \in A(X)$, and let $a_1, \ldots, a_k$ be ω-homogeneous elements such that they do not depend on x_n and

$$\frac{\partial h}{\partial x_n} = \sum_{i=1}^{k} m_i\, a_i. \tag{5.5}$$

Suppose also that $\omega(a_1) \leq \omega(a_i)$ for all $i = 2, \ldots, k$, and there exists an associative word b_0 such that b_0 is contained in a_1 with a nonzero coefficient and the elements $a_2, \ldots, a_k$ do not contain words with the ending b_0. Then $m_1 \in \frac{\partial}{\partial x_n} L_X$.

PROOF We use the induction on the length $\ell(b_0)$ of b_0. The basis of induction is the case when $b_0 = 1$. By ω-homogeneity of a_1 there exists $0 \neq \alpha \in K$ such that $a_1 = \alpha \cdot 1$. One can easily see now that $a_2 = \cdots = a_k = 0$, and

$$m_1 = \frac{1}{\alpha} \sum_{i=1}^{k} m_i\, a_i - \frac{1}{\alpha} \frac{\partial h}{\partial x_n} \in \frac{\partial}{\partial x_n} L_X.$$

Suppose now that the assertion of our lemma is valid for all words b_0 with $\ell(b_0) < t$. Let $\ell(b_0) = t$, $b_0 = b_1\, x_r$, $x_r \in X$, $\ell(b_1) = t - 1$.

Using the inequalities $\omega(a_i) \geq \omega(a_1) > 0$ for all $i = 1, \ldots, k$, and the fact that elements $a_1, \ldots, a_k$ do not depend on x_n, we conclude that

$$\frac{\partial}{\partial x_n} h \frac{\partial}{\partial x_n} = \sum_{i=1}^{k} m_i \cdot \left(a_i \frac{\partial}{\partial x_n} \right) = 0.$$

By Corollary 25.2, there exist $h_j \in L_X$, $\alpha_i \in K$, such that

$$h = \sum_{j=1}^{n-1} [h_j, x_j] + \sum_{i=1}^{n} \alpha_i\, x_i.$$

From the condition $\omega(a_i) > 0$ for all i it follows now that for $r \neq n$ we have

$$\begin{aligned} \frac{\partial h_r}{\partial x_n} &= \left(\sum_{j=1}^{n-1} \frac{\partial h_j}{\partial x_n} x_j \right) \frac{\partial}{\partial x_r} = \frac{\partial}{\partial x_n} h \frac{\partial}{\partial x_r} \\ &= \left(\sum_{i=1}^{k} m_i\, a_i \right) \frac{\partial}{\partial x_r} = \sum_{i=1}^{k} m_i \left(a_i \frac{\partial}{\partial x_r} \right). \end{aligned}$$

The elements $a_i \frac{\partial}{\partial x_r}$ are ω-homogeneous and do not depend on x_n. The coefficient at b_1 in $a_1 \frac{\partial}{\partial x_r}$ is the same as the coefficient at b_0 in a_1, i.e. this coefficient is not equal to zero. If a word of the form $b_2\, b_1$ is contained in $a_i \frac{\partial}{\partial x_r}$, $i = 2, \ldots, k$, then the word $b_2\, b_1\, x_r = b_2\, b_0$ is contained in a_i, and we have come to a contradiction.

Finally, for all $i = 1, \ldots, k$

$$\omega\left(a_i \frac{\partial}{\partial x_r}\right) = \omega(a_i) - \omega(x_r) \geq \omega(a_1) - \omega(x_r) = \omega\left(a_1 \frac{\partial}{\partial x_r}\right).$$

By the induction hypothesis $m_1 \in \frac{\partial}{\partial x_n} L_X$. ■

Lemma 25.7 ([414], [418])
Any set $\{a_1, \ldots, a_k\}$ of ω-homogeneous elements, which do not depend on x_n, is admissible.

PROOF We use the induction on k. In the case when $k = 1$ consider the following two possibilities: $a_1 = 0$ or $a_1 \neq 0$. If $a_1 = 0$ then we get our assertion from Lemma 25.5, (i). If $a_1 \neq 0$ and for some $h \in L_X$, $m_1 \in A(X)$

$$\frac{\partial h}{\partial x_n} = m_1\, a_1$$

then $m_1 \in \frac{\partial}{\partial x_n} L_X$ by Lemma 25.6.

Let now $k > 1$. We may assume that $a_i \neq 0$, $i = 1, \ldots, k$ (otherwise our assertion immediately follows from our induction hypothesis and Lemma 25.5, (ii), (iii)). By Point (iii) of Lemma 25.5 it is sufficient to consider only the case when $\omega(a_1) \leq \omega(a_i)$ for all $i = 2, \ldots, k$.

Denote by b the lexicographically leading word in a_1. By Points (iii), (iv) of Lemma 25.5, it is possible to suppose that the coefficient of b in a_1 is equal to 1. Each element a_i, $i = 2, \ldots, k$, can be rewritten in the form

$$a_i = b_i \cdot b + a_i',$$

where b_i, a_i' are ω-homogeneous elements of $A(X)$ such that they do not depend on x_n,

$$\omega(a_i) = \omega(a_i') = \omega(b_i) + \omega(b),$$

and all words in a_i' do not contain the ending b.

By Points (iii), (v) of Lemma 25.5, the set $\{a_1, \ldots, a_k\}$ is admissible if and only if the set $\{a_1, a_2 - b_2\, a_1, \ldots, a_k - b_k\, a_1\}$ is admissible. It follows from the ω-homogeneity of a_1 that the element $a_1 - b$ does not contain any word ended by b. The same property is valid for the elements

$$a_i - b_i\, a_1 = b_i\, b + a_i' - b_i\, a_1 = -b_i(a_1 - b) + a_i'.$$

Therefore it is sufficient to prove our lemma only in the case when all words in $a_2, \ldots, a_k$ do not have the ending b where b is the leading word in a_1. By Lemma 25.6 there exists $h_1 \in L_X$ such that

$$m_1 = \frac{\partial h_1}{\partial x_n}.$$

Hence,

$$\frac{\partial h}{\partial x_n} = \frac{\partial h_1}{\partial x_n} a_1 + \sum_{i=2}^{k} m_i\, a_i,$$

and by Lemma 23.3

$$\frac{\partial h}{\partial x_n} - \frac{\partial h_1}{\partial x_n} a_1 = \frac{\partial}{\partial x_n}(h - h_1 \cdot \mathrm{Ad}\,(a_1)) = \sum_{i=2}^{k} m_i\, a_i.$$

By induction hypothesis the set $\{a_2, \ldots, a_k\}$ is admissible, therefore, there exist elements $h_2, \ldots, h_k \in L_X$ such that

$$\frac{\partial h}{\partial x_n} - \frac{\partial h_1}{\partial x_n} a_1 = \sum_{i=2}^{k} \frac{\partial h_i}{\partial x_n} a_i.$$

Thus,

$$\frac{\partial h}{\partial x_n} = \sum_{i=1}^{k} \frac{\partial h_i}{\partial x_n} a_i,$$

that is the set $\{a_1, \ldots, a_k\}$ is admissible. ■

Lemma 25.8
Let $h \in L_X$, and let $a_1, \ldots, a_k$ be ω-homogeneous left independent elements of $A(X)$ such that $a_1, \ldots, a_k$ do not depend on x_n. Suppose that there exist $m_1, \ldots, m_k \in A(X)$ such that (5.5) is fulfilled. Then $m_1, \ldots, m_k \in \frac{\partial}{\partial x_n} L_X$.

PROOF By Lemma 25.7, there exist $h_1, \ldots, h_k \in L_X$ such that

$$\frac{\partial h}{\partial x_n} = \sum_{i=1}^{k} \frac{\partial h_i}{\partial x_n} a_i.$$

Therefore,

$$\sum_{i=1}^{k} \frac{\partial h_i}{\partial x_n} a_i = \frac{\partial h}{\partial x_n} = \sum_{i=1}^{k} m_i\, a_i.$$

Hence

$$\sum_{i=1}^{k} \left(m_i - \frac{\partial h_i}{\partial x_n} \right) a_i = 0.$$

Since the set $\{a_1, \ldots, a_k\}$ is left independent,

$$m_i = \frac{\partial h_i}{\partial x_n}$$

for all $i = 1, \ldots, k$. ■

26 The homogeneous case

In this section we prove the main theorem of the chapter (so called the rank theorem) in the ω-homogeneous case (Theorem 26.6).

In all what follows in this section let ω be a fixed functional. Without additional remarks all considered elements and relations are ω-homogeneous.

Lemma 26.1
Let $m_1, \ldots, m_k \in A(X)$, $\omega(m_i) = \omega(x_i) - \omega(x_n)$ for all $i = 1, \ldots, k$, and let φ be an ω-homogeneous automorphism of L_X such that $\varphi(x_i) = x_i$ for all $i > k$. Then:

(i) *there exist the uniquely determined elements $m'_1, \ldots, m'_k \in A(X)$ such that*

$$\varphi(m_j) + \frac{\partial \varphi(x_j)}{\partial x_n} = \sum_{i=1}^{k} m'_i \frac{\partial \varphi(x_j)}{\partial x_i} \tag{5.6}$$

for all $j = 1, \ldots, k$;

(ii) *the elements $m'_1, \ldots, m'_k$ from (5.6) are ω-homogeneous and $\omega(m'_i) = \omega(m_i)$ for all $i = 1, \ldots, k$.*

PROOF Consider the equalities (5.6) in matrix form. We have

$$A = M J_k(\varphi),$$

where

$$\begin{aligned} A &= \left(\varphi(m_1) + \frac{\partial \varphi(x_1)}{\partial x_n}, \ldots, \varphi(m_k) + \frac{\partial \varphi(x_k)}{\partial x_n}\right), \\ M &= (m'_1, \ldots, m'_k). \end{aligned}$$

By Lemma 24.5 the matrix $J_k(\varphi)$ is invertible. It proves (i).

Let $C = J_k(\varphi)^{-1}$,

$$C = \begin{pmatrix} c_{11} & \cdots & c_{1k} \\ \vdots & \ddots & \vdots \\ c_{k1} & \cdots & c_{kk} \end{pmatrix}.$$

Applying Lemma 24.5 we get that c_{ij} are ω-homogeneous and $\omega(c_{ij}) = \omega(x_j) - \omega(x_i)$. Therefore, taking into account that for all $i, j = 1, \ldots, k$

$$\begin{aligned} \omega(\varphi(m_i)\, c_{ij}) &= \omega\left(\frac{\partial \varphi(x_i)}{\partial x_n} c_{ij}\right) \\ &= \omega(x_i) - \omega(x_n) + \omega(x_j) - \omega(x_i) = \omega(x_j) - \omega(x_n) \end{aligned}$$

and

$$m'_j = \sum_{i=1}^{k} \left(\varphi(m_i) + \frac{\partial \varphi(x_i)}{\partial x_n}\right) c_{ij},$$

one can easily see that elements m'_j are ω-homogeneous and $\omega(m'_j) = \omega(x_j) - \omega(x_n) = \omega(m_j)$. ∎

Lemma 26.2
Let $k < n$, $h \in L_X$, $m_1, \ldots, m_k$ elements of $A(X)$ such that $\omega(m_i) = \omega(x_i) - \omega(x_n)$ for all $i = 1, \ldots, k$, and

$$\frac{\partial h}{\partial x_n} = \sum_{i=1}^{k} m_i \frac{\partial h}{\partial x_i}. \tag{5.7}$$

Let φ be an ω-homogeneous automorphism of L_X such that $\varphi(x_i) = x_i$ for all $i > k$, and $m'_1, \ldots, m'_k$ the elements defined by the equality (5.6). Then

$$\frac{\partial \varphi(h)}{\partial x_n} = \sum_{i=1}^{k} m'_i \frac{\partial \varphi(h)}{\partial x_i}.$$

PROOF Consider the presentation of h in the form

$$h = \sum_{i=1}^{n} x_i \frac{\partial h}{\partial x_i}.$$

Then

$$\begin{aligned} \varphi(h) &= \sum_{i=1}^{n} \varphi(x_i) \varphi\left(\frac{\partial h}{\partial x_i}\right) \\ &= \sum_{i=1}^{k} \varphi(x_i) \varphi\left(\frac{\partial h}{\partial x_i}\right) + \sum_{i=k+1}^{n} x_i \varphi\left(\frac{\partial h}{\partial x_i}\right). \end{aligned}$$

Therefore, for all $j \leq k$ we have

$$\frac{\partial \varphi(h)}{\partial x_j} = \sum_{i=1}^{k} \frac{\partial \varphi(x_i)}{\partial x_j} \varphi\left(\frac{\partial h}{\partial x_i}\right).$$

Moreover,

$$\frac{\partial \varphi(h)}{\partial x_n} = \varphi\left(\frac{\partial h}{\partial x_n}\right) + \sum_{i=1}^{k} \frac{\partial \varphi(x_i)}{\partial x_n} \varphi\left(\frac{\partial h}{\partial x_i}\right).$$

Using the equalities (5.6) and (5.7), we derive that

$$\begin{aligned} &\frac{\partial \varphi(h)}{\partial x_n} - \sum_{j=1}^{k} m'_j \frac{\partial \varphi(h)}{\partial x_j} \\ &= \varphi\left(\frac{\partial h}{\partial x_n}\right) + \sum_{i=1}^{k} \frac{\partial \varphi(x_i)}{\partial x_n} \varphi\left(\frac{\partial h}{\partial x_i}\right) - \sum_{j=1}^{k} m'_j \sum_{i=1}^{k} \frac{\partial \varphi(x_i)}{\partial x_j} \varphi\left(\frac{\partial h}{\partial x_i}\right) \\ &= \sum_{i=1}^{k} \left(\varphi(m_i) + \frac{\partial \varphi(x_i)}{\partial x_n} - \sum_{j=1}^{k} m'_j \frac{\partial \varphi(x_i)}{\partial x_j} \right) \varphi\left(\frac{\partial h}{\partial x_i}\right) = 0. \end{aligned}$$

■

We introduce now $\mathbb{Z}$-gradings of algebras $A(X)$ and L_X by the n-th component of multidegree, i.e. by the degree on $x_n \in X$:

$$A(X) = \bigoplus_{k \in \mathbb{Z}_+} A(X)_k, \qquad L_X = \bigoplus_{k \in \mathbb{Z}_+} (L_X)_k.$$

Denote by $\zeta_k: A(X) \to A(X)_k$ the projection of $A(X)$ onto $A(X)_k$ (the restriction of ζ_k onto L_X is the projection onto $(L_X)_k$).

Lemma 26.3

Let $h \in L_X$ and let $m_1, \ldots, m_k$ be elements of $A(X)$ satisfying (5.7). Suppose that there exists s, $s \leq k$, such that elements $\frac{\partial}{\partial x_1}\zeta_0(h), \ldots, \frac{\partial}{\partial x_s}\zeta_0(h)$ are left independent and

$$\frac{\partial}{\partial x_{s+1}}\zeta_0(h) = \cdots = \frac{\partial}{\partial x_k}\zeta_0(h) = 0.$$

Then for any $r > 0$ the following conditions are equivalent:

(i) $\zeta_1(h) = \zeta_2(h) = \cdots = \zeta_r(h) = 0$;

(ii) $\zeta_j(m_i) = 0$ *for all* $i = 1, \ldots, s$ *and* $j = 0, 1, \ldots, r-1$.

PROOF Applying the operator ζ_j, $j = 0, 1, \ldots, r-1$, to the relation (5.7), we have

$$\begin{aligned}\frac{\partial}{\partial x_n}\zeta_{j+1}(h) &= \zeta_j\left(\frac{\partial h}{\partial x_n}\right) = \sum_{i=1}^{k} \zeta_j\left(m_i \frac{\partial h}{\partial x_i}\right) \\ &= \sum_{i=1}^{k}\sum_{l=0}^{j} \zeta_l(m_i)\zeta_{j-l}\left(\frac{\partial h}{\partial x_i}\right) \\ &= \sum_{i=1}^{k}\sum_{l=0}^{j} \zeta_l(m_i)\frac{\partial}{\partial x_i}\zeta_{j-l}(h). \end{aligned} \tag{5.8}$$

Assume now that (i) is valid. Then one can rewrite (5.8) in the following form:

$$0 = \sum_{i=1}^{k} \zeta_j(m_i)\frac{\partial}{\partial x_i}\zeta_0(h) = \sum_{i=1}^{s} \zeta_j(m_i)\frac{\partial}{\partial x_i}\zeta_0(h).$$

Since $\frac{\partial}{\partial x_1}\zeta_0(h), \ldots, \frac{\partial}{\partial x_s}\zeta_0(h)$ are left independent, we see that $\zeta_j(m_i) = 0$ for all $j = 0, 1, \ldots, r-1$ and all $i = 1, \ldots, s$, in other words, (ii) takes place.

Suppose that the property (ii) is valid. Then the equality (5.8) can be rewritten in the form

$$\frac{\partial}{\partial x_n}\zeta_{j+1}(h) = \sum_{i=s+1}^{k}\sum_{l=0}^{j-1} \zeta_l(m_i)\frac{\partial}{\partial x_i}\zeta_{j-l}(h).$$

Using the induction on t we prove that $\zeta_t(h) = 0$ for all $t = 1, \ldots, r$. Starting with $j = 0$ we derive that $\frac{\partial}{\partial x_n}\zeta_1(h) = 0$, hence $\zeta_1(h) = 0$.

If $\zeta_1(h) = \cdots = \zeta_t(h) = 0$ and $t < r$, then for $j = t$ one can see that

$$\frac{\partial}{\partial x_n}\zeta_{t+1}(h) = \sum_{i=s+1}^{k}\sum_{l=0}^{t-1}\zeta_l(m_i)\frac{\partial}{\partial x_i}\zeta_{t-l}(h) = 0.$$

Thus, $\zeta_{t+1}(h) = 0$. Applying the induction hypothesis we complete the proof. ■

Lemma 26.4

Let $h \in L_X$ and let $m_1, \ldots, m_k$ be elements of $A(X)$ satisfying (5.7). Suppose that there exist s, $s \leq k$, and $r \geq 0$ such that elements $\frac{\partial}{\partial x_1}\zeta_0(h), \ldots, \frac{\partial}{\partial x_s}\zeta_0(h)$ are left independent,

$$\frac{\partial}{\partial x_{s+1}}\zeta_0(h) = \cdots = \frac{\partial}{\partial x_k}\zeta_0(h) = 0$$

and $\zeta_1(h) = \cdots = \zeta_r(h) = 0$. Then there exists an ω-homogeneous automorphism φ of L_X with $\varphi(x_i) = x_i$ for all $i > k$, $\zeta_0(\varphi(h)) = \zeta_0(h)$ and

$$\zeta_1(\varphi(h)) = \cdots = \zeta_r(\varphi(h)) = \zeta_{r+1}(\varphi(h)) = 0.$$

PROOF Applying the operator ζ_r to the equality (5.7) we have

$$\begin{aligned}
\frac{\partial}{\partial x_n}\zeta_{r+1}(h) &= \zeta_r\left(\frac{\partial h}{\partial x_n}\right) = \sum_{i=1}^{k}\zeta_r\left(m_i\frac{\partial h}{\partial x_i}\right) \\
&= \sum_{i=1}^{k}\sum_{l=0}^{r}\zeta_l(m_i)\zeta_{r-l}\left(\frac{\partial h}{\partial x_i}\right) = \sum_{i=1}^{k}\sum_{l=0}^{r}\zeta_l(m_i)\frac{\partial}{\partial x_i}\zeta_{r-l}(h) \\
&= \sum_{i=1}^{k}\zeta_r(m_i)\frac{\partial}{\partial x_i}\zeta_0(h) = \sum_{i=1}^{s}\zeta_r(m_i)\frac{\partial}{\partial x_i}\zeta_0(h).
\end{aligned}$$

Since elements $\frac{\partial}{\partial x_1}\zeta_0(h), \ldots, \frac{\partial}{\partial x_s}\zeta_0(h)$ are ω-homogeneous, left independent and do not depend on x_n, by Lemma 25.8 there exist $g_1, \ldots, g_s \in L_X$ such that $\zeta_r(m_i) = \frac{\partial g_i}{\partial x_n}$ for all $i = 1, \ldots, s$. For any i the element $\zeta_r(m_i)$ is ω-homogeneous, and its degree on x_n is equal to r. Consequently we may assume that for any i the element g_i is ω-homogeneous, and that its degree on x_n is equal to $r + 1$. Therefore

$$\begin{aligned}
\omega(g_i) &= \omega(\zeta_r(m_i)) + \omega(x_n) = \omega(m_i) + \omega(x_n) \\
&= \omega(x_i) - \omega(x_n) + \omega(x_n) = \omega(x_i).
\end{aligned}$$

Consider the mapping $\varphi\colon X \to L_X$ given by

$$\varphi(x_i) = \begin{cases} x_i - g_i, & \text{if } i \le s; \\ x_i, & \text{otherwise.} \end{cases}$$

Without loss of generality we may assume that $\omega(x_1) \ge \cdots \ge \omega(x_s)$. Then for all $i = 1, \ldots, s$ and $j \le i$

$$\omega(m_i) = \omega(x_i) - \omega(x_n) < \omega(x_j).$$

Therefore m_i does not depend on x_j, and also g_i does not depend on x_j. Hence for all $i, j = 1, \ldots, n$ with $i \ge j$,

$$\frac{\partial \varphi(x_i)}{\partial x_j} = \delta_{ij}.$$

It means that the matrix $J_n(\varphi)$ is triangular with unity elements on the diagonal. It is clear that such matrix is invertible. By Theorem 24.2, φ can be extended to an automorphism of L_X.

Consider the ideal I of $A(X)$,

$$I = \bigoplus_{k>r} A(X)_k,$$

and the quotient algebra $A(X)/I$. Let $\mu\colon A(X) \to A(X)/I$ be the natural homomorphism of $A(X)$ onto $A(X)/I$.

Since $g_i \in (L_X)_{r+1} \subseteq I$ for all $i = 1, \ldots, s$, we have $\mu(\varphi(x_i)) = \mu(x_i)$ for all $i = 1, \ldots, n$. Hence $\mu(\varphi(a)) = \mu(a)$ for all $a \in A(X)$. Therefore, $\mu(\varphi(h)) = \mu(h)$. Thus $\zeta_0(\varphi(h)) = \zeta_0(h)$. It proves the first part of the lemma.

Since $\zeta_1(h) = \cdots = \zeta_r(h) = 0$, by Lemma 26.3 $\zeta_j(m_i) = 0$ for all $j = 0, 1, \ldots, r-1$ and $i = 1, \ldots, s$. Hence $\mu(\varphi(m_i)) = \mu(m_i) = \mu(\zeta_r(m_i))$ for all $i = 1, \ldots, s$.

By Lemma 26.1, there exist elements $m'_1, \ldots, m'_k \in A(X)$ such that they satisfy the relations (5.6). Applying the natural homomorphism μ to the relations (5.6), we get

$$\mu(\varphi(m_i)) + \mu\left(\frac{\partial \varphi(x_i)}{\partial x_n}\right) = \sum_{j=1}^{k} \mu(m'_j)\mu\left(\frac{\partial \varphi(x_i)}{\partial x_j}\right).$$

On the other hand, since $\dfrac{\partial g_i}{\partial x_j} \in A(X)_{r+1}$ for all $i = 1, \ldots, s$ and $j = 1, \ldots, k$,

$$\begin{aligned} \mu(\varphi(m_i)) + \mu\left(\frac{\partial \varphi(x_i)}{\partial x_n}\right) &= \mu(\zeta_r(m_i)) + \mu\left(\frac{\partial (x_i - g_i)}{\partial x_n}\right) \\ &= \mu\left(\zeta_r(m_i) - \frac{\partial g_i}{\partial x_n}\right) = 0 \end{aligned}$$

and

$$\sum_{j=1}^{k} \mu(m'_j)\mu\left(\frac{\partial\varphi(x_i)}{\partial x_j}\right) = \sum_{j=1}^{k} \mu(m'_j)\mu\left(\frac{\partial(x_i - g_i)}{\partial x_j}\right) = \mu(m'_i).$$

Finally, we derive that $\mu(m'_1) = \cdots = \mu(m'_s) = 0$, in other words, $\zeta_j(m'_i) = 0$ for all $i = 1, \ldots, s$ and $j = 0, \ldots, r$. Since $\zeta_0(\varphi(h)) = \zeta_0(h)$, the elements $\frac{\partial\zeta_0(\varphi(h))}{\partial x_1}, \ldots, \frac{\partial\zeta_0(\varphi(h))}{\partial x_s}$ are left independent and

$$\frac{\partial\zeta_0(\varphi(h))}{\partial x_{s+1}} = \cdots = \frac{\partial\zeta_0(\varphi(h))}{\partial x_k} = 0.$$

By Lemma 26.3,

$$\zeta_1(\varphi(h)) = \cdots = \zeta_r(\varphi(h)) = \zeta_{r+1}(\varphi(h)) = 0.$$

■

Lemma 26.5

Let $h \in L_X$, and let $m_1, \ldots, m_k$ be elements of $A(X)$ satisfying (5.7). Suppose that there exists s, $s \leq k$, such that the elements $\frac{\partial}{\partial x_1}\zeta_0(h), \ldots, \frac{\partial}{\partial x_s}\zeta_0(h)$ are left independent and

$$\frac{\partial}{\partial x_{s+1}}\zeta_0(h) = \cdots = \frac{\partial}{\partial x_k}\zeta_0(h) = 0.$$

Then there exists an ω-homogeneous automorphism φ of L_X with $\varphi(x_i) = x_i$ for $i > k$ and $\frac{\partial\varphi(h)}{\partial x_n} = 0$.

PROOF Assume that for some automorphism φ and $r \geq 0$ the element $\zeta_r(\varphi(h))$ is not equal to zero. Since x_n is contained in each monomial $\zeta_r(\varphi(h))$ exactly r times, one can derive that

$$\omega(x_n) \cdot r \leq \omega(\zeta_r(\varphi(h))) = \omega(\varphi(h)) = \omega(h),$$

and

$$r \leq \frac{\omega(h)}{\omega(x_n)}.$$

Hence it is sufficient to prove for all $r \geq 0$ the existence of ω-homogeneous automorphism φ_r of L_X such that $\zeta_0(\varphi_r(h)) = \zeta_0(h)$,

$$\zeta_1(\varphi_r(h)) = \cdots = \zeta_r(\varphi_r(h)) = 0$$

and $\varphi_r(x_i) = x_i$ for all $i > k$. Now we argue by induction on r. The basis of induction is $r = 0$. In this case φ_0 is equal to the identity mapping.

Suppose that we constructed an ω-homogeneous automorphism φ_r of L_X such that $\zeta_0(\varphi_r(h)) = \zeta_0(h)$, $\zeta_1(\varphi_r(h)) = \cdots = \zeta_r(\varphi_r(h)) = 0$ and $\varphi_r(x_i) = x_i$ for all $i > k$. Then by Lemma 26.1 there exist ω-homogeneous elements $m'_1, \ldots, m'_k \in A(X)$ such that

$$\frac{\partial \varphi_r(h)}{\partial x_n} = \sum_{i=1}^{k} m'_i \frac{\partial \varphi_r(h)}{\partial x_i}.$$

Since the elements $\frac{\partial}{\partial x_1}\zeta_0(\varphi_r(h)), \ldots, \frac{\partial}{\partial x_s}\zeta_0(\varphi_r(h))$ are left independent and

$$\frac{\partial}{\partial x_{s+1}}\zeta_0(\varphi_r(h)) = \cdots = \frac{\partial}{\partial x_k}\zeta_0(\varphi_r(h)) = 0,$$

by Lemma 26.4 there exists an ω-homogeneous automorphism φ'_r of L_X with $\varphi'_r(x_i) = x_i$ for all $i > k$, $\zeta_0(\varphi'_r\varphi_r(h)) = \zeta_0(\varphi_r(h))$ and

$$\zeta_1(\varphi'_r\varphi_r(h)) = \cdots = \zeta_r(\varphi'_r\varphi_r(h)) = \zeta_{r+1}(\varphi'_r\varphi_r(h)) = 0.$$

Setting $\varphi_{r+1} = \varphi'_r\varphi_r$ we complete the proof. ■

Theorem 26.6 ([414], [418])
Let $h \in L_X$, and let $m_1, \ldots, m_k$ be elements of $A(X)$ satisfying (5.7). Then there exists an ω-homogeneous automorphism φ of L_X such that $\varphi(x_i) = x_i$ for $i > k$ and $\frac{\partial \varphi(h)}{\partial x_n} = 0$.

PROOF We proceed by induction on k. The basis of induction is $k = 0$. In this case $\frac{\partial h}{\partial x_n} = 0$ and φ is the identity automorphism.

Assume that we have proved our statement for all numbers which are less than k. Consider the element $g = \zeta_0(h)$. It does not depend on x_n. We may assume also that $\frac{\partial g}{\partial x_1}, \cdots, \frac{\partial g}{\partial x_r}$ are nonzero and $\frac{\partial g}{\partial x_{r+1}} = \cdots = \frac{\partial g}{\partial x_k} = 0$. If the elements $\frac{\partial g}{\partial x_1}, \ldots, \frac{\partial g}{\partial x_r}$ are not left independent then by Theorem 12.2 we get

$$\frac{\partial g}{\partial x_s} = \sum_{\substack{i=1 \\ i \neq s}}^{r} m_i \frac{\partial g}{\partial x_i}$$

for some $s = 1, \ldots, r$ and $m_1, \ldots, m_{s-1}, m_{s+1}, \ldots, m_r \in A(X)$. Without loss of generality we may assume now that $s = r$. By induction hypothesis there exists an ω-homogeneous automorphism ψ of L_X such that $\psi(x_i) = x_i$ for $i \geq r$ and $\frac{\partial \psi(g)}{\partial x_r} = 0$.

Let $Y = \{x_{r+1}, \ldots, x_k, x_n\}$. Since g does not depend on generators from Y, $g \in L_{X\setminus Y}$, therefore it is possible to construct the automorphism ψ as follows. Firstly we construct ψ as an automorphism of $L_{X\setminus Y}$ and then extend it to an automorphism of L_X setting that ψ acts identically on Y. Hence we may suppose that $\psi(g)$ does not depend on $x_{r+1}, \ldots, x_k, x_n$. Consequently, $\frac{\partial\psi(g)}{\partial x_r} = \cdots = \frac{\partial\psi(g)}{\partial x_k} = 0$. Since $\psi(x_n) = x_n$, we see that $\zeta_j(\psi(h)) = \psi(\zeta_j(h))$ for all $j \geq 0$. After a finite number of steps in the above mentioned procedure we get an ω-homogeneous automorphism ψ' for which $\psi'(x_i) = x_i$ for all $i > k$, and there exists $s \leq k$ such that the elements $\frac{\partial}{\partial x_1}\zeta_0(\psi'(h)), \ldots, \frac{\partial}{\partial x_s}\zeta_0(\psi'(h))$ are left independent and

$$\frac{\partial}{\partial x_{s+1}}\zeta_0(\psi'(h)) = \cdots = \frac{\partial}{\partial x_k}\zeta_0(\psi'(h)) = 0.$$

By Lemmas 26.1 and 26.2, there exist $m_1', \ldots, m_k' \in A(X)$ with

$$\frac{\partial\psi'(h)}{\partial x_n} = \sum_{i=1}^{k} m_i' \frac{\partial\psi'(h)}{\partial x_i}.$$

Now by Lemma 26.5 there exists an ω-homogeneous automorphism φ' of L_X such that $\varphi'(x_i) = x_i$ for $i > k$ and $\frac{\partial\varphi'(\psi'(h))}{\partial x_n} = 0$. Finally, $\varphi = \varphi'\psi'$ is the desired automorphism. It completes the proof of the main result in ω-homogeneous case. ■

27 Rank theorems

In this section we prove the main rank theorem (Theorem 27.9). As a corollary we obtain the theorem on the rank of a system of elements (Theorem 27.11).

Rank of an element

Definition 27.1
We say that an element h of $L(X)$ (of $L^p(X)$) has *the rank* k if k is the least number of generators of X on which $\varphi(h)$ depends on, where φ is an automorphism of $L(X)$ (of $L^p(X)$). We use the notation $\operatorname{rank}(h) = k$.

Lemma 27.2
Let $h \in L_X$. Then $\operatorname{rank}(M_h) \leq \operatorname{rank}(h)$.

PROOF Denote by k the rank of h. Then there exists an automorphism φ of L_X such that $\varphi(h)$ depends only on $x_1, \ldots, x_k$. Therefore $\frac{\partial\varphi(h)}{\partial x_i} = 0$ for all $i > k$. Hence the left $A(X)$-module $M_{\varphi(h)}$ is generated by the elements $\frac{\partial\varphi(h)}{\partial x_1}, \ldots, \frac{\partial\varphi(h)}{\partial x_k}$. In other words, $\operatorname{rank}(M_{\varphi(h)}) = \operatorname{rank}(M_h) \leq k = \operatorname{rank}(h)$. ■

Let $a \in A(X)$, $\alpha \in \mathbb{Q}$ and let ω be a functional. For the functional ω we have $\mathbb{Q}$-gradings of $A(X)$ and L_X,

$$A(X) = \bigoplus_{\alpha \in \mathbb{Q}_+} A^\alpha(X), \qquad L_X = \bigoplus_{\alpha \in \mathbb{Q}_+} L_X^\alpha,$$

such that $x_i \in L_X^{\alpha_i} \subseteq A^{\alpha_i}(X)$, $\alpha_i = \omega(x_i)$, for all $x_i \in X$. Denote by $\xi_\alpha^\omega(a)$ the projection of a onto the component $A^\alpha(X)$ of $A(X)$ relative to the functional ω (i.e., $\xi_\alpha^\omega(a) \in A^\alpha(X)$ is such an element that $a - \xi_\alpha^\omega(a) \in \bigoplus_{\beta \neq \alpha} A^\beta(X)$).

For any functional ω we use also the notation $V(\omega)$ for the set of all points a of the lattice $\mathbb{Z}_+^n$ such that $\omega(a) \leq 1$. We say that an element h of L_X is *ω-bounded* if $\xi_\alpha^\omega(h) = 0$ for all $\alpha > 1$. Consider a partial ordering $\prec$ on the set of all functionals setting $\omega_1 \prec \omega_2$ if $V(\omega_1) \subseteq V(\omega_2)$ and $V(\omega_1) \neq V(\omega_2)$.

Lemma 27.3
Let ω be a functional, h an ω-bounded element of $L(X)$ (of $L^p(X)$) depending on $x_1, \ldots, x_n \in X$, $g = \widehat{h}$ the leading component of h relative to ω, and let $\operatorname{rank}(M_h) < n$. *Then* $\operatorname{rank}(M_g) < n$.

PROOF Since $\operatorname{rank}(h) < n$ there exist elements $m_1, \ldots, m_n \in A(X)$ such that

$$\sum_{i=1}^n m_i \frac{\partial h}{\partial x_i} = 0. \tag{5.9}$$

We denote $\alpha_i = \max\{\alpha \in \mathbb{Q} \mid \xi_\alpha^\omega(m_i) \neq 0\}$, $i = 1, \ldots, n$ (if $m_i = 0$ then we set $\alpha_i = -1$) and

$$\beta = \max\{\alpha_i - \omega(x_i) + 1 \mid i = 1, \ldots, n\}.$$

Since not all of elements $\xi_\beta^\omega\left(m_i \dfrac{\partial h}{\partial x_i}\right)$, $i = 1, \ldots, n$, are equal to zero, we deduce that $\beta \geq 0$. Applying the operator ξ_β^ω to the equality (5.9) we get

$$\sum_{i=1}^n \sum_{\alpha \in \mathbb{Q}} \xi_{\beta-\alpha}^\omega(m_i)\, \xi_\alpha^\omega\left(\frac{\partial h}{\partial x_i}\right)$$
$$= \sum_{i=1}^n \sum_{\alpha \in \mathbb{Q}} \xi_{\beta-\alpha}^\omega(m_i) \left(\frac{\partial}{\partial x_i}\left(\xi_{\alpha+\omega(x_i)}^\omega(h)\right)\right) = 0. \tag{5.10}$$

If $\alpha + \omega(x_i) > 1$, then $\xi_{\alpha+\omega(x_i)}^\omega(h) = 0$ by ω-boundedness of h. On the other hand, if $\alpha + \omega(x_i) < 1$ then $\beta - \alpha \geq \alpha_i - \omega(x_i) + 1 - \alpha > \alpha_i$, that is $\xi_{\beta-\alpha}^\omega(m_i) = 0$. Therefore one can rewrite (5.10) in the following form:

$$\sum_{i=1}^n \xi_{\beta-1+\omega(x_i)}^\omega(m_i) \left(\frac{\partial}{\partial x_i} \xi_1^\omega(h)\right) = 0.$$

It is clear that there exists i such that

$$\begin{aligned}\beta - 1 + \omega(x_i) &= (\alpha_i - \omega(x_i) + 1) - 1 + \omega(x_i) \\ &= \alpha_i \geq -1 + \omega(x_i).\end{aligned}$$

By the definition of α_i, $\xi^{\omega}_{\alpha_i}(m_i) \neq 0$. Hence not all of elements $\xi^{\omega}_{\alpha}(m_i)$ are equal to zero. ∎

Lemma 27.4
Let h be an ω-bounded element of L_X, and $m_1, \ldots, m_n$ not all equal to zero elements of $A(X)$ such that

$$\sum_{i=1}^{n} m_i \frac{\partial h}{\partial x_i} = 0.$$

Then there exists an ω-homogeneous automorphism φ of L_X such that the element $\xi^{\omega}_1(\varphi(h))$ depends on not more than $n-1$ generators from X.

PROOF If $\xi^{\omega}_1(h)$ depends on less than n generators from X then by Lemma 27.3 we have a nontrivial relation between elements

$$\frac{\partial}{\partial x_1}\xi^{\omega}_1(h), \ldots, \frac{\partial}{\partial x_n}\xi^{\omega}_1(h).$$

Since these elements are ω-homogeneous, there exists j such that the element $\frac{\partial}{\partial x_j}\xi^{\omega}_1(h)$ lies in the left $A(X)$-submodule of $A(X)$ generated by elements $\frac{\partial}{\partial x_i}\xi^{\omega}_1(h)$, $i \neq j$ (see Theorem 12.2). At the other hand, $\xi^{\omega}_1(h)$ is also ω-homogeneous. Therefore by Lemma 26.6 there exists an ω-homogeneous automorphism φ of L_X such that the element $\varphi(\xi^{\omega}_1(h))$ does not depend on x_j. Since φ is ω-homogeneous, $\xi^{\omega}_1(\varphi(h)) = \varphi(\xi^{\omega}_1(h))$. ∎

Lemma 27.5
Let h be an ω-bounded element of L_X. Suppose that h depends on n generators from X. Then there exists a functional ω_0 such that $V(\omega_0) \subseteq V(\omega)$, h is ω_0-bounded, and $\xi^{\omega_0}_1(h)$ depends on n generators from X.

PROOF We may assume that $\xi^{\omega}_1(h)$ depends on $x_1, \ldots, x_k \in X$ and does not depend on other generators from X. It is sufficient to construct a functional ω_0 such that h is ω_0-bounded, $V(\omega_0) \subseteq V(\omega)$, and $\xi^{\omega_0}_1(h)$ depends at least on $x_1, \ldots, x_k, x_{k+1}$.

For $t \in \mathbb{Q}$, $t > 0$, denote by $\omega(t)$ the functional with $\omega(t)(x_i) = \omega(x_i)$ for $i \neq k+1$, and $\omega(t)(x_{k+1}) = \omega(x_{k+1}) + t$. It is clear that $V(\omega(t)) \subseteq V(\omega)$ and $\omega(t)(\xi^{\omega}_1(h)) = 1$ for any t. The element h is $\omega(t)$-bounded if and only if for each word b appearing in h with nonzero coefficient and having nonzero degree r on x_{k+1}, the following inequalities are fulfilled: $\omega(b) + rt \leq 1$

($\omega(b) < 1$ for all b depending on x_{k+1}). Therefore the property to be $\omega(t)$-bounded for h can be defined by a finite system of inequalities as follows: $t \leq \alpha_i$, where $\alpha_i \in \mathbb{Q}$ and $\alpha_i > 0$. Hence there exists a maximal $t_0 \in \mathbb{Q}$ with this property. Consequently, there exists a word b with the nonzero coefficient in h such that its degree on x_{k+1} is equal to $r > 0$ and $\omega(t_0)(b) = \omega(b) + rt_0 = 1$. Thus $\xi_1^{\omega(t_0)}(h)$ depends on x_{k+1}. Since $\omega(t_0)(\xi_1^{\omega}(h)) = 1$, the element $\xi_1^{\omega(t_0)}(h)$ depends on $x_1, \ldots, x_k$. Setting $\omega_0 = \omega(t_0)$, we complete the proof. ■

Lemma 27.6
Let ω be a functional, h an ω-bounded element of L_X with rank$(h) = n$, *and $m_1, \ldots, m_n$ not all equal to zero elements of $A(X)$ satisfying* (5.9). *Then there exist an automorphism φ of L_X and an functional $\omega_0 \prec \omega$ such that $\varphi(h)$ is ω_0-bounded.*

PROOF By Lemma 27.5 we may assume that $\xi_1^{\omega}(h)$ depends on n generators. By Lemma 27.4 there exists an ω-homogeneous automorphism φ such that $\xi_1^{\omega}(\varphi(h))$ does not depend at least on one generator from X. We may assume that x_n is a such generator.

The element $\xi_1^{\omega}(h)$ depends on x_n. Hence one can find a word b with nonzero coefficient in $\xi_1^{\omega}(h)$ such that b depends on x_n and $\omega(b) = 1$. Denote by $m(b) = (a_1, \ldots, a_n)$ the multidegree of b, $a_n \neq 0$.

As in Lemma 27.5, we consider a functional $\omega(t)$ with the following properties: $\omega(t)(x_i) = \omega(x_i)$ for all $i \neq n$ and $\omega(t)(x_n) = \omega(x_n)+t$. It was shown above that for $\varphi(h)$ such that $\xi_1^{\omega}(\varphi(h))$ does not depend on x_n, the property to be $\omega(t)$-bounded can be defined by a finite system of inequalities $t \leq \alpha_i$, where $\alpha_i \in \mathbb{Q}$, $\alpha_i > 0$. Therefore there exists a sufficiently small $t_0 \in \mathbb{Q}$ such that $\varphi(h)$ is $\omega(t_0)$-bounded.

Let $\omega_0 = \omega(t_0)$. As above, $V(\omega_0) \subseteq V(\omega)$. For the word b we get

$$\omega_0(b) = \omega(b) + a_n \cdot t_0 = 1 + a_n \cdot t_0 > 1.$$

In other words, $m(b) \in V(\omega)$ and $m(b) \notin V(\omega_0)$. Thus $\omega_0 \prec \omega$. ■

Lemma 27.7
Let $X_0 = \{x_1, \ldots, x_k\} \subseteq X$, and let h be an element of L_{X_0} such that the rank of the left $A(X_0)$-submodule M_h of $A(X_0)$ is equal to k. Then the rank of the left $A(X)$-module $A(X)M_h$ is equal to k.

PROOF Suppose that there exist not all equal to zero $m_1, \ldots, m_k \in A(X)$ such that

$$\sum_{i=1}^{k} m_i \frac{\partial h}{\partial x_i} = 0.$$

It is necessary to show that there exist not all equal to zero $m_1', \ldots, m_k' \in A(X_0)$ such that

$$\sum_{i=1}^{k} m_i' \frac{\partial h}{\partial x_i} = 0.$$

Denote by φ_0 the homomorphism of L_X onto L_{X_0} given by $\varphi_0(x_i) = x_i$ for all $x_i \in X_0$, and $\varphi_0(x_i) = 0$ otherwise. We extend it to a homomorphism of $A(X)$ onto $A(X_0)$.

We argue by induction on $l = \max\{\ell(m_i) \mid i = 1, \ldots, k\}$. The basis of induction is $l = 0$. In this case $m_1, \ldots, m_k \in K \cdot 1 \subseteq A(X_0)$.

If for some m_j the element $\varphi_0(m_j)$ is nonzero, then one can set $m_i' = \varphi_0(m_i)$, $i = 1, \ldots, k$. Since h does not depend on $x_{k+1}, \ldots, x_n$, we get

$$0 = \varphi_0(0) = \sum_{i=1}^{k} \varphi_0(m_i)\, \varphi_0\left(\frac{\partial h}{\partial x_i}\right) = \sum_{i=1}^{k} m_i' \frac{\partial h}{\partial x_i}.$$

Let now $\varphi_0(m_1) = \cdots = \varphi_0(m_k) = 0$. Then for all $x_j \in X$

$$0 = \frac{\partial}{\partial x_j} \sum_{i=1}^{k} m_i \frac{\partial h}{\partial x_i} = \sum_{i=1}^{k} \frac{\partial m_i}{\partial x_j} \frac{\partial h}{\partial x_i}.$$

There exists $x_j \in X$, such that not all $\frac{\partial m_i}{\partial x_j}$, $i = 1, \ldots, k$, are equal to zero. It is obvious that

$$\max\left\{\ell\left(\frac{\partial m_i}{\partial x_j}\right)\right\} < \max\{\ell(m_i)\}.$$

Applying the induction hypothesis we complete the proof. ■

Lemma 27.8
Let $h \in L_X$. Then rank$(h) \leq$ rank(M_h).

PROOF Let $k =$ rank(h), and let φ be an automorphism of L_X such that $\varphi(h) \in L_{X_0} \subseteq L_X$, $X_0 = \{x_1, \ldots, x_k\}$. By Corollary 23.6, rank$(M_h) =$ rank$(M_{\varphi(h)})$. By Lemma 27.7, we may assume that $h \in L_{X_0}$, and $X_0 = X$.

Denote $d = \ell(h)$ and consider the functional $\omega\colon \mathbb{Z}_+^n \to \mathbb{Q}$ given by $\omega(x_i) = 1/d$ for all $x_i \in X$. It is clear that h is ω-bounded.

For a pair (h, ω), where $h \in L_X$, rank$(h) = n$, and h is ω-bounded, we would like to prove that the rank of the $A(X)$-submodule M_h of $A(X)$ is equal to n. We use the induction on the cardinality of $V(\omega)$ (in any case this is a finite number).

The starting point is $|V(\omega)| = 1$, i.e. $V(\omega) = \{(0, \ldots, 0)\}$. Since $1 \notin L_X$, there are no such pairs (h, ω).

Let $|V(\omega)| > 1$. By the way of contradiction suppose that $\operatorname{rank}(M_h) < n$. Then for some not all equal to zero elements $m_1, \dots, m_n \in A(X)$ we have the relation

$$\sum_{i=1}^{n} m_i \frac{\partial h}{\partial x_i} = 0.$$

By Lemma 27.6, there exist an automorphism φ and a functional $\omega_0 \prec \omega$ such that $\varphi(h)$ is ω_0-bounded. Now we see that $\operatorname{rank}(\varphi(h)) = \operatorname{rank}(h) = n$ and $|V(\omega_0)| < |V(\omega)|$. By induction hypothesis, $\operatorname{rank}(M_{\varphi(h)}) = n$. Therefore $\operatorname{rank}(M_h) = n$, and we have come to a contradiction. It shows that for any element $h \in L_X$ with $\operatorname{rank}(h) = n$ the rank of the left $A(X)$-submodule M_h of the $A(X)$-module $A(X)$ is equal to n. ■

Theorem 27.9 ([411], [413], [414], [415], [416], [418])
Let h be a G-homogeneous element of L_X. Then $\operatorname{rank}(h)$ *is equal to the rank of the left ideal of $A(X)$ generated by the elements* $\left\{ \frac{\partial h}{\partial x} \,\middle|\, x \in X \right\}$ *(as a free left $A(X)$-module).*

PROOF The assertion of Theorem 27.9 follows immediately from Lemmas 27.2 and 27.8. ■

Rank of a system of elements

Definition 27.10
A system of elements $\{h_1, \dots, h_s\}$ of $L(X)$ (of $L^p(X)$) has *the rank k* ($\operatorname{rank}(\{h_1, \dots, h_s\}) = k$) if k is the least number of generators of X on which the system $\{h_1, \dots, h_s\}$ depends on, where φ is an automorphism of $L(X)$ (of $L^p(X)$).

Theorem 27.11 ([411], [413], [414], [415], [416], [418])
Let $h_1, \dots, h_k$ be G-homogeneous elements of L_X. Then the rank of the system $\{h_1, \dots, h_k\}$ is equal to the rank of the left $A(X)$-submodule of $A(X)$-module $A(X)^k = A(X)e_1 \oplus \cdots \oplus A(X)e_k$ generated by the elements

$$\left\{ \sum_{i=1}^{k} \frac{\partial h_i}{\partial x} e_i \,\middle|\, x \in X \right\}.$$

PROOF Let $h_1, \dots, h_k \in L_X$. Denote by r the rank of the system $\{h_1, \dots, h_k\}$ and by R the rank of the $A(X)$-module generated by the set of elements $\left\{ \frac{\partial h_i}{\partial x} e_i \,\middle|\, x \in X \right\}$.

Let $x_l \in X$, and let φ be an automorphism of L_X such that

$$\frac{\partial \varphi(h_1)}{\partial x_l} = \cdots = \frac{\partial \varphi(h_1)}{\partial x_l} = 0.$$

For all $i = 1, \dots, k$ we can write down

$$h_i = \sum_{j=1}^{n} x_j \frac{\partial h_i}{\partial x_j},$$

and

$$\varphi(h_i) = \sum_{j=1}^{n} \varphi(x_j)\, \varphi\left(\frac{\partial h_i}{\partial x_j}\right).$$

Hence

$$\frac{\partial \varphi(h_i)}{\partial x_l} = \sum_{j=1}^{n} \frac{\partial \varphi(x_j)}{\partial x_l}\, \varphi\left(\frac{\partial h_i}{\partial x_j}\right) = 0.$$

Applying the automorphism φ^{-1}, we get

$$\sum_{j=1}^{n} \varphi^{-1}\left(\frac{\partial \varphi(x_j)}{\partial x_l}\right) \frac{\partial h_i}{\partial x_j} = 0.$$

Therefore $r \geq R$.

On the other hand, suppose that for all $i = 1, \dots, k$

$$\sum_{j=1}^{n} m_j \frac{\partial h_i}{\partial x_j} = 0,$$

where $m_j \in A(X)$.

Consider a G-graded set $Y = \{y_1, \dots, y_k\}$ such that $d(y_i) = -d(h_i)$ for all $i = 1, \dots, k$. We extend the set X to $X' = X \cup Y$. Then the element $h = [h_1, y_1] + \cdots + [h_k, y_k]$ is G-homogeneous,

$$\frac{\partial h}{\partial x_j} = \sum_{i=1}^{k} \frac{\partial h_i}{\partial x_j} y_i,$$

and

$$\sum_{j=1}^{n} m_j \frac{\partial h}{\partial x_j} = \sum_{i=1}^{k} \left(\sum_{j=1}^{n} m_j \frac{\partial h_i}{\partial x_j}\right) y_i = 0. \tag{5.11}$$

As it was shown in the proof of Lemma 27.8, there exist an automorphism φ and $x_l \in X$ such that $\varphi(h)$ does not depend on x_l. Since elements $m_1, \dots, m_n$ do not depend on $y_1, \dots, y_k$, elements $\varphi(x_1), \dots, \varphi(x_n)$ also do not depend on $y_1, \dots, y_k$. Bearing in mind that in (5.11) we use Fox derivatives only by $x_1, \dots, x_n$, we obtain that $\varphi(y_1) = y_1, \dots, \varphi(y_k) = y_k$. Therefore

$$\varphi(h) = [\varphi(h_1), y_1] + \cdots + [\varphi(h_k), y_k].$$

Thus the system $\{\varphi(h_1), \dots, \varphi(h_k)\}$ does not depend on x_l. ∎

At the end of this section we pose the following problem.

Problem 27.12
Are the assertions of Theorems 27.9 and 27.11 valid for free color Lie superalgebra over $\mathbb{Z}$?

Note that the assertion of the Theorem 27.9 does not take place for free Lie algebras over a field of prime characteristic (see Theorem 28.10).

28 Automorphisms and primitive elements

In this section we study primitive elements of the free color Lie (p-)superalgebra L_X. We present a criterion for an element to be primitive (Theorem 28.2). As a corollary we obtain a criterion for a system of elements to be primitive (Theorem 28.3). At the same time Theorem 28.18 shows that an endomorphism of L_X preserving primitivity of elements is an automorphism. We consider also images of linear combinations of elements of the free generating set X under endomorphisms (Theorem 28.20).

In all what follows in this section $X = \{x_1, \dots, x_n\}$, G is an Abelian group.

Criteria of primitivity

Definition 28.1
We say that a G-homogeneous element h of $L(X)$ (of $L^p(X)$) is *primitive* if it can be embedded in a set of free generators of $L(X)$ (respectively, of $L^p(X)$).

We say that a system of G-homogeneous elements $S \subseteq L(X)$ ($S \subseteq L^p(X)$) is *primitive* if S is a subset of some set of free generators of $L(X)$ (of $L^p(X)$).

Theorem 28.2 ([411], [413], [414], [415], [416], [418])
Let h be a G-homogeneous element of L_X. Then the following conditions are equivalent:

(i) *the element h is primitive;*

(ii) *there exist elements $m_1, \dots, m_n \in A(X)$ such that*

$$\sum_{i=1}^{n} m_i \frac{\partial h}{\partial x_i} = 1.$$

PROOF We note that for $h \in L_X$ there exist elements $m_1, \dots, m_n \in A(X)$ such that

$$\sum_{i=1}^{n} m_i \frac{\partial h}{\partial x_i} = 1$$

if and only if $M_h = A(X)$.

Let h be a primitive element. Then there exist an automorphism φ of L_X and $x \in X$ such that $h = \varphi(x)$. Therefore

$$M_h = M_{\varphi(x)} = \varphi(M_x) = \varphi(A(X)) = A(X).$$

Suppose now that $M_h = A(X)$. Then $\operatorname{rank}(M_h) = 1$. By Theorem 27.9 there exists an automorphism φ of L_X such that $\varphi(h)$ depends only on one generator from X. If $\ell(\varphi(h)) > 1$ then $\frac{\partial \varphi(h)}{\partial x} \notin K \cdot 1$ and

$$M_{\varphi(h)} = A(X)\frac{\partial \varphi(h)}{\partial x} \neq A(X).$$

Therefore $\varphi(h) = \alpha x_j$ for some $x_j \in X$, $0 \neq \alpha \in K$. Consider the automorphism φ_0 given by $\varphi_0(x_i) = x_i$ for $i \neq j$ and $\varphi_0(x_j) = (1/\alpha)\, x_j$. Then $\varphi_0\varphi(h) = x_j$. ■

Theorem 28.3 ([411], [413], [414], [415], [416], [418])
Let $h_1, \ldots, h_k$ be G-homogeneous elements of L_X. Then the following conditions are equivalent:

(i) *the system $\{h_1, \ldots, h_k\}$ is primitive;*

(ii) *there exist elements $m_{ij} \in A(X)$, $i = 1, \ldots, k$, $j = 1, \ldots, n$, such that for all $r, s = 1, \ldots, k$*

$$\sum_{i=1}^{n} m_{ri}\frac{\partial h_s}{\partial x_i} = \delta_{rs},$$

i.e. $(n \times k)$-matrix $\left(\frac{\partial h_j}{\partial x_i}\right)$ is left invertible.

PROOF Let $\{h_1, \ldots, h_k\} \in L_X$ be a system of primitive elements. Then there exists an automorphism φ of L_X such that $\varphi(h_i) = x_i$, $i = 1, \ldots, k$, and $\frac{\partial \varphi(h_l)}{\partial x_j} = \delta_{jl}$ for all $l = 1, \ldots, k$, $j = 1, \ldots, n$. At the same time

$$\frac{\partial \varphi(h_l)}{\partial x_j} = \sum_{i=1}^{n} \frac{\partial \varphi(x_i)}{\partial x_j}\varphi\left(\frac{\partial h_l}{\partial x_i}\right) = \delta_{jl}.$$

Applying the automorphism φ^{-1}, we get

$$\sum_{i=1}^{n} \varphi^{-1}\left(\frac{\partial \varphi(x_i)}{\partial x_j}\right)\frac{\partial h_l}{\partial x_i} = \delta_{jl},$$

in other words, $m_{ij} = \varphi^{-1}\left(\frac{\partial \varphi(x_j)}{\partial x_i}\right)$.

Let now for some $h_1, \ldots, h_k \in L_X$ and $m_{ij} \in A(X)$ we have

$$\sum_{j=1}^{n} m_{ij} \frac{\partial h_l}{\partial x_j} = \delta_{il}$$

for all $i = 1, \ldots, n$ and $l = 1, \ldots, k$. Let also $\{e_i \mid 1 \leq i \leq k\}$ be the standard basis of the left $A(X)$-module $A(X)^k$. Then

$$\sum_{j=1}^{n} m_{ij} \cdot \sum_{l=1}^{k} \frac{\partial h_l}{\partial x_j} e_l = \sum_{l=1}^{k} \left(\sum_{j=1}^{n} m_{ij} \frac{\partial h_l}{\partial x_j} \right) e_l = \sum_{l=1}^{k} \delta_{il}\, e_l = e_i,$$

that is the set

$$\left\{ \sum_{i=1}^{k} \frac{\partial h_i}{\partial x_j} e_i \,\middle|\, x_j \in X \right\}$$

generates the $A(X)$-module $A(X)^k$. By Theorem 27.11 the rank of the system $\{h_1, \ldots, h_k\}$ is equal to k. Therefore without loss of generality we may assume that elements $h_1, \ldots, h_k$ depend only on generators $x_1, \ldots, x_k$. Consider the endomorphism φ of L_X given by $\varphi(x_i) = h_i$ with $i = 1, \ldots, k$, and $\varphi(x_i) = x_i$ with $i = k+1, \ldots, n$. The Jacobian matrix $J_n(\varphi)$ is invertible. Thus φ is an automorphism, and $\{h_1, \ldots, h_k\}$ is a system of primitive elements. ■

In connection with Theorems 28.2 and 28.3 we pose some problems.

Problem 28.4
Is it true that in the case when $|X| = n$ a system of elements $\{a_1, \ldots, a_k\}$ of $L = L(X)$ (of $L = L^p(X)$) is primitive if and only if the factor algebra $L/\mathrm{id}(a_1, \ldots, a_k)$ is the free algebra of the rank $n - k$? As for the case when L is the free Lie algebra and $k = 1$ see [288].

Problem 28.5
Let $K[t]$ be the ring of polynomials in one variable, a, b primitive elements of L_X such that the element $a + tb$ is a primitive element of $K[t] \otimes_K L_X$. Is it follows that the system $\{a, b\}$ is primitive in L_X?

Problem 28.6
Do the assertions of Theorems 28.2 and 28.3 are valid for free color Lie superalgebra over $\mathbb{Z}$?

Remark 28.7
As Theorem 28.10 shows the assertion of Theorem 28.2 does not take place in the case when the characteristic of K is equal to $p > 2$ and $L(X)$ is the free Lie algebra. At the same time Theorem 28.10 gives a counterexample to the hypothesis that the analog of Stallings-Swan theorem (see [538], [547]) is valid for free Lie algebras.

Cohomological dimension 1

Let K be a field, L a Lie algebra over K, M an L-module. Consider the set $C^n(L,M)$ of all n-dimensional cochains (i.e., of all K-multilinear skew symmetric mappings of the direct product of n copies of L into M).

The operator of taking the coboundary

$$\partial_n : C^n(L,M) \to C^{n+1}(L,M)$$

is given by

$$\begin{aligned}(\partial_n f)(v_1,\dots,v_{n+1}) &= \sum_{i=1}^{n} v_i f(v_1,\dots,v_{i-1},v_{i+1},\dots,v_n) \\ &+ \sum_{i<j}(-1)^{i+j}([v_i,v_j],v_1,\dots,v_{i-1},v_{i+1},\dots,v_{j-1},v_{j+1},\dots,v_{n+1}).\end{aligned}$$

The set $Z^n(L,M)$ of all n-dimensional cocycles is the set of all $f \in C^n(L,M)$ such that $\partial_n f = 0$. The set $B^n(L,M)$ of all n-dimensional coboundaries is the set of all $f \in C^n(L,M)$ such that there exists $h \in C^{n-1}(L,M)$ with $f = \partial_{n-1} h$.

One can see that $\partial_{n+1}\partial_n = 0$. *The space (Abelian group) of n-dimensional cohomologies* of L relatively M is given by

$$H^n(L,M) = Z^n(L,M)/B^n(L,M).$$

This space can be calculated by the following way. Consider K as a trivial L-module. Let $U(L)$ be the universal enveloping algebra of L. We can consider all L modules as right $U(L)$-modules. Let

$$\cdots \longrightarrow P_n \xrightarrow{d_n} P_{n-1} \longrightarrow \cdots \longrightarrow P_2 \xrightarrow{d_2} P_1 \xrightarrow{d_1} P_0 \xrightarrow{\varepsilon} K \longrightarrow 0$$

be a projective resolution of the module K (i.e., an exact sequence of right $U(L)$-modules with projective P_i, $i \geq 0$). Then we have the induced sequence

$$\operatorname{Hom}_{U(L)}(P_{n-1},M) \xrightarrow{d_n^*} \operatorname{Hom}_{U(L)}(P_n,M) \xrightarrow{d_{n+1}^*} \operatorname{Hom}_{U(L)}(P_{n+1},M)$$

with

$$d_{n+1}^* \cdot d_n^* = 0,$$

and

$$H^n(L,M) = \operatorname{Ker} d_{n+1}^* / \operatorname{Im} d_n^*.$$

Definition 28.8
A Lie algebra L over K has *the cohomological dimension s* (notation: $\operatorname{cd}(L) = s$) if for any L-module M we have $H^{s+1}(L,M) = 0$, and at the same time there exists an L-module M_0 such that $H^s(L,M_0) \neq 0$.

For a Lie algebra L we denote by $I(L)$ the right ideal $L \cdot U(L)$ of $U(L)$ generated by L. By Poincaré-Birkhoff-Witt theorem, $U(L)/I(L) \cong K$.

Remark 28.9
If $I(L)$ is a projective right $U(L)$-module, then $\mathrm{cd}(L) \leq 1$. Indeed, for any L-module M we have the resolution

$$0 \longrightarrow 0 \xrightarrow{d_3} 0 \xrightarrow{d_2} I(L) \xrightarrow{\varepsilon} K \longrightarrow 0,$$

and therefore

$$H^2(L,M) = \mathrm{Ker}\, d_3^* / \mathrm{Im}\, d_2^* = 0.$$

Theorem 28.10 ([407], [408], [415])
Let K be a field, $\mathrm{char}\, K = p > 2$, $X = \{x,y,z\}$, $L(X)$ the free Lie algebra, $h = x + [y,z] + (\mathrm{ad}\, x)^p(z) \in L(X)$. Then:

(i) *h is not a primitive element of $L(X)$;*

(ii) *the rank of the left ideal of $A(X)$ generated by the elements $\dfrac{\partial h}{\partial x}$, $\dfrac{\partial h}{\partial y}$, $\dfrac{\partial h}{\partial z}$ is equal to 1 (h is a primitive element of $L^p(X)$, but not of $L(X)$);*

(iii) *the algebra $L = L(X)/\mathrm{id}(h)$ is not a free Lie algebra;*

(iv) *the universal enveloping algebra of the Lie algebra $L = L(X)/\mathrm{id}(h)$ is the rank two free associative algebra;*

(v) *the cohomological dimension of $L = L(X)/\mathrm{id}(h)$ is equal to one.*

PROOF (i) By the way of contradiction assume that h is a primitive element of $L(X)$. Then there exist primitive elements f_1, f_2 of $L(X)$ such that $\{f_1, f_2, h\}$ is a primitive system of elements, $L(X) = L(\{f_1, f_2, h\})$. By Lemma 14.3 and Theorem 14.21 we may assume that $\ell(f_1) \leq \ell(f_2) < \ell(h)$, and that the leading part $\widehat{h}$ of h is a Lie polynomial on leading parts $\widehat{f_1}$, $\widehat{f_2}$ of elements f_1, f_2, respectively. Since $\widehat{h} = (\mathrm{ad}\, x)^p(z)$, we may assume $\widehat{f_1} = \alpha x$, $0 \neq \alpha \in K$. Therefore $f_1 = \alpha x$, and

$$h - \frac{1}{\alpha} f_1 = [y,z] + (\mathrm{ad}\, x)^p(z)$$

is a primitive element. It gives a contradiction with the fact that any set of free generators of $L(X)$ forms a bases of $L(X)$ modulo $[L(X), L(X)]$.

(ii) Consider the partial derivatives of h:

$$\frac{\partial h}{\partial x} = 1 + x^{p-1} z, \quad \frac{\partial h}{\partial y} = z, \quad \frac{\partial h}{\partial z} = -y - x^p.$$

Setting $m_1 = 1$, $m_2 = -x^{p-1}$, $m_3 = 0$ we get the equality

$$m_1 \frac{\partial h}{\partial x} + m_2 \frac{\partial h}{\partial y} + m_3 \frac{\partial h}{\partial z} = 1.$$

(iii) We prove that for $a \in L(X)$ the Lie algebra $L_0 = L(X)/\mathrm{id}(a)$ is free if and only if a is a primitive element of $L(X)$ (and then our statement follows from (i)). It is obvious that if a is a primitive element of $L(X)$ then L_0 is a free Lie algebra.

Let now L_0 be a free Lie algebra. One can easily see that $\dim(L_0/[L_0, L_0]) \geq n-1$. Therefore rank $(L_0) = n-1$. Denote $x_i' = x_i + \mathrm{id}(a)$. Using a finite number of elementary transformations one can reduce the set $\{x_1', \ldots, x_n'\}$ to the set $\{z_1, \ldots, z_{n-1}, 0\}$, where $\{z_1, \ldots, z_{n-1}\}$ is a set of free generators of L_0. Applying these elementary transformations to $X = \{x_1, \ldots, x_n\}$, we get the set $Y = \{y_1, \ldots, y_n\}$ of free generators of $L(X)$.

Consider the canonical epimorphism $\varphi\colon L(X) \to L_0$, $\varphi(y_i) = z_i, 1 \leq i \leq n-1, \varphi(y_n) = 0$. Then $\mathrm{id}(y_n) \subseteq \mathrm{id}(a)$, $L(X)/\mathrm{id}(y_n)$ is a free Lie algebra and

$$(L(X)/\mathrm{id}(y_n))/(\mathrm{id}(a)/\mathrm{id}(y_n)) \cong L(X)/\mathrm{id}(a) = L_0.$$

Since $L(X)/\mathrm{id}(y_n)$ and L_0 are free Lie algebras of the rank $n-1$, by Theorem 15.1, $\mathrm{id}(a)/\mathrm{id}(y_n) = 0$. Hence $\mathrm{id}(a) = \mathrm{id}(y_n)$, therefore $y = \alpha a$, $0 \neq \alpha \in K$ (see Remark 20.6). Thus, a is a primitive element of $L(X)$.

Since h is not a primitive element of $L(X)$, the Lie algebra L is not free.

(iv) It is clear that $U(L) \cong A(X)/\mathrm{Id}(h)$ where $\mathrm{Id}(h)$ is the ideal of $A(X)$ generated by h. Consider the elements $u = y + x^p$ and $v = z$. One can easily see that the set $\{h, u, v\}$ is a set of free generators of $A(X)$. Therefore, $U(L)$ is a free associative algebra with the set $\{u + \mathrm{Id}(h), v + \mathrm{Id}(h)\}$ of free generators.

(v) Consider the right ideal $I(L)$ of $U(L)$ generated by L, $I(L) = L{\cdot}U(L)$. We have

$$I(L) = (u + \mathrm{Id}(h))U(L) \oplus (v + \mathrm{Id}(h))U(L).$$

Since $U(L)$ is the free associative algebra on $\{u + \mathrm{Id}(h), v + \mathrm{Id}(h)\}$, $I(L)$ is a free right $U(L)$-module. Therefore, by Remark 28.9 we get $\mathrm{cd}(L) \leq 1$. Consider the group $H^1(L, K)$, where

$$H^1(L, K) = \mathrm{Ker}\, d_2^*/\mathrm{Im}\, d_1^*,$$

$$\mathrm{Hom}_{U(L)}(U(L), K) \xrightarrow{d_1^*} \mathrm{Hom}_{U(L)}(I(L), K) \xrightarrow{d_2^*} \mathrm{Hom}_{U(L)}(0, K) = 0.$$

Since $I(L)$ is a free $U(L)$-module of the rank 2, we get

$$\dim_K \mathrm{Hom}_{U(L)}(I(L), K) = 2.$$

Taking into account that

$$\mathrm{Ker}\, d_2^* = \mathrm{Hom}_{U(L)}(I(L), K)$$

and

$$\dim_K \mathrm{Hom}_{U(L)}(U(L), K) = 1,$$

we see that $H^1(L, K) \neq 0$. Thus, $\mathrm{cd}(L) = 1$. It completes the proof. ∎

Problem 28.11
Let char $K = 0$, *and let* L *be a color Lie superalgebra. Suppose that the cohomological dimension of* L *is equal to one. Is it true that* L *is a free color Lie superalgebra? In the case when* char $K = p > 2$ *the same question for a color Lie p-superalgebra is also open.*

Note that G. L. Feldman proved in [178] *that any two generated subalgebra of Lie algebra of cohomological dimension one was free.*

Problem 28.12
Let L *be a Lie algebra over a field,* $\mathrm{cd}(L) = 1$. *Is it true that the universal enveloping algebra* $U(L)$ *of* L *is a free associative algebra?*

Problem 28.13 ([156], Problem 3.47)
Is it true that a Lie algebra has the cohomological dimension not greater than two if and only if its relation module is free?

The same question is also interesting for color Lie p-superalgebras.

Problem 28.14
Let L *be a free color Lie superalgebra,* $a \in L$. *Suppose that* $L/\mathrm{id}(a)$ *is also a free color Lie superalgebra. Is it true that* a *is a primitive element of* L*? The same problem for an element of the free color Lie p-superalgebras.*

Images of primitive elements

In all what follows in this section we study images of primitive elements under endomorphisms of free color Lie (p-)superalgebras.

Lemma 28.15
Let h *be a primitive element of* L_X. *If* h *is a linear combination of monomials such that each of them depends on* $x \in X$ *then* $h = \alpha x$, $0 \neq \alpha \in K$.

PROOF We may assume that $x = x_n$. On the way of contradiction assume that $\ell(h) > 1$. Consider a functional ω such that h is ω-bounded and $\omega(x_n) \leq \omega(x_i)$ for all $i = 1, \dots, n-1$. We use induction on $|V(\omega)|$ to prove that h is not primitive. One can suppose that $\omega(x_1) \geq \omega(x_2) \geq \cdots \geq \omega(x_n)$. Let h^0 be the component of h of the weight 1 relative to ω. Since $\mathrm{rank}\,(h) = 1 < n$, $\mathrm{rank}\,(h^0) < n$ by Lemma 27.3. Since $\omega(x_i) \geq \omega(x_j)$ for all $i \leq j$, for some k and $a_i \in A(X)$, $\omega(a_i) = \omega(x_i) - \omega(x_k)$, we have the relation

$$\frac{\partial h^0}{\partial x_k} = \sum_{i<k} a_i \frac{\partial h^0}{\partial x_i}. \tag{5.12}$$

The derivative $\dfrac{\partial h^0}{\partial x_n}$ could not appear at the right side of (5.12). By Lemma 26.6 there exists an ω-homogeneous automorphism φ such that $\varphi(x_n) = x_n$. Raising the weight of the generator x_n we construct a functional ω_0 such that $|V(\omega_0)| < |V(\omega)|$ and $\varphi(h)$ is ω-bounded. Since $\varphi(x_n) = x_n$, $\varphi(h)$ is a linear combination of monomials such that each of them depends on x_n. It is clear that $\varphi(h) \notin K \cdot x_n$. Applying inductive assumption we have come to a contradiction. ■

Lemma 28.16
Suppose that $d(x_n)$ and 0 belong to the submonoid of G generated by $d(x_1)$, $\ldots, d(x_k)$, $k \geq 2$, $r > 0$. Then there exists a multihomogeneous element $h(x_1, \ldots, x_k)$ of L_X such that $d(h) = d(x_n)$, $\operatorname{rank}(h) = k$ and $\ell(h) > r$.

PROOF Let

$$d(x_n) = t_1\, d(x_1) + \cdots + t_k\, d(x_k), \qquad 0 = t_1'\, d(x_1) + \cdots + t_k'\, d(x_k).$$

Denote $s_1 = t_1 + rt_1', \ldots, s_k = t_k + rt_k'$. It follows from Theorem 8.24 that there exists a regular monomial $u \in L_X$ of multidegree $(s_1, \ldots, s_k)$. Then

$$\begin{aligned} \ell(u) &= s_1 + \cdots + s_k = t_1 + \cdots + t_k + r(t_1' + \cdots + t_k') > r, \\ d(u) &= t_1\, d(x_1) + \cdots + t_k\, d(x_k) + r(t_1'\, d(x_1) + \cdots + t_k'\, d(x_k)). \end{aligned}$$

Consider the elements $\dfrac{\partial u}{\partial x_1}, \ldots, \dfrac{\partial u}{\partial x_k}$. These elements are multihomogeneous with multidegrees

$$(s_1 - 1, s_2, \ldots, s_k), \quad \ldots, \quad (s_1, \ldots, s_{k-1}, s_k - 1),$$

respectively. Therefore there are no $m_1, \ldots, m_k \in A(X)$ such that

$$m_1 \frac{\partial u}{\partial x_1} + \cdots + m_k \frac{\partial u}{\partial x_k} = 0,$$

that is $\operatorname{rank}(u) = k$. ■

Lemma 28.17
Let φ be an endomorphism of L_X, $\varphi(x_i) = a_i$, $i = 1, \ldots, n$, such that for any primitive element u of L_X the element $\varphi(u)$ is primitive, and let ω be a functional. Suppose that the set $\{a_1, \ldots, a_{n-1}\}$ is reduced and that 0 and $d(x_n)$ belong to the submonoid of G generated by the elements $d(x_1), \ldots, d(x_{n-1})$. Then $\operatorname{rank}(\{\widehat{a_1}, \ldots, \widehat{a_{n-1}}\}) < n$.

PROOF If $n = 1$, then the assertion is trivial. If $n = 2$ then $\operatorname{rank}(a_1) = 1 < n$, and $\operatorname{rank}(\widehat{a_1}) < n$ by Lemma 27.3. For $n > 2$ consider a multihomogeneous element $h(x_1, \ldots, x_{n-1})$ of L_X such that $\operatorname{rank}(h(x_1, \ldots, x_{n-1})) = n - 1$, $\omega(h(\widehat{a}_1, \ldots, \widehat{a_{n-1}})) > \omega(\widehat{a_n})$ and $d(h) = d(x_n)$ (there exists such element by Lemma 28.16). The element $x_n + h$ is primitive. Therefore the element $f = \varphi(x_n + h)$ is primitive, and for some $m_i \in A(X)$, $1 \le i \le n$, we have

$$1 = \sum_{i=1}^{n} m_i \frac{\partial f}{\partial x_i} = \sum_{i=1}^{n} m_i \frac{\partial a_n}{\partial x_i} + \sum_{j=1}^{n-1} \left(\sum_{i=1}^{n} m_i \frac{\partial a_j}{\partial x_i} \right) \frac{\partial h}{\partial x_j}\Bigg|_{\substack{x_1 = a_1 \\ \cdots \\ x_{n-1} = a_{n-1}}} . \quad (5.13)$$

Consider in (5.13) the component of the weight $r + \omega(h(\widehat{a}_1, \dots, \widehat{a_{n-1}}))$ relative to ω, where $r = \max_i\{\omega(m_i) - \omega(x_i)\}$. Since

$$\omega(\widehat{a_n}) < \omega(h(\widehat{a}_1, \dots, \widehat{a_{n-1}}))$$

and

$$\omega\left(\frac{\partial h}{\partial x_j}\Bigg|_{\substack{x_1 = \widehat{a_1} \\ \cdots \\ x_{n-1} = \widehat{a_{n-1}}}} \right) = \omega(h(\widehat{a}_1, \dots, \widehat{a_{n-1}})) - \omega(\widehat{a}_j),$$

we see that

$$\sum_{j=1}^{n-1} \left(\sum_{i=1}^{n} m_i' \frac{\partial \widehat{a}_j}{\partial x_i} \right) \frac{\partial h}{\partial x_j}\Bigg|_{\substack{x_1 = \widehat{a_1} \\ \cdots \\ x_{n-1} = \widehat{a_{n-1}}}} = 0, \quad (5.14)$$

where m_i' is the components of the weight $r + \omega(x_i)$ relative to ω of the element m_i. Since $\operatorname{rank}(h) = n - 1$, it follows from Theorem 27.9 that elements $\frac{\partial h}{\partial x_1}, \dots, \frac{\partial h}{\partial x_{n-1}}$ are left independent. By Theorems 14.1 and 14.2 for any reduced set of elements $u_1, \dots, u_{n-1} \in L_X$ the images of elements $\frac{\partial h}{\partial x_1}, \dots, \frac{\partial h}{\partial x_{n-1}}$ under the endomorphism $x_1 \to u_1, \dots, x_{n-1} \to u_{n-1}$, are left independent. But the system $\{\widehat{a}_1, \dots, \widehat{a_{n-1}}\}$ is reduced. Therefore it follows from (5.14) that

$$\sum_{i=1}^{n} m_i' \frac{\partial \widehat{a}_j}{\partial x_i} = 0, \qquad 1 \le j \le n-1.$$

By Theorem 27.11, $\operatorname{rank}(\{\widehat{a}_1, \dots, \widehat{a_{n-1}}\}) < n$. ∎

Theorem 28.18 ([410], [419])
Let $X = \{x_1, \dots, x_n\}$, and let φ be an endomorphism of L_X such that for any primitive element u of L_X the element $\varphi(u)$ is also primitive. Then φ is an automorphism of L_X.

PROOF If $n = 1$ then the assertion of the theorem is obvious.

Denote $a_i = \varphi(x_i)$, $i = 1, \dots, n$. Since the image under φ of any nonzero G-homogeneous linear combination of elements of X is a primitive element, it has the nonzero linear part. Hence we may assume that $a_i = x_i + b_i$, and all monomials in b_i are of the length greater than 1, $i = 1, \dots, n$.

At first suppose that in L_X there are no G-homogeneous elements of the zero degree. Consider the monoid $G_0 \subseteq G$ generated by the elements $d(x_1), \dots, d(x_n)$. Then $0 \notin G_0$ by Theorem 8.24. Hence it is possible to introduce the partial ordering $\triangleleft$ on G_0 by the following way: $g_1 \triangleleft g_2$ if and only if there exists $h \in G_0$ such that $g_2 = g_1 + h$. Without loss of generality one may assume that for all $i, j = 1, \dots, n$ it follows from $d(x_i) \triangleleft d(x_j)$ that $i < j$.

Note that each primitive element is G-homogeneous. If b_i depends on x_j, then $d(x_j) \triangleleft d(x_i)$, and therefore $j < i$. Hence the Jacobian matrix $J(\varphi)$ is triangular with unity element on the diagonal. Thus $J(\varphi)$ is invertible, and φ is an automorphism by Theorem 24.2.

Therefore we may assume that for some not all equal to zero integers $k_1, \dots, k_n \geq 0$ we have the equality

$$k_1 d(x_1) + \cdots + k_n d(x_n) = 0. \tag{5.15}$$

If for some $j = 1, \dots, n$ the element $d(x_j)$ of G does not belong to the submonoid of G generated by the elements

$$d(x_1), \dots, d(x_{j-1}), d(x_{j+1}), \dots, d(x_n),$$

then in the presentation of the G-homogeneous primitive element b_j there are no monomials depending only on $x_1, \dots, x_{j-1}, x_{j+1}, \dots, x_n$, and do not depending on x_j. By Lemma 28.15, $a_j = \alpha_j x_j$, $0 \neq \alpha_j \in K$. In the case when $a_j = \alpha_j x_j$, $0 \neq \alpha_j \in K$, for all $j = 1, \dots, n$, φ is an automorphism. Therefore one can assume that there exists j such that $d(x_j)$ belongs to the submonoid generated by the elements $d(x_i)$, $i \neq j$. Without loss of generality we set $j = n$.

We have proved that there exist not all equal to zero $k'_1, \dots, k'_{n-1} \in \mathbb{Z}_+$ such that

$$d(x_n) = k'_1 d(x_1) + \cdots + k'_{n-1} d(x_{n-1}). \tag{5.16}$$

Substituting (5.16) into (5.15), we get

$$(k_1 + k_n k'_1) d(x_1) + (k_2 + k_n k'_2) d(x_2) + \cdots + (k_{n-1} + k_n k'_{n-1}) d(x_{n-1}) = 0,$$

with not all equal to zero $k_i + k_n k'_i$, $i = 1, \dots, n-1$. In other words, the zero element of G belongs to the submonoid of the monoid G generated by $d(x_1), \dots, d(x_{n-1})$.

Let $u_1, \dots, u_k \in L_X$ and

$$W(u_1, \dots, u_k) = \min_{\omega} \{ |V(\omega)| \bigm| u_1, \dots, u_k \text{ are } \omega\text{-bounded} \}.$$

We show by induction on $W(a_1, \dots, a_{n-1})$ that there exists an automorphism ψ such that the elements $\psi(a_1), \dots, \psi(a_{n-1})$ do not depend on x_n.

Let ω be a functional such that elements a_i, $i = 1, \ldots, n-1$, are ω-bounded, and $|V(\omega)| = W(a_1, \ldots, a_{n-1})$.

We reduce the set $\{a_1, \ldots, a_{n-1}\}$ to a reduced set $\{a'_1, \ldots, a'_{n-1}\}$. Then there exists an automorphism θ of L_X such that $\varphi\theta(x_n) = a_n$ and $\varphi\theta(x_i) = a'_i$ for all $i = 1, \ldots, n-1$. Hence we may assume that $\{a_1, \ldots, a_{n-1}\}$ is a reduced set.

Since 0 and $d(x_n)$ belong to the submonoid of G, generated by the elements $d(x_1), \ldots, d(x_{n-1})$, $\operatorname{rank}(\{\widehat{a}_1, \ldots, \widehat{a_{n-1}}\}) < n$ by Lemma 28.17. Hence there exists an ω-homogeneous automorphism ξ of L_X such that the elements $\xi(\widehat{a}_i) = (\widehat{\xi(a_i)})$, $1 \leq i \leq n-1$, do not depend on x_j for some j. For each $t \in \mathbb{Q}_+$ consider the functional ω_t given by $\omega_t(x_i) = \omega(x_i)$ with $i \neq j$, and $\omega_t(x_j) = \omega(x_j) + t$. For sufficiently small t the elements $\xi(a_1), \ldots, \xi(a_{n-1})$ are ω_t-bounded, $V(\omega_t) \subseteq V(\omega)$ and $V(\omega_t) \neq V(\omega)$, that is $|V(\omega_t)| < W(a_1, \ldots, a_{n-1})$. By our inductive hypothesis there exists an automorphism ψ of L_X such that elements $\psi\xi(a_i)$, $1 \leq i \leq n-1$, do not depend on some variable. It is possible to suppose that it is x_n.

The image of any primitive element under the endomorphism $\psi\xi\varphi$ is a primitive element. By induction on $|X|$ we suppose that the restriction $\varphi' = \psi\xi\varphi|_{L_{X_0}}$, $X_0 = X \backslash \{x_n\}$, is an automorphism. The case $|X| = 1$ is trivial. Consider the extension of the automorphism $\mu' = (\varphi')^{-1}$ to the automorphism μ of L_X given by $\mu(x_n) = x_n$. Then the image of any primitive element of L_X under the endomorphism $\nu = \mu\psi\xi\varphi$ is a primitive element,

$$\nu(x_n) = a(x_n) + b(x_1, \ldots, x_{n-1}), \qquad \nu(x_i) = x_i, \quad 1 \leq i \leq n-1,$$

all monomials in $a(x_n)$ depend on x_n, $b(x_1, \ldots, x_{n-1})$ does not depend on x_n. It is clear that the element $f = x_n - b(x_1, \ldots, x_{n-1})$ is primitive. The element $\nu(f) = a(x_n)$ is also primitive. By Lemma 28.15, $a(x_n) = \alpha x_n$, $0 \neq \alpha \in K$, therefore ν is an automorphism. Thus φ is an automorphism of L_X. ∎

Problem 28.19

Is the assertion of Theorem 28.18 valid in the case when $\operatorname{char} K = p > 2$ *and* $L = L(X)$ *is the free color Lie superalgebra? The main case is the case when* $|X| = 2$.

Now we consider the images of linear combinations of free generators under an endomorphism and pose some problems. In Theorem 28.18 it is impossible to consider only images of such nonzero G-homogeneous linear combinations as the following example shows. Let $X = \{x, y, z\}$, $G = \mathbb{Z}_2$, $d(x) = 0$, $d(y) = d(z) = 1$, $L = L(X)$ the free Lie superalgebra, φ the endomorphism of L given by

$$\varphi(x) = x + [y, z] + [x, z, z], \qquad \varphi(y) = y, \qquad \varphi(z) = z.$$

Then it is clear that the image under φ of any $\mathbb{Z}_2$-homogeneous nonzero linear combinations of x, y and z is a primitive element, and at the same time φ is not an automorphism of L by Theorem 15.2.

Moreover, even in the case of free Lie algebras it is impossible in Theorem 28.18 to consider only images of linear combinations of elements of the free generating set. Indeed, the following theorem takes place (for the case of algebraically closed field see Problem 28.22).

Theorem 28.20 ([419])
Suppose that $n \geq 3$, $X = \{x_1, \ldots, x_n\}$, the field K is not algebraically closed, $L = L(X)$ ($L = L^p(X)$) is the free Lie (p-)algebra. Then there exists an endomorphism φ of L such that φ is not an automorphism of L and at the same time for any nonzero linear combination a of elements of X the image $\varphi(a)$ is a primitive element.

PROOF Since K is not algebraically closed, there exist $\alpha_0, \alpha_1, \ldots, \alpha_r \in K$, $r \geq 0$, $\alpha_r \neq 0$, such that the polynomial

$$q(t) = 1 + t\sum_{i=0}^{r} \alpha_i t^i$$

has no roots in K. Consider the endomorphism φ of L given by

$$\begin{aligned} \varphi(x_1) &= x_1 + \sum_{i=0}^{r} \alpha_i x_2 (\operatorname{Ad} x_3)^i, \\ \varphi(x_2) &= x_2 - [x_1, x_3], \\ \varphi(x_j) &= x_j \qquad \text{with } j \geq 3. \end{aligned}$$

Since the system $\{x_2(\operatorname{Ad} x_3)^r, [x_1, x_3], x_3, \ldots, x_n\}$ is reduced, φ is not an automorphism by Theorem 15.2.

For not all equal to zero $\beta_1, \ldots \beta_n \in K$ the element

$$h = \varphi(\beta_1 x_1 + \cdots + \beta_n x_n) = \sum_{j=1}^{n} \beta_j x_j + \beta_1 \sum_{i=0}^{r} \alpha_i x_2 (\operatorname{Ad} x_3)^i - \beta_2 [x_1, x_3]$$

is primitive. Indeed, if either $\beta_i \neq 0$ for some $4 \leq i \leq n$, or $\beta_1 = 0$, or $\beta_2 = 0$, then our statement is evident. Let now $\beta_4 = \cdots = \beta_n = 0$ and $\beta_1, \beta_2 \neq 0$. One may assume that $\beta_2 = 1$. Then

$$\begin{aligned} h &= \beta_1 x_1 + x_2 + \beta_3 x_3 - [x_1, x_3] + \beta_1 \sum_{i=0}^{r} \alpha_i x_2 (\operatorname{Ad} x_3)^i \\ &= \beta_1 x_1 + \beta_3 x_3 - [x_1, x_3] + x_2(\operatorname{Ad}(q_0(x_3))), \end{aligned}$$

where

$$q_0(x_3) = 1 + \beta_1 \sum_{i=0}^{r} \alpha_i x_3^i \in K[x_3] \subseteq A(X).$$

Since $q(t)$ has no roots in K,

$$q_0(\beta_1) = 1 + \beta_1 \sum_{i=0}^{r} \alpha_i \beta_1^i = q(\beta_1) \neq 0.$$

We use the induction on the degree of the polynomial q_0 to prove that it follows from $q_0(\beta_1) \neq 0$ that the element

$$h = \beta_1 x_1 + \beta_3 x_3 - [x_1, x_3] + x_2(\mathrm{Ad}\,(q_0(x_3)))$$

is primitive. The basis of the induction is the case $q_0 \in K$, $q_0 \neq 0$, in which the assertion is clear. Let $q_0(x_3) = q_0'(x_3) + \alpha x_3^r$, where the degree of q_0' is less than the degree of q_0. Consider the automorphism ψ given by

$$\begin{aligned} \psi(x_1) &= x_1 + \alpha x_2(\mathrm{Ad}\, x_3)^{r-1}, \\ \psi(x_i) &= x_i \qquad \text{with } i > 1. \end{aligned}$$

Then

$$\begin{aligned} \psi(h) &= \beta_1 x_1 + \alpha\beta_1 x_2(\mathrm{Ad}\, x_3)^{r-1} + \beta_3 x_3 - [x_1, x_3] - \alpha x_2(\mathrm{Ad}\, x_3)^r \\ &\quad + x_2(\mathrm{Ad}\,(q_0'(x_3))) + \alpha x_2(\mathrm{Ad}\, x_3)^r \\ &= \beta_1 x_1 + \beta_3 x_3 - [x_1, x_3] + x_2(\mathrm{Ad}\,(q_1(x_3))), \end{aligned}$$

where $q_1(x_3) = \alpha\beta_1 x_3^{r-1} + q_0'(x_3)$ has the degree less than r, and $q_1(\beta_1) = q_0'(\beta_1) + \alpha\beta_1^r = q_0(\beta_1) \neq 0$. By the induction hypothesis, $\psi(h)$ is a primitive element. Hence h is a primitive element. ■

Problem 28.21
Let K be an algebraically closed field, φ an endomorphism of the free Lie algebra L of finite rank such that for any nonzero linear combination a of elements of X the image $\varphi(a)$ is a primitive element. Is it true that φ is an automorphism of L?

Problem 28.22
Let K be an algebraically closed field, φ an endomorphism of the free color Lie (p-)superalgebra L of finite rank such that for any pairwise distinct elements $g_1, \ldots, g_k \in G$ and for any linear independent G-homogeneous system $\{a_1, \ldots, a_k\}$ of linear combinations of elements of X such that $d(a_i) = g_i$, the system $\{\varphi(a_1), \ldots, \varphi(a_k)\}$ is primitive. Is it true that φ is an automorphism of L?

E. C. Turner in [558] proves that if F is the free group of rank n, φ is a monomorphism of F, $u \in F$, $\mathrm{rank}\,(u) = n$, $\varphi(u) = u$, then φ is an automorphism of F. Hence it is interesting to consider the following problems.

Problem 28.23*

Let K be a field, L the free color Lie (p-)superalgebra of the rank n, u a G-homogeneous element of L, $\operatorname{rank}(u) = n$, φ is a monomorphism of L, $\varphi(u) = \alpha u$, $0 \neq \alpha \in K$. Is it true that φ is an automorphism of L?

Problem 28.24

Describe all test elements of the free Lie (p-)superalgebra L, i.e. such G-homogeneous elements $u \in L$ that for any monomorphism φ of L it follows from $\varphi(u) = \alpha u$, $0 \neq \alpha \in K$, that φ is an automorphism of L.

At the end of this section we formulate some open problems for possible further directions of research.

Problem 28.25

Does the generalized Lyndon property take place for free color Lie (p-)superalgebras, i.e. if $X = \{x_1, \ldots, x_n\}$, $L = L(X)$ ($L = L^p(X)$), $a_1, \ldots, a_m \in L$, $m < n$, then in the factor algebra $L/\operatorname{id}(a_1, \ldots, a_m)$ there exists a free subalgebra of the rank $n - m$. As for free Lie algebras and free solvable Lie algebras see [269].

Problem 28.26

Do Theorems 28.2 and 28.3 take place for $A(X)$ (with matrix elements from $A(X)^e$)?

Problem 28.27**

Do the assertions of Theorems 27.9 and 27.11 take place for the free group F (instead of $A(X)$ we consider the group algebra of F over a field K; this algebra is also fir, see [126])? As for algorithms to construct a basis of a free left KF-module see, for example, [23], [332]. A characterization of primitive systems of elements of the free group (as Theorems 28.2 and 28.3) was given in [568], [570].

Problem 28.28

In this chapter we considered the minimal number of free generators for the orbit of an element under action of the group of the automorphisms. What can be said on the minimal length in this situation?

Problem 28.29

In [612] I. A. Yunus gave a solution of Fox problem for free Lie algebras: if L is the free Lie algebra over a field, R is an ideal of L, Δ is the ideal of $U(L) = A(X)$ generated by L, $\operatorname{id}(R)$ is the ideal of $U(L)$ generated by R, $n \in \mathbb{N}$, then $L \cap \operatorname{id}(R) \cdot \Delta^n = \sum [R_{i_1}, \ldots, R_{i_k}]$, where $2 \leq k \leq n+1$,

---

*Note added in the proof: Positive answer on the question of Problem 28.23 was given by A. A. Mikhalev and A. A. Zolotykh in [424].

**Note added in the proof: Problem 28.27 was solved by U. U. Umirbaev in [573].

$R_i = R \cap L^i$, $(\sum_{j=1}^{k} i_j) - i_l \geq n$ for all $l = 1, \dots, k$ (L^i is the i-th Lie power of L). What is the exact form of such intersection in the case when L is the free color Lie (p-)superalgebra?

29 Algorithms and examples

In this section, we present the following series of algorithms: an algorithm to decide whether an endomorphism of L_X is an automorphism, or not (Algorithm 29.1); an algorithm to construct an element with the given partial derivative (Algorithm 29.2); an algorithm to find the rank of an element of L_X and an algorithm to decide whether an element h of L_X is primitive, or not (Algorithm 29.3), an algorithm for solving the same problems for a system of elements (Algorithm 29.4); algorithms to construct automorphisms which realize the rank and the primitivity of a system of elements (Algorithms 29.5–29.9); an algorithm to construct a complement of a primitive system of elements with respect to a set of free generators of L_X (Algorithm 29.10). We give also some concrete examples of applications of these algorithms.

Algorithms

Algorithm 29.1
Let φ be an endomorphism of $L(X)$ (of $L^p(X)$). The problem is to decide whether φ is an automorphism, or not. We present here the algorithm which is different from Algorithm 17.6. By Theorem 24.2, φ is an automorphism if and only if the Jacobian matrix $J(\varphi)$ is invertible over $A(X)$.

Consider the $A(X)$-submodule M of the left $A(X)$-module

$$A(X)e_1 \oplus \cdots \oplus A(X)e_n$$

generated by elements

$$\left\{ \sum_{j=1}^{n} \frac{\partial \varphi(x_j)}{\partial x_i} e_j \,\middle|\, i = 1, \dots, n \right\}.$$

It is clear that $J(\varphi)$ is invertible if and only if $M = A(X)e_1 \oplus \cdots \oplus A(X)e_n$. Then one can use Algorithm 22.4 to find the Gröbner basis of M.

Algorithm 29.2
Let a be a G-homogeneous element of $A(X)$, $x \in X$. The problem is to decide whether there exists an element u of $L(X)$ (of $L^p(X)$) such that $a = \dfrac{\partial u}{\partial x}$, or not. In the case of positive answer it is necessary to find all such elements u.

Assume x to be the largest letter in X. We consider the element xa and the leading term $\overline{xa}$. Then there exists $\alpha \in K$, $\alpha \neq 0$, such that $\alpha\overline{xa} \in S(X)$. If $a = \frac{\partial u}{\partial x}$ for some $u \in L(X)$ ($u \in L^p(X)$), then $\alpha\overline{xa}$ is an s-regular word or p^l-th power of some s-regular word v with $d(v) \in G_+$. There exists the unique arrangement of brackets $[\alpha\overline{xa}]$ on $\alpha\overline{xa}$ such that $[\alpha\overline{xa}]$ is a s-regular (ps-regular) monomial. Hence there exists $\beta \in K$, $\beta \neq 0$, such that $\overline{b} < \overline{a}$ for $b = a - \beta[\alpha\overline{xa}]$. After a finite number of such steps one can decide whether $a = \frac{\partial u}{\partial x}$, or not. In the case of positive answer we construct the element u of $L(X)$ (of $L^p(X)$) with the following properties: $\frac{\partial u}{\partial x} = a$; all words in associative presentation of u depend on x. On the other hand, if $u, u' \in L(X)$ ($L^p(X)$) and $\frac{\partial u}{\partial x} = \frac{\partial u'}{\partial x}$, then $\frac{\partial(u'-u)}{\partial x} = 0$, and by Lemma 23.4, $u = u' + w$, $w \in L(X)$ ($L^p(X)$), w does not depend on x.

Algorithm 29.3
The problem is to find the rank of a G-homogeneous element h of L_X and to decide whether h is a primitive element, or not.

We consider the left ideal M_h of $A(X)$ generated by the set of partial derivatives $\left\{\frac{\partial h}{\partial x} \,\middle|\, x \in X\right\}$. Using Algorithm 22.3, we construct the Gröbner basis of M_h. By Theorem 27.9, the cardinality of this basis is exactly the rank of h. The element h is primitive if and only if $\operatorname{rank}(h) = 1$ and the Gröbner basis of M_h consists only of the unity element.

Algorithm 29.4
The problem is to find the rank of a system $\{g_1, \ldots, g_r\}$ of G-homogeneous elements of L_X and to decide whether this system is primitive, or not.

We consider the left $A(X)$-submodule M of the $A(X)$-module

$$A(X)e_1 \oplus \cdots \oplus A(X)e_r$$

generated by the elements

$$\left\{\sum_{j=1}^{n} \frac{\partial g_j}{\partial x_i} e_j \,\middle|\, i = 1, \ldots, n\right\}.$$

Using Algorithm 22.4 we construct the Gröbner basis of M. By Theorem 27.11, the cardinality of this basis is equal to the rank of the system $\{g_1, \ldots, g_r\}$. By Theorem 28.3 our system is primitive if and only if $M = A(X)e_1 \oplus \cdots \oplus A(X)e_r$.

Algorithm 29.5
Let ω be a functional, h an ω-homogeneous element of L_X, $\operatorname{rank}(h) = k$.

The problem is to construct an ω-homogeneous automorphism φ of L_X, such that $\varphi(h)$ depends only on k generators from X.

The algorithm is inductive. More precisely, we assume that we are able to construct desirable automorphisms for all ω-homogeneous elements $h' \in L_X$ depending on less generators than h depends on (the inductive assumption).

If h depends on more than k generators then by Lemma 27.3 there exist a generator $x_j \in X$ and not all equal to zero ω-homogeneous elements $m_1, \dots, m_{j-1}, m_{j+1}, \dots, m_n \in A(X)$ such that

$$\frac{\partial h}{\partial x_j} = \sum_{\substack{i=1 \\ i \neq j}}^{n} m_i \frac{\partial h}{\partial x_i} \tag{5.17}$$

and $\omega(m_i) = \omega(x_i) - \omega(x_j)$ for all $i \neq j$. The generator $x_j \in X$ and the elements $m_i \in A(X)$ can be found using Algorithm 22.5. Without loss of generality we may assume that $j = n$.

Consider the ω-homogeneous element $\zeta_0(h)$ (see the notations in the Section 26). Since $\zeta_0(h)$ does not depend on x_n, by inductive assumption we are able to construct an ω-homogeneous automorphism φ' of L_X, such that $\varphi'(\zeta_0(h)) = \zeta_0(\varphi'(h))$ depends only on $\operatorname{rank}(\zeta_0(h))$ generators from X. Therefore, it is sufficient to describe the algorithm in the case when the elements $\frac{\partial}{\partial x_1}\zeta_0(h), \dots, \frac{\partial}{\partial x_l}\zeta_0(h)$ are left independent and

$$\frac{\partial}{\partial x_{l+1}}\zeta_0(h) = \cdots = \frac{\partial}{\partial x_n}\zeta_0(h) = 0.$$

Using the equality (5.17) we construct an ω-homogeneous automorphism φ'' of L_X such that $\varphi''(h)$ does not depend on x_n. By inductive assumption we are able to construct an ω-homogeneous automorphism φ''' of L_X, such that $\varphi'''(\varphi''(h))$ depends exactly on k generators from X. The automorphism $\varphi = \varphi''' \circ \varphi''$ is desirable.

To construct an automorphism φ'' we determine two sequences: the sequence of ω-homogeneous automorphisms $\varphi_1, \dots, \varphi_N$ of L_X and the sequence of ω-homogeneous elements $h_0, h_1, \dots, h_N \in L_X$, such that $h_0 = h$ and $h_i = \varphi_i(h_{i-1})$ for all $i = 1, \dots, N$. For each s, $s = 1, \dots, N$, we have $\zeta_1(h_s) = \cdots = \zeta_s(h_s) = 0$. If $s > \omega(h)/\omega(x_n)$, then h_s does not depend on x_n, and the resulting automorphism φ'' is equal to $\varphi_N \circ \cdots \circ \varphi_1$ with any $N > \omega(h)/\omega(x_n)$.

For each $s \geq 1$ we find $m_i^{(s)} \in A(X)$, such that

$$\frac{\partial h_{s-1}}{\partial x_n} = \sum_{i=1}^{n-1} m_i^{(s)} \frac{\partial h_{s-1}}{\partial x_i}$$

and $\omega(m_i^{(s)}) = \omega(x_i) - \omega(x_n)$ for all $i = 1, \dots, n-1$ (see Algorithm 22.5). We have proved that $\zeta_{s-1}(m_1^{(s)}), \dots, \zeta_{s-1}(m_l^{(s)}) \in \frac{\partial}{\partial x_n} L_X$. Using Algorithm 29.2 we find elements $g_1^{(s)}, \dots, g_l^{(s)} \in L_X$ with $\frac{\partial}{\partial x_n} g_i^{(s)} = m_i^{(s)}$, $i = 1, \dots, l$. The mapping $\psi_s : X \to L_X$, such that $\psi_s(x_i) = x_i - g_i^{(s)}$ for all $i = 1, \dots, l$ and $\psi_s(x_i) = x_i$ otherwise, can be extended to the above mentioned ω-homogeneous automorphism φ_s of L_X.

Algorithm 29.6
Let ω be a functional, $\{g_1, \dots, g_r\}$ a system of elements of L_X such that $g_1, \dots, g_r$ are ω-homogeneous and $\omega(g_1) = \dots = \omega(g_r)$. The problem is to construct an ω-homogeneous automorphism φ of L_X such that $\varphi(g_1), \dots,$ $\varphi(g_r)$ depend only on rank$(\{g_1, \dots, g_r\})$ generators from X.

We define

$$\begin{aligned}
\frac{\partial(a_1e_1 + \dots + a_re_r)}{\partial x_i} &= \frac{\partial a_1}{\partial x_i}e_1 + \dots + \frac{\partial a_r}{\partial x_i}e_r, \\
\zeta_j(a_1e_1 + \dots + a_re_r) &= \zeta_j(a_1)e_1 + \dots + \zeta_j(a_r)e_r, \\
\varphi(a_1e_1 + \dots + a_re_r) &= \varphi(a_1)e_1 + \dots + \varphi(a_r)e_r,
\end{aligned}$$

where $a_1, \dots, a_r \in L_X$, $j \geq 0$, $x_i \in X$, φ is an endomorphism of L_X. Then we set $h = g_1e_1 + \dots + g_re_r$, and in all what follows this algorithm works as Algorithm 29.5.

Algorithm 29.7
Let h be an element of L_X, rank$(h) = k$. The problem is to construct an automorphism φ of L_X such that $\varphi(h)$ depends exactly on k generators from X. If h is primitive then $\varphi(h)$ depends on one generator, therefore $\varphi(h) \in X$.

We determine the sequence of automorphisms $\varphi_1, \dots, \varphi_N$ of L_X, the sequence of elements $h_0, h_1, \dots, h_N \in L_X$ and the sequence of functionals $\omega_0, \omega_1, \dots, \omega_N$, such that $h_0 = h$, $\omega_0(x_1) = 1/\ell(h)$ for all $x_i \in X$, $h_s = \varphi_s(h_{s-1})$ and h_s is ω_s-bounded, $s = 1, \dots, N$. The resulting automorphism φ is equal to $\varphi_N \circ \dots \circ \varphi_1$, and the element $h_N = \varphi(h)$ depends only on k generators from X.

The algorithm consists of N steps. On the s-th step, $s = 1, \dots, N$, we determine the functional ω_s, the automorphism φ_s and the element h_s. We finish the procedure in the case when h_s depends on k generators from X $(s = N)$.

Now we describe the s-th step, $s = 1, \dots, N$. Let h'_{s-1} be the ω_{s-1}-homogeneous component of h_{s-1} of the weight 1. If h_{s-1} and h'_{s-1} depend on the same set of generators, then we set $\omega_s = \omega_{s-1}$ and determine the ω_s-homogeneous automorphism φ_s of L_X such that $\varphi_s(h'_{s-1})$ depends only on rank(h'_{s-1}) generators from X, $h_s = \varphi_s(h_{s-1})$ (see Algorithm 29.5). Otherwise, there exists $x_j \in X$ such that h_{s-1} depends on x_j, but h'_{s-1} does not depend on this generator. Consider the set of functionals $\{\omega_{s-1}(t) \mid t \in \mathbb{Q}_+\}$, such that $\omega_{s-1}(t)(x_i) = \omega_{s-1}(x_i) + \delta_{ij} \cdot t$ for all $x_i \in X$. We set

$h_s = h_{s-1}$, $\varphi_s = 1$ and $\omega_s = \omega_{s-1}(t_0)$, where $t_0 \in \mathbb{Q}_+$ is the maximal number t, such that the element h_s is $\omega_{s-1}(t)$-bounded.

Algorithm 29.8
Let $g_1, \ldots, g_r \in L_X$ be G-homogeneous elements. The problem is to construct an automorphism φ of L_X which realizes the rank of this system.

We define

$$(a_1 e_1 + \cdots + a_r e_r)' = a_1' e_1 + \cdots + a_r' e_r,$$

and set $h_0 = g_1 e_1 + \cdots + g_r e_r$, $\omega_0(x_i) = 1/\max\{\ell(g_1), \ldots, \ell(g_r)\}$ for all $x_i \in X$. Further this algorithm works as Algorithm 29.7.

Algorithm 29.9
Let $\{g_1, \ldots, g_r\} \subseteq L_X$ be a primitive set of elements. The problem is to construct an automorphism φ of L_X such that $\varphi(g_i) \in X$ for all $i = 1, \ldots, r$.

The algorithm is inductive. Suppose that we are able to construct such automorphism for any primitive set consisting of less than r elements. Denote $X' = \{x_1, \ldots, x_r\} \subseteq X$. Using Algorithm 29.5 we construct an automorphism φ_1 of L_X, such that the set $\{\varphi_1(g_1), \ldots, \varphi_1(g_r)\}$ depends on rank $(\{g_1, \ldots, g_r\}) = r$ generators. Hence we may assume that $\varphi_1(g_i) \in L_{X'}$ for all $i = 1, \ldots, r$.

By the inductive assumption we can construct an automorphism φ_2 of L_X, such that $\varphi_2(\varphi_1(g_i)) = x_i$ for all $i = 1, \ldots, r-1$. Then $\varphi_2(\varphi_1(g_r)) = \alpha x_r + u$, where $0 \neq \alpha \in K$, $u \in L_{X'}$ and u does not depend on x_r (because the set $\{x_1, \ldots, x_{r-1}, \varphi_2(\varphi_1(g_r))\}$ is primitive). Consider the automorphism φ_3 of L_X, given by $\varphi_3(x_i) = x_i$ for all $i \neq r$ and $\varphi_3(x_r) = x_r/\alpha - u$. Setting $\varphi = \varphi_3 \circ \varphi_2 \circ \varphi_1$ we get a solution of the problem.

Algorithm 29.10
Let $\{g_1, \ldots, g_r\} \subseteq L_X$ be a primitive set of elements. The problem is to construct a complement of this system with respect to a set of free generators of L_X.

Using Algorithm 29.7 we construct an automorphism φ of L_X as a product of elementary automorphisms and such that $\varphi(g_i) = x_i$, $i = 1, \ldots, r$. Therefore it is possible to construct effectively φ^{-1}. The set

$$\{\varphi^{-1}(x_j) \mid j = r+1, \ldots, n\}$$

is a complement of our primitive system with respect to a set of free generators of L_X.

Examples

Example 29.11
Let $X = \{x, y, z\}$, and let φ be an endomorphism of $L(X)$ defined by

$$\varphi(x) = x + [y, z] + 3[x, z, z],$$

$$\begin{aligned}\varphi(y) &= -y+2z+3[z,x],\\ \varphi(z) &= y+z+3[x,z].\end{aligned}$$

We consider the problem to decide whether φ is an automorphism of $L(X)$, or not. The Jacobian matrix $J(\varphi)$ has the following form

$$\begin{pmatrix} 1+3z^2 & z & -y-6xz+3zx \\ -3z & -1 & 2+3x \\ 3z & 1 & 1-3x \end{pmatrix}.$$

We use elementary transformations of rows in the left $A(X)$-module $A(X)^3$:

$$\begin{pmatrix} 1+3z^2 & z & -y-6xz+3zx \\ -3z & -1 & 2+3x \\ 0 & 0 & 3 \end{pmatrix}, \quad \begin{pmatrix} 1+3z^2 & z & -y \\ -3z & -1 & 2+3x \\ 0 & 0 & 3 \end{pmatrix},$$

$$\begin{pmatrix} 1 & 0 & -y \\ -3z & -1 & 2+3x \\ 0 & 0 & 3 \end{pmatrix}, \quad \begin{pmatrix} 1 & 0 & -y \\ 0 & -1 & 2+3x-3zx \\ 0 & 0 & 3 \end{pmatrix}.$$

If $\operatorname{char} K = 3$ then $J(\varphi)$ is not invertible, and therefore φ is not an automorphism. Otherwise the last matrix is triangular with nonzero elements of K at the diagonal. Hence φ is an automorphism. □

Example 29.12

Let g be any G-homogeneous element of L_X, g does not depend on x_1 and $d(g) = d(x_1)$. Consider two G-homogeneous elements $h_1 = x_1$ and $h_2 = x_1 + g$ of L_X. Since $\frac{\partial h_1}{\partial x_1} = \frac{\partial h_2}{\partial x_1} = 1$, these elements are primitive, but the rank of the system $\{h_1, h_2\}$ is equal to k, where $k = \operatorname{rank}(g) + 1$.

Indeed, we may assume that g depends on $\operatorname{rank}(g)$ generators from X, i.e. the derivatives $\frac{\partial g}{\partial x_2}, \ldots, \frac{\partial g}{\partial x_k}$ are left independent. Consider the $A(X)$-submodule of $A(X)^2$ generated by the set

$$\left\{ \frac{\partial h_1}{\partial x_i} e_1 + \frac{\partial h_2}{\partial x_i} e_2 \,\middle|\, i = 1, \ldots, n \right\}.$$

Since

$$\begin{aligned}\frac{\partial h_1}{\partial x_1} e_1 + \frac{\partial h_2}{\partial x_1} e_2 &= e_1 + e_2,\\ \frac{\partial h_1}{\partial x_i} e_1 + \frac{\partial h_2}{\partial x_i} e_2 &= \frac{\partial g}{\partial x_i} e_2 \qquad \text{for } i = 2, \ldots, k,\end{aligned}$$

one can easily see that the set

$$\left\{ \frac{\partial h_1}{\partial x_i} e_1 + \frac{\partial h_2}{\partial x_i} e_2 \,\middle|\, i = 1, \ldots, n \right\}$$

is a set of free generators of this submodule.

We have shown that the set of two primitive elements could have an arbitrary large rank. □

Example 29.13

Let $X = \{x, y, z, t\}$, and let $h \in L_X$ be defined by

$$\begin{aligned} h &= [z, y] + [y, x] + [z, x] - [x, t, z] + [x, t, x] - [t, y, x, y, y] \\ &\quad + [z, x, y, x, z] - [z, x, y, x, x] - [t, y, x, y, x] + [[t, y, x, y], [t, x]] \\ &\quad - [z, x, y, y, x, y, y] - [z, x, y, y, x, y, x] + [[z, x, y, y, x, y], [t, x]] \\ &\quad + [[t, y, x, y], [z, x, y, x]] + [[z, x, y, y, x, y], [z, x, y, x]]. \end{aligned}$$

We construct an automorphism φ such that $\varphi(h)$ depends on rank(h) generators from X.

Since $\ell(h) = 10$, we set $h_0 = h$ and $\omega_0(x) = \omega_0(y) = \omega_0(z) = \omega_0(t) = 1/10$.

Step 1. Since $h_0' = [[z, x, y, y, x, y], [z, x, y, x]]$ does not depend on t and h_0 depends on t, we raise the weight of t and set

$$h_1 = h_0, \qquad \omega_1(x) = \omega_1(y) = \omega_1(z) = 1/10, \quad \omega_1(t) = 3/10.$$

Step 2. For

$$\begin{aligned} h_1' &= [[t, y, x, y], [t, x]] + [[z, x, y, y, x, y], [t, x]] \\ &\quad + [[t, y, x, y], [z, x, y, x]] + [[z, x, y, y, x, y], [z, x, y, x]] \end{aligned}$$

we have

$$\begin{aligned} \frac{\partial h_1'}{\partial z} &= xy^2xy[t, x] - xyx[t, y, x, y] + xy^2xy[z, x, y, x] \\ &\quad - xyx[z, x, y, y, x, y], \\ \frac{\partial h_1'}{\partial t} &= yxy[t, x] - x[t, y, x, y] - x[z, x, y, y, x, y] + yxy[z, x, y, x], \end{aligned}$$

and

$$\frac{\partial h_1'}{\partial z} = xy\frac{\partial h_1'}{\partial t}.$$

Since $\dfrac{\partial [z, x, y]}{\partial z} = xy$, we set

$$\begin{gathered} \omega_2 = \omega_1, \\ \varphi_2(x) = x, \quad \varphi_2(y) = y, \quad \varphi_2(z) = z, \quad \varphi_2(t) = t - [z, x, y], \\ h_2 = \varphi_2(h_1) = [z, y] + [y, x] + [z, x] - [x, t, z] + [x, t, x] \\ - [t, y, x, y, y] - [t, y, x, y, x] + [[t, y, x, y], [t, x]]. \end{gathered}$$

Step 3. Since $h_2' = \big[[t,y,x,y],[t,x]\big]$ does not depend on z and h_2 depends on z, we raise the weight of z and set

$$h_3 = h_2, \qquad \omega_3(x) = \omega_3(y) = 1/10, \quad \omega_3(z) = 6/10, \quad \omega_3(t) = 3/10.$$

Step 4. For $h_3' = -[x,t,z] + \big[[t,y,x,y],[t,x]\big]$ we have

$$\begin{aligned} \frac{\partial h_3'}{\partial y} &= -txy[t,x] - [t,y,x][t,x], \\ \frac{\partial h_3'}{\partial z} &= [x,t], \end{aligned}$$

and

$$\frac{\partial h_3'}{\partial y} = (txy + [t,y,x])\frac{\partial h_3'}{\partial z}.$$

Since $\dfrac{\partial [y,t,x,y]}{\partial y} = txy + [t,y,x]$, we set

$$\begin{gathered} \omega_4 = \omega_3, \\ \varphi_4(x) = x, \quad \varphi_4(y) = y, \quad \varphi_4(z) = z - [y,t,x,y], \quad \varphi_4(t) = t, \\ h_4 = \varphi_4(h_3) = [z,y] + [y,x] + [z,x] - [x,t,z] + [x,t,x]. \end{gathered}$$

Step 5. Since $h_4' = -[x,t,z]$ does not depend on y and h_4 depends on y, we raise the weight of y and set

$$h_5 = h_4, \quad \omega_5(x) = 1/10, \quad \omega_5(y) = 4/10, \quad \omega_5(z) = 6/10, \quad \omega_5(t) = 3/10.$$

Step 6. For $h_5' = [z,y] - [x,t,z]$ we have

$$\frac{\partial h_5'}{\partial x} = -tz, \qquad \frac{\partial h_5'}{\partial y} = -z,$$

and

$$\frac{\partial h_5'}{\partial x} = t\frac{\partial h_5'}{\partial y}.$$

Since $\dfrac{\partial [x,t]}{\partial x} = t$, we set

$$\begin{gathered} \omega_6 = \omega_5, \\ \varphi_6(x) = x, \quad \varphi_6(y) = y - [x,t], \quad \varphi_6(z) = z, \quad \varphi_6(t) = t, \\ h_6 = \varphi_6(h_5) = [z,y] + [y,x] + [z,x]. \end{gathered}$$

Step 7. Since $h_6' = [z,y]$ does not depend on x and h_6 depends on x, we raise the weight of x and set

$$h_7 = h_6, \qquad \omega_7(x) = \omega_7(y) = 4/10, \quad \omega_7(z) = 6/10, \quad \omega_7(t) = 3/10.$$

Step 8. For $h_7' = [z,y] + [z,x]$ we have

$$\frac{\partial h_7'}{\partial x} = -z, \qquad \frac{\partial h_7'}{\partial y} = -z,$$

and

$$\frac{\partial h_7'}{\partial y} = \frac{\partial h_7'}{\partial x}.$$

Since $\frac{\partial y}{\partial y} = 1$, we set

$$\omega_8 = \omega_7,$$
$$\varphi_8(x) = x - y, \quad \varphi_8(y) = y, \quad \varphi_8(z) = z, \quad \varphi_8(t) = t,$$
$$h_8 = \varphi_8(h_7) = [y,x] + [z,x].$$

Step 9. Since $h_8' = [z,x]$ does not depend on y and h_8 depends on y, we raise the weight of y and set

$$h_9 = h_8, \qquad \omega_9(x) = 4/10, \quad \omega_9(y) = \omega_9(z) = 6/10, \quad \omega_9(t) = 3/10.$$

Step 10. For $h_9' = h_9 = [y,x] + [z,x]$ we have

$$\frac{\partial h_9}{\partial y} = x, \qquad \frac{\partial h_9}{\partial z} = x,$$

and

$$\frac{\partial h_9}{\partial z} = \frac{\partial h_9}{\partial y}.$$

Since $\frac{\partial z}{\partial z} = 1$, we set

$$\omega_{10} = \omega_9,$$
$$\varphi_{10}(x) = x, \quad \varphi_{10}(y) = y - z, \quad \varphi_{10}(z) = z, \quad \varphi_{10}(t) = t,$$
$$h_{10} = \varphi_{10}(h_9) = [y,x].$$

The elements $\frac{\partial [y,x]}{\partial x} = -y$ and $\frac{\partial [y,x]}{\partial y} = x$ are left independent. Hence the desirable automorphism φ is equal to $\varphi_{10} \circ \varphi_8 \circ \varphi_6 \circ \varphi_4 \circ \varphi_2$, and $\varphi(h) = [y,x]$, $\operatorname{rank}(h) = \operatorname{rank}([y,x]) = 2$. □

Example 29.14
Let $X = \{x,y,z,t\}$ and

$$\begin{aligned} h &= x - [y,z] + [t,x,z] + [x,y,x,z] - [y,z,z,x,z], \\ g &= y + [x,t] + [y,x,x] + [y,z,z,x]. \end{aligned}$$

We find the rank of the system $\{h, g\}$ in L_X. Consider the $A(X)$-submodule M of the left $A(X)$-module $A(X)e_1 \oplus A(X)e_2$ generated by the elements

$$\begin{aligned}
a_1 = \frac{\partial h}{\partial x}e_1 + \frac{\partial g}{\partial x}e_2 &= (1 - tz + yxz - [x,y]z + [y,z,z]z)e_1 \\
&\quad + (t - 2yx + xy - [y,z,z])e_2, \\
a_2 = \frac{\partial h}{\partial y}e_1 + \frac{\partial g}{\partial y}e_2 &= (-z - x^2 z - z^2 xz)e_1 + (1 + x^2 + z^2 x)e_2, \\
a_3 = \frac{\partial h}{\partial z}e_1 + \frac{\partial g}{\partial z}e_2 &= (y - [t,x] - [x,y,x] + yzxz + [y,z]xz \\
&\quad + [y,z,z,x])e_1 + (-yzx - [y,z]x)e_2, \\
a_4 = \frac{\partial h}{\partial t}e_1 + \frac{\partial g}{\partial t}e_2 &= xz\, e_1 - x\, e_2.
\end{aligned}$$

The rank of the system $\{h, g\}$ is equal to the rank of the left $A(X)$-module M. Since the elements

$$\begin{aligned}
a_5 &= a_2 + (x + z^2)a_4 = -z\, e_1 + e_2, \\
a_6 &= a_1 - (t - 2yx + xy - [y,z,z])a_5 = e_1, \\
a_7 &= a_5 + z a_6 = e_2
\end{aligned}$$

belong to M, the rank of M is equal to 2. Moreover, since $M = A(X)e_1 \oplus A(X)e_2$, the system $\{h, g\}$ is primitive. □

Example 29.15

For the primitive system $\{h, g\}$ from Example 29.14 we construct an automorphism φ such that $\varphi(h) = x$ and $\varphi(g) = y$.

Since $\ell(h) = 5$, $\ell(g) = 4$, we set $a_0 = he_1 + ge_2$ and $\omega_0(x) = \omega_0(y) = \omega_0(z) = \omega_0(t) = 1/5$.

Step 1. Since $a_0' = -[y,z,z,x,z]e_1$ does not depend on t and a_0 depends on t, we raise the weight of t and set

$$a_1 = a_0, \qquad \omega_1(x) = \omega_1(y) = \omega_1(z) = 1/5, \quad \omega_1(t) = 3/5.$$

Step 2. For $a_1' = ([t,x,z] - [y,z,z,x,z])e_1$ we have

$$\frac{\partial a_1'}{\partial y} = -z^2 xz\, e_1, \qquad \frac{\partial a_1'}{\partial t} = xz\, e_1,$$

and

$$\frac{\partial a_1'}{\partial y} = -z^2 \frac{\partial a_1'}{\partial t}.$$

Since $\dfrac{\partial(-[y,z,z])}{\partial y} = -z^2$, we set

$$\begin{gathered}
\omega_2 = \omega_1, \\
\varphi_2(x) = x, \quad \varphi_2(y) = y, \quad \varphi_2(z) = z, \quad \varphi_2(t) = t + [y,z,z], \\
a_2 = \varphi_2(a_1) = (x - [y,z] + [t,x,z] + [x,y,x,z])e_1 \\
+ (y + [x,t] + [y,x,x])e_2.
\end{gathered}$$

Step 3. Since $a_2' = [t,x,z]e_1$ does not depend on y and a_2 depends on y, we raise the weight of y and set

$$a_3 = a_2, \qquad \omega_3(x) = \omega_3(z) = 1/5, \quad \omega_3(y) = 2/5, \quad \omega_3(t) = 3/5.$$

Step 4. For $a_3' = ([t,x,z] + [x,y,x,z])e_1$ we get

$$\frac{\partial a_3'}{\partial y} = -x^2 z\, e_1, \qquad \frac{\partial a_3'}{\partial t} = xz\, e_1,$$

and therefore

$$\frac{\partial a_3'}{\partial y} = -x \frac{\partial a_3'}{\partial t}.$$

Since $\dfrac{\partial[x,y]}{\partial y} = -x$, we set

$$\omega_4 = \omega_3,$$
$$\varphi_4(x) = x, \quad \varphi_4(y) = y, \quad \varphi_4(z) = z, \quad \varphi_4(t) = t - [x,y],$$
$$a_4 = \varphi_4(a_3) = (x - [y,z] + [t,x,z])e_1 + (y + [x,t])e_2.$$

Step 5. Since $a_4' = [t,x,z]e_1$ does not depend on y and a_4 depends on y, we raise the weight of y and set

$$a_5 = a_4, \qquad \omega_5(x) = \omega_5(z) = 1/5, \quad \omega_5(y) = 4/5, \quad \omega_5(t) = 3/5.$$

Step 6. For $a_5' = (-[y,z] + [t,x,z])e_1$ we have

$$\frac{\partial a_5'}{\partial y} = -z\, e_1, \qquad \frac{\partial a_5'}{\partial t} = xz\, e_1,$$

and therefore

$$\frac{\partial a_5'}{\partial t} = -x \frac{\partial a_5'}{\partial y}.$$

Since $\dfrac{\partial[x,t]}{\partial t} = -x$, we set

$$\omega_6 = \omega_5,$$
$$\varphi_6(x) = x, \quad \varphi_6(y) = y - [x,t], \quad \varphi_6(z) = z, \quad \varphi_6(t) = t,$$
$$a_6 = \varphi_6(a_5) = (x - [y,z])e_1 + y\, e_2.$$

Step 7. Since $a_6' = -[y,z]e_1$ does not depend on x and a_6 depends on x, we raise the weight of x and set

$$a_7 = a_6, \qquad \omega_7(x) = 1, \quad \omega_7(y) = 4/5, \quad \omega_7(z) = 1/5, \quad \omega_7(t) = 3/5.$$

Step 8. For $a_7' = a_7 = (x - [y,z])e_1 + y\, e_2$ we have

$$\frac{\partial a_7}{\partial x} = e_1, \qquad \frac{\partial a_7}{\partial y} = -z\, e_1.$$

Hence

$$\frac{\partial a_7}{\partial y} = -z\frac{\partial a_7}{\partial x}.$$

Since $\dfrac{\partial[z,y]}{\partial y} = -z$, we set

$$\omega_8 = \omega_7$$

$$\varphi_8(x) = x - [z,y], \quad \varphi_8(y) = y, \quad \varphi_8(z) = z, \quad \varphi_8(t) = t,$$

$$a_8 = \varphi_8(a_7) = x\,e_1 + y\,e_2.$$

We have constructed the automorphism $\varphi = \varphi_8 \circ \varphi_6 \circ \varphi_4 \circ \varphi_2$ such that $\varphi(a) = x\,e_1 + y\,e_2$. Thus $\varphi(h) = x$, $\varphi(g) = y$. □

Example 29.16

For the primitive system $\{h, g\}$ from Example 29.14 we construct a complement with respect to a set of free generators of L_X.

Using the automorphism φ from Example 29.15 we see that the system

$$\{\varphi^{-1}(x), \varphi^{-1}(y), \varphi^{-1}(z), \varphi^{-1}(t)\}$$

is primitive. Since $\varphi^{-1}(x) = h$, $\varphi^{-1}(y) = g$, the problem is to find the elements $\varphi^{-1}(z)$ and $\varphi^{-1}(t)$. Since

$$\varphi^{-1} = \varphi_2^{-1} \circ \varphi_4^{-1} \circ \varphi_6^{-1} \circ \varphi_8^{-1}$$

and

$$\begin{gathered}
\varphi_2^{-1}(x) = x, \quad \varphi_2^{-1}(y) = y, \quad \varphi_2^{-1}(z) = z, \quad \varphi_2^{-1}(t) = t - [y,z,z], \\
\varphi_4^{-1}(x) = x, \quad \varphi_4^{-1}(y) = y, \quad \varphi_4^{-1}(z) = z, \quad \varphi_4^{-1}(t) = t + [x,y], \\
\varphi_6^{-1}(x) = x, \quad \varphi_6^{-1}(y) = y + [x,t], \quad \varphi_6^{-1}(z) = z, \quad \varphi_6^{-1}(t) = t, \\
\varphi_8^{-1}(x) = x + [z,y], \quad \varphi_8^{-1}(y) = y, \quad \varphi_8^{-1}(z) = z, \quad \varphi_8^{-1}(t) = t,
\end{gathered}$$

we get

$$\begin{aligned}
\varphi^{-1}(z) &= \varphi_2^{-1} \circ \varphi_4^{-1} \circ \varphi_6^{-1} \circ \varphi_8^{-1}(z) = \varphi_2^{-1} \circ \varphi_4^{-1} \circ \varphi_6^{-1}(z) \\
&= \varphi_2^{-1} \circ \varphi_4^{-1}(z) = \varphi_2^{-1}(z) = z, \\
\varphi^{-1}(t) &= \varphi_2^{-1} \circ \varphi_4^{-1} \circ \varphi_6^{-1} \circ \varphi_8^{-1}(t) = \varphi_2^{-1} \circ \varphi_4^{-1} \circ \varphi_6^{-1}(t) \\
&= \varphi_2^{-1} \circ \varphi_4^{-1}(t) = \varphi_2^{-1}(t + [x,y]) = t + [x,y] - [y,z,z].
\end{aligned}$$

Thus the system $\{h, g, z, t + [x,y] - [y,z,z]\}$ is primitive, and $\{z, t + [x,y] - [y,z,z]\}$ is a desirable complement. □

Comments

Free differential calculus for free groups was introduced by R. H. Fox in [184]. J. Birman in [69] gave a matrix criterion for an endomorphism of free group to be an automorphism. In [564] U. U. Umirbaev proved an analog of this result for free Lie algebras (Ch. Reutenauer in [479] also obtained such criterion, this theorem for free Lie algebras over commutative rings was proved by A. A. Mikhalev and A. A. Zolotykh in [423]; more general case was considered by V. Shpilrain in [523] and by U. U. Umirbaev in [566]). In [524] V. Shpilrain considered necessary and sufficient conditions for an endomorphism of the free group to be an automorphism. The general Jacobian problem for varieties of algebras was considered by A. V. Yagzhev, he solved it in [609] for the varieties of all algebras and of all anticommutative algebras. The Jacobian problem for free associative algebras was considered by W. Dicks and J. Lewin in [154] and by A. H. Schofield in [493] (A. A. Mikhalev and A. A. Zolotykh solved this problem for free associative algebras over commutative rings in [421]). Automorphisms and Jacobian matrices in varieties of associative algebras were considered by R. M. Bryant and V. Drensky in [100]. Free differentials were studied also in [93] by A. Borowiec, V. K. Kharchenko, and Z. Oziewicz. O. G. Kharlampovich in [269] used Fox differential calculus in connection with Lyndon condition for Lie algebras. As for universal derivations of relatively free algebras see for example [242], [508], [572]. Universal derivations of associative algebras are under consideration in articles [65] and [66] of G. M. Bergman and W. Dicks.

U. U. Umirbaev in [568], [570] gave a criterion for a system of elements of the free group to be primitive. In [525] V. Shpilrain defined the rank of an element of the free Lie algebra and gave a description of the rank of a homogeneous (relative to the length) element (the rank of an element a of the length m is equal to the dimension of the linear subspace spanned by all its right Fox partial derivatives of degree $m-1$).

The main results of this chapter are due to A. A. Mikhalev and A. A. Zolotykh (see [396], [407], [408], [410], [414], [415], [418], [419]). In Sections 23, 24 we used some ideas of U. U. Umirbaev's articles [569], [572]. The example of a nonfree Lie algebra with cohomological dimension one in Theorem 28.10 goes back to A. A. Mikhalev, U. U. Umirbaev and A. A. Zolotykh (see [407]). In the proof of Point (iv) of Theorem 28.10 we expose in fact G. P. Kukin's criterion for an element of the free Lie algebra $L(X)$ to be primitive (see [288]).

Appendix
Program Description

The program **CALS** (**C**ombinatorial **A**spects of **L**ie **S**uperalgebras) is intended for performance of **CAL**culation**S** in the free Lie (p-)superalgebras. The program expresses elements of the free Lie (p-)superalgebras in terms of the Lyndon-Shirshov basis and elements of the free associative algebra in terms of the basis of words. The regular arrangement of brackets on s-regular (ps-regular) words can be also performed.

How the program works

The program **CALS** is an interactive program. At first, you should specify the initial conditions for the calculations (the type of the main ring (field) of coefficients, the type of the free algebra and the set of generators). These initial conditions we call *options*. The options can be viewed, saved into a file and loaded from a file. The options which were saved into the file `OPTIONS.CNF` are loaded automatically at the beginning of the work. When the wanted calculations have been completed you can return to selection of options and start the next series of the calculations.

In operation you type the expression consisting of constants and of operators and the program calculates this expression and shows the result. The value of any expression can be assigned to a variable, and then you can use this variable typing other expressions.

The name of any variable is the sequence of letters (Greek or Roman, capital or small) and digits, possibly followed by a sequence of digits in subscript. The first character at the name of a variable must be a letter. The maximal length of the name of variable (considering the subscripts) is equal to 20. The variable with the name `prev` contains the value of the previous calculation. The following names have been used already for other aims: `add`, `ass`, `brack`, `div`, `lie`, `mul`, `neg`, `p`, `power`, `sub`. None of these names can be used as names of variables. In addition, you cannot use the names of the indeterminates of the polynomial ring and the names of the generators of the free algebra as the name of variables.

The meaning of the keys

The following keys are used in the program. For example, the notation "the keys `Shift` + `Tab`" means that these keys should be pushed simultaneously).

- the key `Left`: Go to the previous position;
- the key `Right`: Go to the next position;
- the key `Up`: Go to the previous line;
- the key `Down`: Go to the next line;
- the key `Home`: Go to the beginning of the current line;
- the key `End`: Go to the end of the current line;
- the keys `Ctrl` + `Home`: Go to the beginning of the expression;
- the keys `Ctrl` + `End`: Go to the end of the expression;
- the key `Tab`: The right tabulation;
- the keys `Shift` + `Tab`: The left tabulation;
- the key `Del`: Delete the current symbol;
- the key `BackSpace`: Delete the previous symbol;
- the keys `Alt` + `D`: Delete the current line;
- the keys `Alt` + `I`: Insert a new line;
- the key `Ins`: The Insert mode/Replace mode switch;
- the key `F1`: Show the help window;
- the key `F2`: Show the previous result;
- the key `F3`: Show the list of the variables;
- the key `F4`: Go to the subscript line;
- the key `F5`: Go to the base line;
- the key `F6`: Go to the superscript line;
- the key `F7`: Draw a Greek letter;
- the key `F8`: Clear the window;
- the key `F9` or `Enter`: Enter the written expression;
- the key `F10` or `Esc`: Return to the previous window.

The main field (ring) of coefficients

The following fields (rings) can be selected as the main ring of coefficients using the corresponding menu: the ring of integers $\mathbb{Z}$, the field of rational numbers $\mathbb{Q}$, the ring of residue classes $\mathbb{Z}_n$ (which is a field in the case when n is prime), the ring of polynomials over integers, rational numbers or residue classes. The maximal number of indeterminates of polynomials is 20. The maximal residue class module is equal to $2147483647 = 2^{31} - 1$.

You can also select the main field (ring) of coefficients editing the corresponding window (line `Edit` of the menu). You should type `Z` for the ring of integers, `Q` for the ring of rational numbers, $\mathtt{Z}_{37}$ for the field of residue classes modulo 37, `Q[t]` for the ring of polynomials over rational numbers with a indeterminate t, $\mathtt{Z}_{1111111}[\mathtt{aa}_1, \mathtt{bb}_2, \mathtt{cc}_3]$ for the ring of polynomials over the ring $\mathbb{Z}_{1111111}$ of residue classes modulo 1111111 with the set of indeterminates $\{aa_1, bb_2, cc_3\}$.

Remember that the more complicated type of the main ring you choose the more time and memory for the calculations of the coefficients you need. Thus, try to use a type of ring which is not more complicated than you really need.

The type of the free algebra

The program **CALS** deals with elements of one of the following type:

- free Lie algebra;
- free Lie superalgebra;
- free Lie p-algebra;
- free Lie p-superalgebra,

and, in each case, with elements of associative envelope (i.e., with elements of the free associative algebra). The last two types of free algebras can be chosen only if the main ring of coefficients has the prime characteristic.

The selection of the type of the free algebra and of generators of the free algebra can be done with the use of the corresponding menus. The maximal number of generators of the free Lie algebra and of the free Lie p-algebra is equal to 20. The free Lie superalgebra and the free Lie p-superalgebra can not have more that 20 odd and 20 even generators.

Operands and operators

The expressions in the program are constructed from constants and variables using the operators.

The following operators are allowed:

- the operator of addition:

```
<operand>+<operand>
add(<operand>,<operand>,...)
```

- the operator of subtraction:

```
<operand>-<operand>
sub(<operand>,<operand>)
```

- two unary sign operators:

```
+<operand>
-<operand>
neg(<operand>)
```

- the operator of multiplication:

```
<operand>*<operand>
<operand><operand>
mul(<operand>,<operand>,...)
```

- the operator of division:

```
<operand>/<operand>
div(<operand>,<operand>)
```

- the power operator:

<operand>$^{\langle\mathtt{operand}\rangle}$

```
<operand>^<operand>
power(<operand>,<operand>)
```

- the unary operator of arrangement of brackets:

```
[<operand>]
brack(<operand>)
```

- the operator of bracket multiplication:

```
[<operand>,<operand>,...]
brack(<operand>,<operand>,...)
```

- the operator of conversion of element into Lie form:

```
lie(<operand>)
```

- the operator of conversion of element into associative form:

  `ass(<operand>)`

- the operator of assignment:

  `<variable>=<operand>`

CALS and TEX

A few ideas connected with the document preparation system TEX are realized as a matter of convenience in **CALS**.

The first of these ideas is to use the control sequences for typing of Greek letters. In any window of the program where Greek letters are admissible you can print `\alpha`, `\beta`, `\gamma`, `\delta`, `\Gamma`, `\Delta` and so on instead of pushing the key `F7`.

The second observation is connected with the way of typing subscripts. You can print `a_1` to have a_1, `bb_2` to have bb_2, and so on. Note that curly brackets is not allowed in subscripts, so you should print arbitrary long subscript without brackets, for example, `abc_1234567890`. The subscript sequence terminates when the first nondigit symbol appears after the underbar character `_`.

TEX's superscript character `^` is used for the notation of the power function similarly the character `+` is used for the notation of the function of addition or the character `/` is used for the notation of the function of division.

Note that TEX's curly brackets `{}` play the same role in the program as the usual brackets do.

Examples of CALS applications

Example A.1

We would like to select the free Lie p-algebra with the set of generators $\{x_1, x_2\}$ over the ring $\mathbb{Z}_5[\alpha]$ and to save these options.

- Call the program `CALS.EXE`;
- push any key to continue;
- push the key `E` to select options;
- push the key `P` to select polynomials as the main ring of coefficient;
- push the key `S` to select polynomials over residue classes;
- type `5` and push the key `Enter` to select the residue class module;

- type 1 and push the key Enter to select the number of indeterminates of the polynomial ring;
- push the keys F7, Enter, Enter to select the name of the indeterminate;
- push the key O to finish selection of the main ring of coefficients;
- push the keys Up, Up and Enter to select Lie p-algebra as the type of the free algebra;
- push the key O to finish selection of the type of the free algebra;
- push the key S to begin selection of the generators of the free algebra;
- type 2 and push the key Enter to select the number of generators;
- type x_1 and push the key Enter to select the name of the first generator;
- type x_2 and push the key Enter to select the name of the second generator;
- push the key O to finish selection of the generators of the free algebra;
- push the key S to save current options;
- type RESCL5.CNF and push the key Enter to select the file name;
- push the key Esc to finish the work with the program.

□

Example A.2
We would like to write down the element $(x_1 + \alpha x_2)^{[p]}$ in terms of the standard Lyndon-Shirshov basis of the algebra considered in Example A.1 and then to write down this element as an element of the free associative algebra.

- Call the program CALS.EXE;
- push any key to continue;
- push the key L to load the options from a file;
- type RESCL5.CNF and push the key Enter to select the file name;
- push the key V to view the current options;
- push the key Esc to stop viewing of current options;
- push the key C to begin the calculations;

- type `(x_1+\alpha x_2)^5` and push the key `Enter` to compute $(x_1 + \alpha x_2)^5$ as an element of the free Lie p-algebra;
- push the key `Esc` to stop viewing of the result;
- push the key `F2` to view the previous result;
- push the key `Esc` to stop viewing of the previous result;
- type `ass(prev)` and push the key `Enter` to compute the associative form of the previous result;
- push the key `Esc` to stop viewing of the result;
- push the key `Esc` to finish the calculations;

□

Example A.3
We would like to change the type of our free algebra and to consider now the example of the free Lie algebra instead of the free Lie p-algebra over the same ring and with the same set of generators as in Example A.1 and then to compute $(x_1 + x_2)^5 - x_1^5 - x_2^5$ as an element of the free Lie algebra.

- Push the key `E` to select options;
- push the key `Enter` to finish selection of the main ring of coefficients;
- push the keys `Up`, `Up`, `Up`, `Up` and `Enter` to select Lie algebra as the type of the free algebra;
- push the key `O` to finish selection of the type of the free algebra;
- push the key `Enter` to finish selection of generators of the free algebra;
- push the key `C` to begin the calculations;
- type `a=(x_1+x_2)^5` and push the key `Enter` to compute $(x_1 + x_2)^5$ as an element of the free associative algebra; now the value of the variable `a` is equal to this element;
- push the key `Esc` to stop viewing of the result;
- type `b=x_1^5` and push the key `Enter` to compute x_1^5 as an element of the free associative algebra; now the value of the variable `b` is equal to this element;
- push the key `Esc` to stop viewing of the result;
- type `c=x_2^5` and push the key `Enter` to compute x_2^5 as an element of the free associative algebra; now the value of the variable `c` is equal to this element;

- push the key `Esc` to stop viewing of the result;
- push the key `F3` to view the list of the defined variables;
- push the key `Esc` to stop viewing of the list of variables;
- type `lie(a-b-c)` and push the key `Enter` to compute the desirable element;
- push the key `Esc` to stop viewing of the result;
- push the key `Esc` to finish the calculations;

□

Example A.4
We would like to change the type of our free algebra and to consider now the example of the free Lie superalgebra instead of the free Lie algebra over the same ring as in Example A.1 with the even generators x_1, x_2 and the odd generator y_1. Then we want to compute the s-regular arrangement of brackets on the word $x_1^2 x_2 x_1 y_1 x_1^2 x_2 x_1 y_1$.

- Push the key `E` to select options;
- push the key `Enter` to finish selection of the main ring of coefficients;
- push the keys `Up`, `Up`, `Up` and `Enter` to select the free Lie superalgebra as the type of the free algebra;
- push the key `O` to finish selection of the type of the free algebra;
- push the key `O` to select odd generators of the free algebra;
- type `1` and push the key `Enter` to select the number of odd generators;
- type `y_1` and push the key `Enter` to select the name of the odd generator;
- push the key `K` to finish selection of generators of the free algebra;
- push the key `C` to begin the calculations;
- type `[(x_1^2 x_2 x_1 y_1)^2]` and push the key `Enter` to compute the s-regular arrangement of brackets $[(x_1^2 x_2 x_1 y_1)^2]$;
- push the key `Esc` to stop viewing of the result;
- push the key `Esc` to finish the calculations;
- push the key `Esc` to finish the work with the program.

□

Bibliography

[1] E. Abe, *Hopf Algebras*. Cambridge Univ. Press, Cambridge–New York, 1980.

[2] S. A. Agalakov, On finite separability of groups and Lie algebras. *Algebra i Logika* **22** (1983), no. 4, 363–371.

[3] S. A. Agalakov, *Notes on Residual Finiteness of Lie Algebras with Respect to Entry*. Deposited at VINITI, no. 5096-84.

[4] S. A. Agalakov, A Lie algebra with one defining relation need not be residual finite. *Mat. Zametki* **51** (1992), no. 4, 3–7.

[5] S. A. Agalakov, On entry problem for free product of Lie algebras. *Internat. Conf. on Algebra. Abstracts of Reports*. Krasnoyarsk, 1993, 4–5.

[6] V. K. Agrawala, Invariants of generalized Lie algebras. *Hadronic J.* **4** (1981), 444–496.

[7] R. Amayo and I. Stewart, *Infinite Dimensional Lie Algebras*. Leyden, Noordhoof, 1974.

[8] D. J. Anick, On the cohomology of associative algebras. *Trans. Amer. Math. Soc.* **296** (1986), 641–659.

[9] J. Apel and W. Lassner, Computation and simplification in Lie fields. *Lecture Notes in Comput. Sci.* **378** (1989), 468–478.

[10] D. Arlet, Zum Poincaré-Birkhoff-Witt Theorem. *Arch. Math. (Basel)* **19** (1968), 291–295.

[11] V. A. Artamonov, Projective metabelian D-groups and Lie superalgebras. *Trudy Sem. Petrovsk.* **15** (1991), 189–195.

[12] V. A. Artamonov, Structure of Hopf algebras. *Algebra Topol. Geom.* **29** (1991), 3–63.

[13] V. A. Artamonov, Algebra of quantum polynomials. *Sovrem. Mat. i Prilozh.*, in press.

[14] M. Artin, *Algebra*. Prentice-Hall Internat., 1991.

[15] M. Aubri and J.-M. Lemaire, Zero divisors in enveloping algebras of graded Lie superalgebras. *J. Pure Appl. Algebra* **38** (1985), 159–166.

[16] L. L. Avramov, Free Lie subalgebras of the homology of local rings. *Trans. Amer. Math. Soc.* **270** (1982), 589–608.

[17] L. L. Avramov, *Homological Methods in the Theory of Commutative Noetherian Rings*. Thesis of D.Sci., Sofia, 1983.

[18] I. K. Babenko, Analytic properties of Poincaré series of a loop space. *Mat. Zametki* **27** (1980), 751–765.

[19] N. B. Backhouse, The Killing form for graded Lie algebras. *J. Math. Phys.* **18** (1977), no. 2, 239–244.

[20] N. B. Backhouse, Computing the invariants of Lie superalgebras. *J. Phys. A* **12** (1979), no. 1, 21–30.

[21] R. Baer, Algebraische Theorie der differentierbaren Funktionenkörper, I. *Sitzungsb. Heidelb. Akad. Wiss. Math.-Natur. Kl.* 1927, 15–32.

[22] J. Bagger and J. Wess, *Supersymmetry and Supergravity*. Princeton Univ. Press, 1983.

[23] Yu. A. Bahturin, On one algorithm of equality. *Sibirsk. Mat. Zh.* **12** (1971), no. 2, 459–462.

[24] Yu. A. Bahturin, On homomorphisms of soluble Lie algebras. *J. London Math. Soc. (2)* **20** (1979), 415–422.

[25] Yu. A. Bahturin, Identities in the universal enveloping algebra for a Lie superalgebra. *Mat. Sb.* **127** (1985), 384–397.

[26] Yu. A. Bahturin, *Identical Relations in Lie Algebras*. VNU Science Press, Utrecht, 1987.

[27] Yu. A. Bahturin and V. S. Drensky, Identities of the soluble color Lie superalgebras. *Algebra i Logika* **26** (1987), 403–418.

[28] Yu. A. Bahturin, D. Fischman, and S. Montgomery, On the generalized Lie structure of associative algebras. *Israel J. Math.*, in press.

[29] Yu. A. Bahturin and A. Giambruno, Identities of commutative subalgebras, in press.

[30] Yu. A. Bahturin, A. A. Mikhalev, M. V. Zaicev, and V. M. Petrogradsky, *Infinite Dimensional Lie Superalgebras*. Walter de Gruyter Publ., Berlin–New York, 1992.

[31] Yu. A. Bahturin and S. Nabiev, Automorphisms and derivations of Abelian extensions of some Lie algebras. *Abh. Math. Sem. Univ. Hamburg* **62** (1992), 43–57.

[32] Yu. A. Bahturin and A. Yu. Ol'shanskii, Identities. *Sovrem. Probl. Mat. Fund. Naprav.* **18** (1988), 117–248.

[33] Yu. A. Bahturin and M. V. Zaicev, Residual finiteness of color Lie superalgebras. *Trans. Amer. Math. Soc.* **337** (1993), 159–180.

[34] H. F. Baker. *Proc. London Math. Soc. (1)* **34** (1902), p. 347.

[35] H. F. Baker, Alternants and continuous groups. *Proc. London Math. Soc. (2)* **3** (1905), 24–47.

[36] A. A. Baranov, Volichenko algebras and nonhomogeneous subalgebras of Lie superalgebras. *Sibirsk. Mat. Zh.*, in press.

[37] A. A. Baranov, Color smash products, universal enveloping algebras and skew polynomial rings. *Izv. Vyssh. Uchebn. Zaved. Mat.*, in press.

[38] H. Barcelo, On the action of the symmetric group on the free Lie algebra and the partition lattice. *J. Combin. Theory Ser. A* **55** (1990), 93–129.

[39] H. Barcelo and N. Bergeron, The Orlik-Solomon algebra on the partition lattice and the free Lie algebra. *J. Combin. Theory Ser. A* **55** (1990), 80–92.

[40] H. Barcelo and S. Sundaram, On some submodules of the action of the symmetric group on the free Lie algebra. *J. Algebra* **154** (1993), 12–26.

[41] H. Bass, E. H. Connell, and D. Wright, The Jacobian conjecture: reduction of degree and formal expansion of the inverse. *Bull. Amer. Math. Soc. (N. S.)* **7** (1982), no. 2, 287–330.

[42] A. Baudisch, *Application of the Theorem of Širšov-Witt in the Theory of Groups and Lie Algebras*. Preprint P-25/80, Zentrallinstitut für Mathematik und Mechanik, Berlin, 1980.

[43] A. Baudisch, *Application of the Theorem of Širšov-Witt in the Theory of Groups and Lie Algebras. Addendum*. Preprint P-30/80, Zentrallinstitut für Mathematik und Mechanik, Berlin, 1980.

[44] A. Baudisch, *On Elementary Properties of Free Lie Algebras*. Preprint P-02/84, Zentrallinstitut für Mathematik und Mechanik, Berlin, 1984.

[45] G. Baumslag, Free Lie algebras and free groups. *J. London Math. Soc. (2)* **4** (1972), 523–532.

[46] P. Beckman and J. Stückrad, The concept of Gröbner algebras. *J. Symbolic Comput.* **10** (1990), 465–479.

[47] E. J. Behr, Enveloping algebras of Lie superalgebras. *Pacific J. Math.* **130** (1987), 9–25.

[48] K. I. Beidar, On radicals of finitely generated algebras. *Uspekhi Mat. Nauk* **36** (1981), no. 6, 203–204.

[49] K. I. Beidar and A. V. Mikhalev, Generalized polynomial identities and rings which are sums of two subrings. *Internat. Conf. "Algebra and Analysis". Abstracts of Reports*. Kazan, 1994, part 1, p. 16.

[50] K. I. Beidar, W. S. Martindale III, and A. V. Mikhalev, *Rings with Generalized Identities*. Marcel Dekker, in press.

[51] A. A. Belavin, A. M. Polyakov, and A. B. Zamolodchikov, Infinite conformal symmetry in two-dimensional Field Theory. *J. Nucl. Phys. B* **241** (1984), 333–380.

[52] A. D. Bell, A criterion for primeness of enveloping algebras of Lie superalgebras. *J. Pure Appl. Algebra* **69** (1990), 111–120.

[53] A. I. Belov, *Periodic Automorphisms of Free Lie Algebras and Their Fixed Points*. Preprint.

[54] A. Ya. Belov, V. V. Borisenko, and V. N. Latyshev, Monomial algebras. *Sovrem. Mat. i Prilozh.*, in press.

[55] F. A. Berezin, *Introduction in Algebra and Analysis with Anticommuting Variables*. Moscow State Univ., Moscow, 1983; *Introduction to Superanalysis*. Reidel, 1987.

[56] J. Bergen and P. Grzeszczuk, Engel-type theorems for Lie color algebras. *Rings, Extensions, and Cohomology*, Marcel Dekker, 1994.

[57] J. Bergen and P. Grzeszczuk, Invariants of Lie superalgebras acting on associative algebras, in press.

[58] J. Bergen and D. S. Passman, Delta methods in enveloping algebras of Lie superalgebras. *Trans. Amer. Math. Soc.* **334** (1992), no. 1, 259–280.

[59] R. Berger, Quantification de l'identité de Jacobi. *C. R. Acad. Sci. Paris Sér. I Math.* **312** (1991), 721–724.

[60] R. Berger, The quantum Poincaré-Birkhoff-Witt theorem. *Comm. Math. Phys.* **143** (1992), 215–234.

[61] F. Bergeron, N. Bergeron, and A. M. Garsia, Idempotents for the free Lie algebras and q-numerations. *IMA Vol. Math. Appl.* **19** (1990), 166–190.

[62] N. Bergeron, An hyperoctahedral analogue of the free Lie algebra. *J. Combin. Theory Ser. A* **58** (1991), 256–278.

[63] G. M. Bergman, The diamond lemma for ring theory. *Adv. Math.* **29** (1978), 178–218.

[64] G. M. Bergman, Everybody knows what a Hopf algebra is. *Contemp. Math.* **43** (1985), 25–48.

[65] G. M. Bergman and W. Dicks, On universal derivations. *J. Algebra* **36** (1975), 193–211.

[66] G. M. Bergman and W. Dicks, Universal derivations and universal ring constructions. *Pacific J. Math.* **79** (1978), no. 2, 293–337.

[67] G. Birkhoff, Representability of Lie algebras and Lie groups by matrices. *Ann. of Math. (2)* **38** (1937), no. 2, 526–532.

[68] G. Birkhoff and T. C. Bartee, *Modern Applied Algebra*. McGraw-Hill Book Company, 1970.

[69] J. S. Birman, An inverse function theorem for free groups. *Proc. Amer. Math. Soc.* **41** (1973), 634–638.

[70] D. Blessenohl and H. Laue, Generalized Jacoby identities. *Note Mat.* **8** (1988), 111–121.

[71] D. Blessenohl and H. Laue, On Witt's dimension formula for free Lie algebras and a theorem of Klyachko. *Bull. Austral. Math. Soc.* **40** (1989), 49–57.

[72] D. Blessenohl and H. Laue, Symmetric groups and free Lie algebras. *Rend. Circ. Mat. Palermo (2) Suppl.* **23** (1990), 201–208.

[73] D. Blessenohl and H. Laue, On the coefficients in the associative expansion of a Lie word. *European J. Combin.* **12** (1991), no. 3, 205–210.

[74] D. Blessenohl and H. Laue, A basis construction for free Lie algebras. *Exposition. Math.* **11** (1993), 145–152.

[75] D. Blessenohl and H. Laue, Algebraic combinatorics related to the free Lie algebra. *Publ. Inst. Rech. Math. Av.* 1993, Actes 29 Séminaire Lotharingien, 1–21.

[76] R. Block, A characterization of Lie elements at prime characteristic. *Contemp. Math.* **110** (1990), p. xv.

[77] L. A. Bokut', Embeddings of Lie algebras into algebraically closed Lie algebras. *Algebra i Logika* **1** (1962), no. 2, 47–53.

[78] L. A. Bokut', A base of free polynilpotent Lie algebras. *Algebra i Logika* **2** (1963), no. 4, 13–19.

[79] L. A. Bokut', Unsolvability of the equality problem and subalgebras of finitely presented Lie algebras. *Izv. Akad. Nauk SSSR Ser. Mat.* **36** (1972), no. 6, 1173–1219.

[80] L. A. Bokut', Embeddings into simple associative algebras. *Algebra i Logika* **15** (1976), 117–142.

[81] L. A. Bokut', *Associative Rings, I*. Novosibirsk State Univ., Novosibirsk, 1977.

[82] L. A. Bokut', On algebraically closed and simple Lie Algebras. *Trudy Mat. Inst. Steklov.* **148** (1978), 30–41.

[83] L. A. Bokut', On one condition of solvability of Lie algebras. *Sibirsk. Mat. Zh.* **22** (1981), no. 3, 15–20.

[84] L. A. Bokut', Lie algebra and its universal enveloping algebra have the same Gröbner-Shirshov basis. *Internat. J. Algebra Comput.*, in press.

[85] L. A. Bokut' and A. A. Klein, Gröbner-Shirshov bases for the classical Lie algebras, I. *Comm. Algebra*, in press.

[86] L. A. Bokut' and G. P. Kukin, Unsolvable algorithmic problems for semigroups, groups, and rings. *Algebra Topol. Geom.* **25** (1987), 3–66.

[87] L. A. Bokut' and G. P. Kukin, *Algorithmic and Combinatorial Algebra*. Kluwer Academic Publ., Dordrecht, 1994.

[88] L. A. Bokut' and L. G. Makar-Limanov, A base of the free metabelian associative algebra. *Sibirsk. Mat. Zh.* **32** (1991), no. 6, 12–18.

[89] P. J. M. Bongaarts, Mathematical aspects of supersymmetry and anticommuting variables. The one-dimensional Wess-Zumino model. *Proc. Semin. Math. Stract. Field Theor. 1986–1987*, Centr. wisk. en inf., Amsterdam, 1990, 55–80.

[90] M. Boral, On ideals of free and free polynilpotent Lie algebras. *Proc. Amer. Math. Soc.* **94** (1985), 23–28.

[91] M. Boral, On subalgebras of free solvable and polynilpotent Lie algebras. *Comm. Algebra* **18** (1990), 1199–1214.

[92] M. Boral, On ideals of free polynilpotent Lie algebras. *Comm. Algebra* **19** (1991), 29–35.

[93] A. Borowiec, V. K. Kharchenko, and Z. Oziewicz, *On Free Differentials on Associative Algebras*. Preprint no. 861, Univ. of Wroclaw, 1993.

[94] A. Bose, Dynkin's method of computing the terms of the Baker-Campbell-Hausdorff series. *J. Math. Phys.* **30** (1989), 2035–2037.

[95] N. Bourbaki, *Algebra I. Chapters 1–3*. Springer-Verlag, Berlin, 1988.

[96] N. Bourbaki, *Lie Groups and Lie Algebras, 1–3*. Springer-Verlag, Berlin, 1988.

[97] A. J. Bracken and H. S. Green, Parastatistics and the quantum model. *J. Math. Phys.* **14** (1973), 1784–1793.

[98] A. Brandt, The free Lie ring and Lie representations of the full linear group. *Trans. Amer. Math. Soc.* **56** (1944), 528–536.

[99] R. M. Bryant, On the fixed points of a finite group acting on a free Lie algebra. *J. London Math. Soc. (2)* **43** (1991), 215–224.

[100] R. M. Bryant and V. Drensky, Obstructions to lifting automorphisms of free algebras. *Comm. Algebra* **21** (1993), 4361–4389.

[101] B. Buchberger, *An Algorithm for Finding a Basis for the Residue Class Ring of a Zero-Dimensional Polynomial Ideals*. Ph.D. Thesis, Univ. of Ihnsbruck, 1965.

[102] B. Buchberger and R. Loos, Algebraic simplification. *Comput. Suppl.* **4** (1982), 11–43; *Computer Algebra. Symbolic and Algebraic Computation. 2nd Ed.* Springer-Verlag, Wien–New York, 1983.

[103] R. G. Burns, On the intersection of finitely generated subgroups of a free group. *Math. Z.* **119** (1971), 121–130.

[104] M. D. Burrow, Invariants of free Lie rings. *Comm. Pure Appl. Math.* **11** (1958), 419–431.

[105] A. R. Calderbank, P. Hanlon, and S. Sundaram, Representations of the symmetric group in deformations of the free Lie algebra. *Trans. Amer. Math. Soc.* **341** (1994), no. 1, 315–333.

[106] J. E. Campbell, On a low of combination of operators, I, II. *J. London Math. Soc. (1)* **28** (1897), 381–390; **29** (1898), 14–32.

[107] J. E. Campbell, *Introductory Treatise on Lie Theory of Finite Continuous Transformation Groups*. Oxford, 1903.

[108] H. Cartan and S. Eilenberg, *Homological Algebra*. Princeton Univ. Press, 1956.

[109] P. Cartier, Démonstration algèbrique de la formule de Hausdorff. *Bull. Soc. Math. France* **84** (1956), 241–249.

[110] P. Cartier, Remarques sur la théorème de Birkhoff-Witt, *Ann. Scuola Norm. Sup. Pisa Cl. Sci. (4)* **12** (1958), 1–4.

[111] P. Cartier and D. Foata, *Problèmes Combinatoires de Commutation et Réarrangements*. Lecture Notes in Math. **85** (1969).

[112] R. Cecchini and M. Tarlini, Symbolic superalgebra manipulation using COMMON LISP. *Comput. Phys. Comm.* **60** (1990), no. 2, 265–270.

[113] R. Cecchini and M. Tarlini, *A Computer Algebra System for Lie and Poisson Superalgebras Manipulation*. Univ. di Firenze, Italy.

[114] M. Chaichian and P. Kulish, Quantum Lie superalgebras and q-oscillators. *Phys. Lett. B* **234** (1990), no. 1–2, 72–80.

[115] A. D. Chanyshev, On asymmetric sequences. *Vestnik Moskov. Univ. Ser. I Mat. Mekh.* 1990, no. 2, 86–88.

[116] A. D. Chanyshev, Regular words and a theorem on sandwich algebras. *Vestnik Moskov. Univ. Ser. I Mat. Mekh.* 1990, no. 5, 84–85.

[117] A. D. Chanyshev and V. A. Ufnarovsky, Regular and semiregular words. *6th All-Union Simpos. on the Theory of Rings, Algebras and Modules. Abstracts of Reports*. Lvov, 1990, p. 132.

[118] K. T. Chen, Integration of paths, geometric invariants and a generalized Baker-Hausdorff formula. *Ann. of Math. (2)* **65** (1957), 163–178.

[119] K. T. Chen, R. H. Fox, and R. C. Lyndon, Free differential calculus. IV. The quotient groups of the lower central series. *Ann. of Math. (2)* **68** (1958), 81–95.

[120] I. V. Chirkov, On generators of automorphism groups of free Lie algebras. *Internat. Conf. on Algebra. Abstracts of Reports*. Barnaul, 1991, p. 140.

[121] P. M. Cohn, *Integral Modules, Lie Rings and Free Groups*. Ph.D. Thesis, Univ. of Cambridge, 1951.

[122] P. M. Cohn, Sur le critère de Friedrichs pour les commutateurs dans une algèbre associative libre. *C. R. Acad. Sci. Paris Sér. I Math.* **239** (1954), 743–745.

[123] P. M. Cohn, On a class of simple rings. *Mathematika* **5** (1958), 103–117.

[124] P. M. Cohn, On a generalization of the Euclidean algorithm. *Proc. Cambridge Philos. Soc.* **57** (1961), 18–30.

[125] P. M. Cohn, A remark on the Birkhoff-Witt theorem. *J. London Math. Soc. (1)* **38** (1963), 197–203.

[126] P. M. Cohn, Free ideal rings. *J. Algebra* **1** (1964), 47–69.

[127] P. M. Cohn, Subalgebras of free associative algebras, *Proc. London Math. Soc. (3)* **14** (1964), 618–632.

[128] P. M. Cohn, *Algebra. Vol. 1.* Wiley, London–New York, 1974.

[129] P. M. Cohn, *Universal Algebra. 2nd Ed.* Reidel, Dordrecht, 1981.

[130] P. M. Cohn, *Free Rings and Their Relations. 2nd Ed.* Academic Press, London–New York, 1985.

[131] P. M. Cohn, *Algebra. Vol. 2.* Wiley, New York, 1989.

[132] A. Colojoara, Sur le théorème de Poincaré-Birkhoff-Witt. *Rev. Roumaine Math. Pures Appl.* **25** (1980), no. 9, 1327–1347.

[133] *Computer Algebra.* Marcel Dekker, New York, 1989.

[134] *Computers in Algebra.* Ed. M. C. Tangora. Marcel Dekker, New York, 1988.

[135] *Computers in Mathematics*, Marcel Dekker, New York, 1990.

[136] *Computers in Nonassociative Rings and Algebras.* Ed. R. E. Beck and B. Kolman. Academic Press, New York, 1977.

[137] L. Corwin, Y. Ne'eman, and S. Sternberg, Graded Lie algebras in mathematics and physics (Bose-Fermi symmetry). *Rev. Modern Phys.* **47** (1975), 573–603.

[138] D. Cox, J. Little, and D. O'Shea, *Ideals, Varieties, and Algorithms.* Springer-Verlag, 1992.

[139] C. W. Curtis and I. Reiner, *Representation Theory of Finite Groups and Associative Algebras.* Wiley, New York, 1962.

[140] A. J. Czerniakiewicz, Automorphisms of a free associative algebra of rank 2, I, II. *Trans. Amer. Math. Soc.* **160** (1971), 393–401; **171** (1972), 309–315.

[141] J. Czyz, On Lie supergroups and superbundles defined via the Baker-Campbell-Hausdorff formula. *J. Geom. Phys.* **6** (1989), no. 4, 595–626.

[142] I. Damiani, A basis of type Poincaré-Birkhoff-Witt for the quantum algebra of $\widehat{\mathrm{sl}}(2)$. *J. Algebra* **161** (1993), no. 2, 291–310.

[143] J. H. Davenport, Y. Siret, and E. Tournier, *Computer Algebra. Systems and Algorithms for Algebraic Computation.* Academic Press, 1988.

[144] J. Demetrovics, E. Knuth, and P. Rado, *Computer-Aided Specification Techniques.* World Scientific, Singapore, 1986.

[145] E. E. Demidov, On multiparametric quantum deformation of the group GL(n). *Uspekhi Mat. Nauk* **46** (1991), no. 4, 147–148.

[146] E. E. Demidov, *Multiparametric Quantum Deformation of General Linear Group.* Ph.D. Thesis, Moscow State Univ., 1992.

[147] E. E. Demidov, On some aspects of the theory of quantum groups. *Uspekhi Mat. Nauk* **48** (1993), no. 6, 39–74.

[148] J. Désarménien, G. Duchamp, D. Krob, and G. Melançon, Quelques remarques sur les super-algèbres de Lie libres. *C. R. Acad. Sci. Paris. Sér. I Math.* **318** (1994), 419–424.

[149] B. DeWitte, *Supermanifolds.* Cambridge Univ. Press, 1984.

[150] C. E. Dicidze, Nonassociative free sums of algebras with the amalgamated subalgebra. *Mat. Sb.* **43** (1957), 373–396.

[151] C. E. Dicidze, Subalgebras of nonassociative free sums with arbitrary amalgamated subalgebras. *Mat. Sb.* **54** (1961), no. 3, 381–384.

[152] W. Dicks, On one-relator associative algebras. *J. London Math. Soc. (2)* **5** (1972), 249–252.

[153] W. Dicks, A commutator test for two elements to generate the free algebra of rank two. *Bull. London Math. Soc.* **14** (1982), 48–51.

[154] W. Dicks and J. Lewin, A Jacobian conjecture for free associative algebras. *Comm. Algebra* **10** (1982), 1285–1306.

[155] J. Dixmier, *Algèbres Enveloppantes*. Hermann, Paris, 1974.

[156] *Dnestrovskaya Tetrad' (Dnestr Notebook). Nonsolvable Problems of the Theory of Rings and Modules. 4th Ed.* Novosibirsk, 1993.

[157] D. Dokovič, A generalization of Witt's formulae. *J. Algebra* **111** (1987), 262–278.

[158] D. Dokovič, On some inner derivations of free Lie algebras over commutative rings. *J. Algebra* **119** (1988), 233–245.

[159] V. Drensky, Endomorphisms and automorphisms of relatively free algebras. *Rend. Circ. Mat. Palermo (2) Suppl.* **31** (1993), 97–132.

[160] V. Drensky, Fixed algebras of residually nilpotent Lie algebras. *Proc. Amer. Math. Soc.* **120** (1994), 1021–1028.

[161] V. Drensky and C. K. Gupta, Automorphisms of free nilpotent Lie algebras. *Canad. J. Math.* **42** (1990), 259–279.

[162] V. G. Drinfeld, Quantum groups. *Proc. of the Internat. Congress of Math., Berkeley, 1986*, part 1, 798–820.

[163] V. G. Drinfeld, On quadratic commuting relations. *Trudy FTINT*, Kharkov.

[164] G. Duchamp, Orthogonal projection onto the free Lie algebra. *Theoret. Comput. Sci.* **79** (1991), 227–239.

[165] G. Duchamp, *Identité Cyclotomique Superalgèbres de Lie et Bicaractères.* Preprint LITP 94.01, Inst. Blaise Pascal, Paris, 1994.

[166] G. Duchamp and D. Krob, Computing with P.B.W. in enveloping algebras. *Lecture Notes in Control and Inform. Sci.* **165** (1991), 223–240.

[167] G. Duchamp and D. Krob, The free partially commutative Lie algebra: bases and ranks. *Adv. Math.* **95** (1992), 92–126.

[168] G. Duchamp and J.-Y. Thibon, Le support de l'algèbre de Lie libre. *Discrete Math.* **76** (1989), 123–129.

[169] J.-P. Duval, Factorizing words over an ordered alphabet. *J. Algorithms* **4** (1983), 363–381.

[170] J.-P. Duval, Gènèration d'une section des classes de conjugaison et arbre des mots de Lyndon de longueur bornée. *Theoret. Comput. Sci.* **60** (1988), 255–283.

[171] E. B. Dynkin, Calculation of the coefficients in the Campbell-Hausdorff formula. *Dokl. Akad. Nauk* **51** (1947), 323–326.

[172] E. B. Dynkin, On the representation by means of commutators of the series $\log(e^x e^y)$ for noncommutative x and y. *Mat. Sb.* **25** (1949), 155–162.

[173] D. I. Eidelkind, Verbal products of Magnus groups. *Mat. Sb.* **85** (1971), no. 4, 504–526.

[174] N. Ekici, Free polynilpotent subalgebras of a free polynilpotent Lie algebra. *Comm. Algebra* **19** (1991), 37–43.

[175] N. Ekici and Y. Ünlü, Quasi-primeness of lower central terms of an ideal in a free Lie algebra. *J. Pure Appl. Algebra* **43** (1986), 299–302.

[176] M. El-Agawani and A. Micali, Superalgèbres de Lie et le théorème de Poincaré-Birkhoff-Witt. *Algebras Groups Geom.* **4** (1987), 1–44.

[177] M. El-Agawani and A. Micali, Le théorème de Poincaré-Birkhoff-Witt pour les algèbre de Lie $\mathbb{Z}(2)$-graduées p-nilpotentes. Cas on $p = 2$ et $p = 3$. *An. Acad. Brasil. Ciênc.* **53** (1981), no. 2, 233–241.

[178] G. L. Feldman, Ends of Lie algebras. *Uspekhi Mat. Nauk* **38** (1983), no. 1, 199–200.

[179] S. Ferrara, *Supersymmetry*. North-Holland, Amsterdam, 1987.

[180] D. Finkelstein, On relations between commutators. *Comm. Pure Appl. Math.* **8** (1955), 245–250.

[181] D. Fischman and S. Montgomery, A Schur double centralizer theorem for cotriangular Hopf algebras and generalized Lie algebras. *J. Algebra*, in press.

[182] R. Floreanini, D. A. Leites, and L. Vinet, On the defining relations of quantum superalgebras. *Lett. Math. Phys.* **23** (1991), no. 2, 127–131.

[183] D. Foata, Etude algèbrique de certains problèmes d'analyse combinatoire et du calcul des probabilités. *Publ. Inst. Statist. Univ. Paris* **14** (1965), 81–241.

[184] R. H. Fox, Free differential calculus. I. Derivations in free group rings. *Ann. of Math. (2)* **57** (1953), 547–560.

[185] K. O. Friedrichs, Mathematical aspects of the quantum theory of fields. V. Fields modified by linear homogeneous forces. *Comm. Pure Appl. Math.* **6** (1953), no. 1, 1–72.

[186] A. Frölicher and A. Nijenhuis, Theory of vector valued differential forms, I. *Indag. Math. (N. S.)* **18** (1956), 338–359.

[187] D. B. Fuchs, *Cohomology of Infinite Dimensional Lie Algebras*. Plenum, New York, 1986.

[188] I. Fujivara, On the evaluation of the operator function $\log(e^x e^y)$. *Progr. Theoret. Phys.* **9** (1953), p. 976.

[189] A. M. Garsia, Combinatorics of the free Lie algebra and the symmetric group. *Analysis*. Academic Press, Boston, 1990, 309–382.

[190] A. M. Garsia and Ch. Reutenauer, A decomposition of Solomon's descent algebra. *Adv. Math.* **77** (1989), 189–262.

[191] T. Gateva-Ivanova and V. N. Latyshev, On recognizing properties of associative algebras. *J. Symbolic Comput.* **6** (1988), 371–388.

[192] K. O. Geddes, S. R. Czapor, and G. Labahn, *Algorithms for Computer Algebra*. Kluwer Academic Publ., 1992.

[193] I. M. Gelfand and Y. L. Daleckii, *Some Formal Differential Structures Connected with Lie Superalgebras*. Preprint no. 85, Inst. Prikl. Mat. Akad. Nauk SSSR, 1994.

[194] S. I. Gelfand and Yu. I. Manin, *Methods of Homologigal Algebra, I*. Nauka, Moscow, 1988.

[195] V. N. Gerasimov, Distributive lattices of subspaces and the word problem for one-relator algebras. *Algebra i Logika* **15** (1976), no. 4, 384–435.

[196] V. P. Gerdt, I. R. Akselrod, V. E. Kovtun, and V. N. Robuk, Construction of a Lie algebra by a subset of generators and commutation relations. *Computer Algebra in Physical Researches*. World Scientific, Singapore, 1991, 306–312.

[197] V. P. Gerdt and W. Lassner, Isomorphism verification for complex and real Lie algebras by Gröbner basis technique. *Modern Group Analysis: Advanced Analytical and Computational Methods in Mathematical Physics*, Kluwer Academic Publ., Dordrecht, 1993, 245–254.

[198] S. M. Gersten, Intersections of finitely generated subgroups of free groups and resolutions of graphs. *Invent. Math.* **71** (1983), 567–591.

[199] K. Goldberg, The formal power series for $\log(e^x e^y)$. *Duke Math. J.* **23** (1956), 13–21.

[200] E. S. Golod, Standard bases and homology. *Lecture Notes in Math.* **1352** (1988), 88–95.

[201] E. S. Golod, Standard bases and homology. II. *Trudy Mat. Inst. Steklov.*, in press.

[202] Y. M. Gorchakov, Commutator subgroups. *Sibirsk. Mat. Zh.* **10** (1969), no. 5, 1023–1033.

[203] M. Goto and F. D. Grosshans, *Semisimple Lie Algebras*. Marcel Dekker, New York–Basel, 1978.

[204] M. D. Gould, Polynomial identities for simple Lie superalgebras, *J. Austral. Math. Soc. Ser. B* **26** (1987), 310–327.

[205] V. E. Govorov, Algebras which are free generated by finite amalgams. *Mat. Sb.* **50** (1960), no. 2, 241–246.

[206] P. K. H. Gragert, Lie algebra computations. *Acta Appl. Math.* **16** (1989), 231–242.

[207] M. Grayson, Models for free nilpotent Lie algebras. *J. Algebra* **135** (1990), 177–191.

[208] E. L. Green, An introduction into noncommutative Gröbner bases. *Comput. Algebra*, Marcel Dekker, 1993, 167–190.

[209] H. S. Green and P. D. Jarvis, Casimir invariants, characteristic identities, and Young diagrams for color algebras and superalgebras. *J. Math. Phys.* **24** (1983), 1681–1687.

[210] M. B. Green, J. H. Schwartz, and E. Witten, *Superstring Theory*. Cambridge Univ. Press, Cambridge, 1987.

[211] G. Griffing, The dual Friedrich's theorem. *J. Algebra* **132** (1990), 446–460.

[212] W. Gröbner, *Modern Algebraic Geometry*. Springer-Verlag, 1949.

[213] J. R. J. Groves, Rewriting systems and homology of groups. *Lecture Notes in Math.* **1456** (1990), 114–141.

[214] N. Gupta, *Free Group Rings*. Contemp. Math. **66** (1987).

[215] D. I. Gurevich, Generalized translation operators on Lie groups. *Soviet J. Contemporary Math. Anal.* **18** (1983), 57–70.

[216] R. M. Hain, *Iterated Integrals and Homotopy Periods*. Amer. Math. Soc., Providence, Rhode Island, 1984.

[217] M. Hall, A basis for free Lie rings and higher commutators in free groups. *Proc. Amer. Math. Soc.* **1** (1950), 575–581.

[218] M. Hall, *Theory of Groups*. MacMillan, New York, 1959.

[219] M. Hall, *Combinatorial Theory*. Wiley, New York, 1986.

[220] P. Hall, A contribution to the theory of groups of a prime power order. *Proc. London Math. Soc. (2)* **36** (1933), 29–95.

[221] Ya. Halluf, *Lie Pseudo-Algebras and Their Representations*. Ph.D. Thesis, Kiev State Univ., 1986.

[222] K. C. Hannabuss, An obstruction to generalized superalgebras. *J. Phys. A* **20** (1987), no. 17, L1135–L1138.

[223] K. C. Hannabuss, Corepresentations of superalgebras and an antilinear Dirac operator. *J. Math. Phys.* **31** (1990), 2520–2524.

[224] F. Hausdorff, Die symbolische Exponentialformel in der Gruppentheorie. *Leipziger Ber.* **58** (1906), 19–48.

[225] G. Havas, M. F. Newman, and M. R. Vaughan-Lee, A nilpotent quotient algorithm for graded Lie rings. *J. Symbolic Comput.* **9** (1990), 653–664.

[226] M. C. Hedges, The Freiheitssatz for graded algebras. *J. London Math. Soc. (2)* **35** (1987), 395–405.

[227] I. R. Hentzel and D. P. Jacobs, A dynamic programming method for building free algebras. *Comput. Math. Appl.* **22** (1991), no. 12, 61–66.

[228] G. Hermann, Die Frage der endlichen vielen Schritte in der Theorie der Polynomideale. *Math. Ann.* **95** (1926), 736–788.

[229] I. N. Herstein, *Abstract Algebra. 2nd Ed.* Macmillan, New York, 1990.

[230] H.-J. Higgins, Baer invariants and the Birkhoff-Witt theorem. *J. Algebra* **11** (1969), 469–482.

[231] P. J. Hilton, On the homotopy groups of the union of spheres. *J. London Math. Soc. (1)* **30** (1955), 154–172.

[232] A. G. Howson, On the intersection of finitely generated free groups. *J. London Math. Soc. (2)* **29** (1954), 428–434.

[233] D. I. Hughes, Symbolic computation with fermions. *J. Symbolic Comput.* **10** (1990), 657–664.

[234] J. W. B. Hughes and J. Van der Jeugt, Unimodal polynomials associated with Lie algebras and superalgebras. *J. Comput. Appl. Math.* **37** (1991), 81–88.

[235] J. E. Humphreys, *Introduction to Lie Algebras and Representation Theory*. Springer-Verlag, Berlin, 1972.

[236] T. C. Hurley, On free polynilpotent groups, their group rings and corresponding Lie algebras. *Proc. Roy. Irish. Acad. Sect. A* **85** (1985), no. 1, 67–80.

[237] C. Itzykson and J. B. Zuber, *Quantum Field Theory*. McGraw-Hill Book Company, New York, 1980.

[238] N. Jacobson, Abstract derivations and Lie algebras. *Trans. Amer. Math. Soc.* **42** (1937), no. 2, 206–224.

[239] N. Jacobson, Restricted Lie algebras of characteristic p. *Trans. Amer. Math. Soc.* **50** (1941), 15–25.

[240] N. Jacobson, *Lecture in Abstract Algebra, I–III*. v. Nostrand, Princeton, 1951, 1953, 1964.

[241] N. Jacobson, *Lie Algebras*. Wiley, New York, 1962.

[242] N. Jacobson, *Structure and Representations of Jordan Algebras*. Amer. Math. Soc., Providence, Rhode Island, 1968.

[243] N. Jacobson, *Basic Algebra I*. Freeman, San Francisco, 1974.

[244] N. Jacobson, *Basic Algebra II. 2nd Ed.* Freeman, New York, 1974.

[245] G. D. James, *The Representation Theory of the Symmetric Groups*. Springer-Verlag, Berlin–New York, 1978.

[246] A. Joseph, Some ring theoretic techniques and open problems in enveloping algebras. *Non-Commutative Rings*, Springer-Verlag, Berlin, 1992, 27–67.

[247] V. G. Kac, A sketch of Lie superalgebras theory. *Comm. Math. Phys.* **53** (1977), 31–64.

[248] V. G. Kac, Classification of simple $\mathbb{Z}$-graded Lie superalgebras and simple Jordan superalgebras. *Comm. Algebra* **13** (1977), 1375–1400.

[249] V. G. Kac, Lie superalgebras. *Adv. Math.* **26** (1977), 8–96.

[250] V. G. Kac, *Infinite Dimensional Lie Algebras*. Birkhaüser, Boston, 1983.

[251] V. G. Kac and J. M. van der Leur, Super boson-fermion correspondense. *Ann. Inst. Fourier (Grenoble)* **37** (1987), no. 4, 99–137.

[252] A. Kandry-Rody and D. Kapur, Algorithms for Gröbner bases of polynomial ideals over various euclidean rings. *Lecture Notes in Comput. Sci.* **174** (1984), 195–205.

[253] A. Kandri-Rody and D. Kapur, Computing a Gröbner basis of a polynomial ideal over a Euclidean ring. *J. Symbolic Comput.* **6** (1988), no. 1, 37–57.

[254] A. Kandri-Rody and V. Weispfenning, Non-commutative Groebner bases in algebras of solvable type. *J. Symbolic Comput.* **9** (1990), no. 1, 1–26.

[255] I. L. Kantor, *An Analogue of E. Witt's Formula for the Dimensions of Homogeneous Components of Free Lie Superalgebras*. Deposited at VINITI, no. 2384-84.

[256] I. L. Kantor, Jordan and Lie superalgebras determined by a Poisson algebra. *Amer. Math. Soc. Transl. Ser. 2* **151** (1992), 55–80.

[257] I. Kaplansky, *Graded Lie Algebras, I, II*. Univ. of Chicago Report, 1976.

[258] I. Kaplansky, Superalgebras. *Pacific J. Math.* **86** (1980), 93–98.

[259] I. Kaplansky, $\mathbb{Z}_2$-graded algebras. *Illinois J. Math.* **35** (1991), no. 1, 85–92.

[260] M. V. Karasev and V. P. Maslov, *Nonlinear Poison Brackets. Geometry and Quantization*. Nauka, Moscow, 1991.

[261] J. Katriel and A. I. Solomon, A q-analoque of the Campbell-Baker-Hausdorff expansion. *J. Phys. A* **24** (1991), L1139–L1142.

[262] M. Katsura and Y. Kobayashi, The shuffle algebra and its derivations. *Theoret. Comput. Sci.* **115** (1993), no. 2, 359–369.

[263] O. H. Kegel, Zur nilpotent gewissen associative Ringe. *Math. Ann.* **149** (1962–63), 258–260.

[264] O. H. Keller, Ganze Cremona Transformationen. *Monatsh. Math.* **47** (1939), 299–306.

[265] A. R. Kemer, A remark on the standard identity. *Mat. Zametki* **23** (1978), 753–757.

[266] A. R. Kemer, Varieties and $\mathbb{Z}_2$-graded algebras. *Izv. Akad. Nauk SSSR Ser. Mat.* **48** (1984), no. 5, 1042–1059.

[267] V. K. Kharchenko, Differential identities of semiprime rings. *Algebra i Logika* **18** (1979), no. 1, 86–119.

[268] V. K. Kharchenko, *Automorphisms and Derivations of Associative Rings*. Kluwer Academic Publ., Dordrecht, 1991.

[269] O. G. Kharlampovich, Lyndon condition for solvable Lie algebras. *Izv. Vyssh. Uchebn. Zaved. Mat.* 1984, no. 9, 50–59.

[270] O. G. Kharlampovich and M. V. Sapir, *Algorithmic Problems in Varieties*, in press.

[271] C.-H. Kim, K. Y. Kim, Y. D. Kim, H. W. Lee, W. S. l'Yi, and Y.-J. Park, Superalgebras for superstring states. *Phys. Rev. D. (3)* **40** (1989), no. 6, 1969–1973.

[272] A. A. Klyachko, Lie elements in the tensor algebra. *Sibirsk. Mat. Zh.* **15** (1974), 914–920.

[273] Y. Kobayashi and S. Nagamachi, Lie groups and Lie algebras with generalized supersymmetric parameters. *J. Math. Phys.* **25** (1984), 3367–3374.

[274] A. T. Kolotov, Free subalgebras of free associative algebras. *Sibirsk. Mat. Zh.* **19** (1978), no. 2, 328–335.

[275] A. V. Kondratiev and A. B. Verevkin, *Package of Programs for Calculation of Groebner Basis, Canonical Elements, Hilbert Series, and to Define the Rationality of the Hilbert Series of Graded Associative Algebras*. Ul'yanovsk Branch of Moscow State Univ., 1994.

[276] A. N. Koryukin, Color Lie superalgebras and Hopf algebras. *Algebra i Logika*, in press.

[277] P.-V. Koseleff, Jeux de mots dans les algèbres de Lie libres: quelques bases et formules. *Theoret. Comput. Sci.* **79** (1991), 241–256.

[278] B. Kostant, Graded manifolds, graded Lie theory and prequantization. *Lecture Notes in Math.* **570** (1977), 177.

[279] V. A. Kostelevsky, M. M. Nieto, and D. R. Truax, Baker-Campbell-Hausdorff relations for the super-groups. *J. Math. Phys.* **27** (1986), 1419–1429.

[280] V. A. Kostelevsky and D. R. Truax, Baker-Campbell-Hausdorff relations for the super-Poincaré group. *J. Math. Phys.* **28** (1987), 2480–2487.

[281] A. I. Kostrikin, *Introduction in Algebra*. Nauka, Moscow, 1977.

[282] A. I. Kostrikin, *Around Burnside*. Springer-Verlag, 1990.

[283] A. I. Kostrikin and E. I. Zelmanov, A theorem on sandwich algebras. *Trudy Mat. Inst. Steklov.* **183** (1988), 142–149.

[284] W. Kraskiewicz and J. Weyman, *Algebra of Coinvariants and the Action of Coxeter Elements*, submitted.

[285] G. R. Krause and T. H. Lenagan, *Growth of Algebras and Gelfand-Kirillov Dimension*. Pitman, London, 1985.

[286] G. V. Kryazhovskih, On generators and defining relations of subalgebras of Lie algebras. *Sibirsk. Mat. Zh.* **24** (1983), no. 6, 80–86.

[287] G. P. Kukin, On Cartesian subalgebra of free sum of Lie algebras. *Algebra i Logika* **9** (1970), no. 6, 701–713.

[288] G. P. Kukin, Primitive elements of free Lie algebras. *Algebra i Logika* **9** (1970), no. 4, 458–472.

[289] G. P. Kukin, On subalgebras of free Lie p-algebras. *Algebra i Logika* **11** (1972), 535–550.

[290] G. P. Kukin, Subalgebras of free sum of Lie algebras with the amalgamated subalgebra. *Algebra i Logika* **11** (1972), no. 1, 59–86.

[291] G. P. Kukin, On free products of restricted Lie algebras. *Mat. Sb.* **95** (1974), no. 1, 53–83.

[292] G. P. Kukin, Embeddings of solvable Lie algebras of countable rank into Lie algebras with two generators. *Algebra i Logika* **14** (1975), no. 4, 414–421.

[293] G. P. Kukin, The intersection of subalgebras of a free Lie algebra. *Algebra i Logika* **16** (1977), no. 5, 577–587.

[294] G. P. Kukin, Algorithmical problems for solvable Lie algebras. *Algebra i Logika* **17** (1978), no. 4, 402–415.

[295] G. P. Kukin, Bases of a free Lie algebra. *Mat. Zametki* **24** (1978), no. 3, 375–382.

[296] G. P. Kukin, On derivations of free Lie p-algebras. *Algebra i Logika* **20** (1981), no. 5, 555–562.

[297] G. P. Kukin, On normal commutant. *Internat. Conf. on Algebra. Abstracts of Reports*. Barnaul, 1991, p. 162.

[298] G. P. Kukin, On co-final subalgebras of a free product of Lie algebras. *Internat. Conf. on Algebra. Abstracts of Reports*. Krasnoyarsk, 1993, p. 195.

[299] P. P. Kulish, Quantum Lie superalgebras and supergroups. *Proc. of the Conf. "Problems of Modern Quantum Field Theory"*. Alushta, 1989, 14–21.

[300] A. G. Kurosh, Die Untergruppen der freien Producte von beliebiegen Gruppen. *Math. Ann.* **109** (1934), 647–660.

[301] A. G. Kurosh, Nonassociative free algebras and free products of algebras. *Mat. Sb.* **20** (1947), no. 2, 239–262.

[302] A. G. Kurosh, Nonassociative free sum of algebras. *Mat. Sb.* **37** (1955), no. 2, 251–264.

[303] A. G. Kurosh, *Group Theory. 3rd Ed.* Nauka, Moscow, 1967.

[304] E. N. Kuzmin, *Color n-Lie Superalgebras*. Preprint, Novosibirsk, 1987.

[305] E. N. Kuzmin, Color n-Lie superalgebras. *19th All-Union Algebraic Conf. Abstracts of Reports*. Lvov, 1987, part 2, p. 154.

[306] K. Kwaśniewski, *Cohomological Description of Generalized Lie Algebras*. Preprint no. 554, Univ. of Wroclaw, 1982.

[307] K. Kwaśniewski, *Explicite Construction of Generalized Lie Algebras*. Preprint no. 585, Univ. of Wroclaw, 1983.

[308] K. Kwaśniewski, *Clifford and Grassmann Like Algebras*. Preprint, Univ. of Wroclaw.

[309] J. P. Labute, Algèbres de Lie et pro-p-groupes définis par une seule relations. *Invent. Math.* **4** (1967), 142–158.

[310] J. P. Labute, Free Lie algebras as modules over their enveloping algebras. *Proc. Amer. Math. Soc.* **68** (1978), 135–139.

[311] S. Lang, *Algebra*. Addison-Wesley, 1965.

[312] W. Lassner, *Operator Ordering and Symbol Representations on Enveloping Algebras*. Preprint, 1990.

[313] V. N. Latyshev, On choosing a basis in one T-ideal. *Sibirsk. Mat. Zh.* **4** (1963), no. 5, 1122–1127.

[314] V. N. Latyshev, Commutatinity of PI-algebras of polynomial type. *Algebra*, Moscow State Univ., 1982, 86–94.

[315] V. N. Latyshev, An equality algorithm for Lie-nilpotent associative algebras. *Vestnik Kiev. Univ. Mat. Mekh.* **27** (1985), 67–69.

[316] V. N. Latyshev, *Combinatorial Theory of Rings. Standard bases*. Moskow State Univ., Moscow, 1988.

[317] V. N. Latyshev, Lie nilpotency: recognition and word problem. *Proc. of the 1st Internat. Workshop Tainan–Moscow–Hong Kong–Stockholm*. Tainan, Taiwan, 1994, in press.

[318] M. Lazard, Sur les algèbres enveloppantes universelles de certaines algèbres de Lie. *Publ. Sci. Univ. Alger. Sér. A* **1** (1954), 281–294.

[319] M. Lazard, Sur les groupes nilpotents et les anneaux de Lie. *Ann. Sci. École Norm. Sup. (3)* **71** (1954), 101–190.

[320] M. Lazard, *Groupes, Anneaux de Lie et Problème de Burnside*. Inst. Math. dell' Univ. Roma, Roma, 1960.

[321] M. Lazard, Quelques calculs concernant la formule de Hausdorff. *Bull. Soc. Math. France* **41** (1963), 435–451.

[322] Ph. Le Chenadec, *Canonical Forms in Finitely Presented Algebras*. Pitman, London–New York, 1986.

[323] P. A. B. Lecomte, Applications of the cohomology of graded Lie algebras to formal deformations of Lie algebras. *Lett. Math. Phys.* **13** (1987), 157–166.

[324] P. A. B. Lecomte, P. W. Michor, and H. Schicketanz, The multigraded Nijenhuis-Richardson algebra, its universal property and applications. *J. Pure Appl. Algebra* **77** (1992), 87–102.

[325] D. A. Leites, Introduction to the theory of supermanifolds. *Uspekhi Mat. Nauk* **35** (1980), 3–57.

[326] D. A. Leites, *The Theory of Supervarieties*. Petrozavodsk, 1983.

[327] D. A. Leites, Lie superalgebras. *Sovrem. Probl. Mat. Fund. Naprav.* **25** (1984), 3–50.

[328] D. A. Leites and V. Serganova, Metasymmetry and Volichenko algebras. *Phys. Lett. B* **252** (1990), no. 1, 91–96.

[329] D. A. Leites and V. Serganova, Defining relations for classical Lie superalgebras. I. Superalgebras with Cartan matrix or Dynkin-type diagram. *Topological and Geometric Methods in Field Theory*, World Scientific, 1992, 194–201.

[330] J. M. Lemaire, *Algèbres Connexes et Homologie des Espaces de Lacets*. Lecture Notes in Math. **422** (1974).

[331] S. Z. Levendorskii, On PBW bases for Yangians. *Lett. Math. Phys.* **27** (1993), no. 1, 37–42.

[332] J. Lewin, Free modules over free algebras and free group algebras: the Schreier technique. *Trans. Amer. Math. Soc.* **145** (1969), 455–465.

[333] J. Lewin and T. Lewin, On ideals of free associative algebras generated by a single element. *J. Algebra* **8** (1968), 248–255.

[334] J.-P. Loday and T. Pirashvili, Universal enveloping algebras of Leibniz algebras and (co)homology. *Math. Ann.* **296** (1993), 139–158.

[335] M. Lothaire, *Combinatorics on Words*. Addison-Wesley, 1983.

[336] J. Lukierski and V. Rittenberg, Color-de Sitter and color-conformal superalgebras. *Phys. Rev. D. (3)* **18** (1978), 385–389.

[337] G. Lusztig, Canonical bases arising from quantized enveloping algebras. *J. Amer. Math. Soc.* **3** (1990), 447–498.

[338] G. Lusztig, Canonical bases arising from quantized enveloping algebras, II. *Progr. Theoret. Phys.* **102** (1990).

[339] R. C. Lyndon, On Burnside's problem, I. *Trans. Amer. Math. Soc.* **77** (1954), 202–215.

[340] R. C. Lyndon, On Burnside's problem, II. *Trans. Amer. Math. Soc.* **78** (1955), 329–332.

[341] R. C. Lyndon, A theorem of Friedrichs. *Michigan Math. J.* **3** (1955), 27–29.

[342] R. C. Lyndon and P. E. Shupp, *Combinatorial Group Theory*. Springer-Verlag, Berlin, 1977.

[343] S. MacLane, *Homology*. Springer-Verlag, Berlin, 1963.

[344] W. Magnus, Über diskontinuierliche Gruppen mit einer definierden Relation (Der Freiheitssatz). *J. Reine Angew. Math.* **163** (1930), 141–165.

[345] W. Magnus, Beziehungen zwischen Gruppen und Idealen in einem speziellen Ring. *Math. Ann.* **111** (1935), 259–280.

[346] W. Magnus, Über Beziehungen zwischen höheren Kommutatoren. *J. Reine Angew. Math.* **177** (1937), 105–115.

[347] W. Magnus, Über Gruppen und zugeordnete Liesche Ringe. *J. Reine Angew. Math.* **182** (1940), 142–149.

[348] W. Magnus, A connection between the Baker-Hausdorff formula and a problem of Burnside. *Ann. Math.* **52** (1950), 111–126.

[349] W. Magnus, *Algebraic Aspects in the Theory of Systems of Linear Differential Equations*. New York Univ. Research Report BR-3, 1953.

[350] W. Magnus, On the exponential solution of differential equations for a linear operator. *Comm. Pure Appl. Math.* **7** (1955), 649–673.

[351] W. Magnus, A. Karrass, and D. Solitar, *Combinatorial Group Theory. 2nd Ed.* Dover, New York, 1976.

[352] L. G. Makar-Limanov, The automorphisms of the free algebra of two generators. *Funktsional. Anal. i Prilozhen.* **4** (1970), no. 3, 107–108.

[353] L. G. Makar-Limanov, Algebras with a single relation. *Uspekhi Mat. Nauk* **30** (1975), no. 2, p. 217.

[354] L. G. Makar-Limanov, Algebraically closed skew fields. *J. Algebra* **93** (1985), 117–135.

[355] L. G. Makar-Limanov, A version of the Poincaré-Birkhoff-Witt theorem. *Bull. London Math. Soc.* **26** (1994), 273–276.

[356] A. I. Malcev, *Algebraic Systems*. Nauka, Moscow, 1970.

[357] F. Maltey, *Calcul des Formules de Baker-Campbell-Hausdorff dans une Base de Lyndon*. Etude no. F. 137, Centre Scientifique IBM, Paris, 1988.

[358] Yu. I. Manin, *Gauge Fields and Complex Geometry*. Springer-Verlag, 1988.

[359] Yu. I. Manin, *Quantum Groups and Non-Commutative Geometry*. Preprint CRM-1561, Montreal, 1988.

[360] Yu. I. Manin, Multiparametric quantum deformation of the general linear superfroup. *Comm. Math. Phys.* **123** (1989), 163–175.

[361] Yu. I. Manin, Multiplicative matrices in Hopf superalgebras. *Vestnik Moskov. Univ. Ser. I Mat. Mekh.* 1989, no. 2, 13–17.

[362] Yu. I. Manin, *Topics in Non-Commutative Geometry*. Princeton Univ. Press, 1991.

[363] W. Marcinek, Generalized Lie algebras and related topics, I, II. *Acta Univ. Wratislav. Mat. Fiz. Astronom.* LV, no. 1170, 3–52.

[364] V. T. Markov, Some examples of finitely generated algebras. *Uspekhi Mat. Nauk* **36** (1981), no. 5, 185–186.

[365] I. M. Masson, A classification of primitive ideals in the enveloping algebras of a classical simple Lie superalgebra. *Adv. Math.* **91** (1992), no. 2, 252–268.

[366] I. M. Masson, Enveloping algebras of Lie superalgebras: a survey. *Contemp. Math.* **124** (1992), 141–149.

[367] J. C. McConnell and J. C. Robson, *Noncommutative Noetherian Rings*. Wiley, New York, 1987.

[368] H. Meier-Wunderli, Note on a basis of P. Hall for the higher commutators in free groups. *Comment. Math. Helv.* **26** (1952), 1–5.

[369] G. Melançon, *Réécritures dans le Groupe Libre, l'Algèbre Libre et l'Algèbre de Lie Libre*. Ph.D. Thesis, Univ. du Québec à Montréal, 1991.

[370] G. Melançon, Combinatorics of Hall trees and Hall words. *J. Combin. Theory Ser. A* **59** (1992), 285–308.

[371] G. Melançon and Ch. Reutenauer, Lyndon words, free algebras and shuffles. *Canad. J. Math.* **41** (1989), 577–591.

[372] G. Melançon and Ch. Reutenauer, Une présentation combinatoire de la super-algèbre de Lie libre. *C. R. Acad. Sci. Paris Sér. I Math.* **315** (1992), 1215–1220.

[373] I. Mendas and P. Milutinović, Anticommutator analogue of the Baker-Hausdorff formula. *J. Phys. A* **22** (1989), L687–L689.

[374] W. Michaelis, The dual Poincaré-Birkhoff-Witt theorem. *Adv. Math.* **57** (1985), 93–162.

[375] W. Michaelis, Properness of Lie algebras and enveloping algebras, II. *Can. Math. Soc. Conf. Proc.* **5** (1986), 265–280.

[376] W. Michaelis, Properness of Lie algebras and enveloping algebras, I. *Proc. Amer. Math. Soc.* **101** (1987), 17–23.

[377] J. Michel, Base des algèbres de Lie libres et séries de Hausdorff. *Séminaire Dubreil* **27** (1974), no. 6, Institut Henri Poincaré, Paris.

[378] J. Michel, Calculs dans les algèbres de Lie libre: la série de Hausdorff et le problème de Burnside. *Astérisque* **38–39** (1976), 139–148.

[379] A. A. Mikhalev, Bases of free color Lie algebras. *Summary of Report at Sci. Res. Seminar of Chair of Higher Algebra of Moscow State Univ. Held Febr. 27, 1984. Vestnik Moskov. Univ. Ser. I Mat. Mekh.* 1984, no. 5, p. 94.

[380] A. A. Mikhalev, The theorem on freeness of subalgebras of free color Lie superalgebra. *Summary of Report at Sci. Res. Seminar of Chair of Higher Algebra of Moscow State Univ. Held May 7, 1984. Vestnik Moskov. Univ. Ser. I Mat. Mekh.* 1984, no. 5, p. 96.

[381] A. A. Mikhalev, Free color Lie superalgebras. *18th All-Union Algebraic Conf. Abstracts of Reports*. Kishinev, 1985, part 2, p. 31.

[382] A. A. Mikhalev, Basis of free product with the amalgamated subalgebra of color Lie superalgebras. *Summary of Report at Sci. Res. Seminar of Chair of Higher Algebra of Moscow State Univ. Held Nov. 19, 1984. Vestnik Moskov. Univ. Ser. I Mat. Mekh.* 1985, no. 3, p. 97.

[383] A. A. Mikhalev, Subalgebras of free color Lie superalgebras. *Mat. Zametki* **37** (1985), no. 5, 653–661.

[384] A. A. Mikhalev, Free Lie p-superalgebras. *Summary of Report at Sci. Res. Seminar of Chair of Higher Algebra of Moscow State Univ. Held May 12, 1986. Vestnik Moskov. Univ. Ser. I Mat. Mekh.* 1986, no. 6, p. 80.

[385] A. A. Mikhalev, Free color Lie superalgebras. *Dokl. Akad. Nauk* **286** (1986), no. 3, 551–554.

[386] A. A. Mikhalev, Subalgebras of free Lie p-superalgebras. *Mat. Zametki* **43** (1988), no. 2, 178–191.

[387] A. A. Mikhalev, *Free Lie Superalgebras*. Ph.D. Thesis, Moscow State Univ., Moscow, 1989.

[388] A. A. Mikhalev, The junction lemma and the equality problem for color Lie superalgebras. *Vestnik Moskov. Univ. Ser. I Mat. Mekh.* 1989, no. 5, 88–91.

[389] A. A. Mikhalev, The junction lemma and the equality problem for color Lie superalgebras, for Lie p-superalgebras and for Lie p-algebras. *Internat. Conf. on Algebra. Abstracts in the Theory of Rings, Algebras and Modules*. Novosibirsk, 1989, p. 90.

[390] A. A. Mikhalev, Embedding of Lie superalgebras of countable rank into a two generated Lie superalgebras. *Uspekhi Mat. Nauk* **45** (1990), no. 6, 139–140.

[391] A. A. Mikhalev, On residual finiteness of color Lie superalgebras and their universal enveloping algebras. *6th All-Union Symposium on the Theory of Rings, Algebras and Modules. Abstracts of Reports*. Lvov, 1990, p. 86.

[392] A. A. Mikhalev, The composition lemma and the equality problem for color Lie superalgebras and for Lie p-superalgebras. *Summary of Report at Sci. Res. Seminar of Chair of Higher Algebra of Moscow State Univ. Held Apr. 3, 1989. Vestnik Moskov. Univ. Ser. I Mat. Mekh.* 1990, no. 1, p. 100.

[393] A. A. Mikhalev, Ado-Iwasawa theorem, graded Hopf algebras and residual finiteness of color Lie (p-)superalgebras and their universal enveloping algebras. *Vestnik Moscov. Univ. Ser. I Mat. Mekh.* 1991, no. 5, 72–74.

[394] A. A. Mikhalev, Free color Lie super-rings. *Izv. Vyssh. Uchebn. Zaved. Mat.* 1991, no. 10, 46–49.

[395] A. A. Mikhalev, On fixed points of a free color Lie superalgebra under the action of a finite group of linear automorphisms. *Uspekhi Mat. Nauk* **47** (1992), no. 4, 205–207.

[396] A. A. Mikhalev, On right ideals of the free associative algebra generated by free color Lie (p-)superalgebras. *Uspekhi Mat. Nauk* **47** (1992), no. 5, 187–188.

[397] A. A. Mikhalev, The composition lemma for color Lie superalgebras and for Lie p-superalgebras. *Contemp. Math.* **131.2** (1992), 91–104.

[398] A. A. Mikhalev, Differential Lie superalgebras. *Internat. Conf. on Algebra. Abstracts of Reports*. Krasnoyarsk, 1993, p. 417.

[399] A. A. Mikhalev, Differential Lie superalgebras. *Uspekhi Mat. Nauk* **48** (1993), no. 5, 179–180.

[400] A. A. Mikhalev, Images of inner derivations of elements of free Lie algebras and superalgebras. *Vestnik Moskov. Univ. Ser. I Mat. Mekh.* 1993, no. 1, 35–38.

[401] A. A. Mikhalev, A. I. Shirshov's composition techniques in Lie superalgebras (non-commutative Gröbner bases). *Trudy Sem. Petrovsk.* **18** (1994).

[402] A. A. Mikhalev, On some properties of super Lie elements in the free associative algebra. *Proc. of the Internat. Conf. on Algebra, Barnaul, 1991, Contemp. Math.*, in press.

[403] A. A. Mikhalev, Some aspects of combinatorial theory of Lie superalgebras. *Proc. of the Internat. School "Algebra and Analysis", Baikal Lake, 1992*, in press.

[404] A. A. Mikhalev, Combinatorial aspects of the theory of Lie superalgebras. *Proc. of the 1st Internat. Workshop Tainan–Moscow–Hong Kong–Stockholm.* Tainan, Taiwan, 1994, in press.

[405] A. A. Mikhalev, Free and differential Lie superalgebras. *Transections of Soros Fellowships*, in press.

[406] A. A. Mikhalev, V. Shpilrain, and A. A. Zolotykh, *Subalgebras of Free Algebras.* Preprint, 1995.

[407] A. A. Mikhalev, U. U. Umirbaev, and A. A. Zolotykh, An example of a nonfree Lie algebra of cohomological dimension one. *Uspekhi Mat. Nauk* **49** (1994), no. 1, 203–204.

[408] A. A. Mikhalev, U. U. Umirbaev, and A. A. Zolotykh, A Lie algebra with cohomological dimension one over a field of prime characteristic is not necessary free. *Proc. of the 1st Internat. Workshop Tainan–Moscow–Hong Kong–Stockholm.* Tainan, Taiwan, 1994, in press.

[409] A. A. Mikhalev and A. A. Zolotykh, On Witt's superformula and Klyachko's theorem. *6th All-Union Conf. on the Theory of Varieties of Algebraic Systems. Abstracts of Reports.* Magnitigorsk, 1990, p. 16.

[410] A. A. Mikhalev and A. A. Zolotykh, Endomorphism of the free Lie algebra preserving primitivity of elements is an automorphism. *Uspekhi Mat. Nauk* **48** (1993), no. 6, 149–150.

[411] A. A. Mikhalev and A. A. Zolotykh, The rank of an element of a free color Lie (p-)superalgebra. *Internat. Conf. on Algebra. Abstracts of Reports.* Krasnoyarsk, 1993, p. 418.

[412] A. A. Mikhalev and A. A. Zolotykh, Free color shuffle algebras. *Internat. Conf. "Algebra and Analysis". Abstracts of Reports.* Kazan, 1994, part 1, p. 38.

[413] A. A. Mikhalev and A. A. Zolotykh, Free differential calculus in free Lie superalgebras. *Internat. Congress of Math. Short Communications.* Zürich, 1994, p. 18.

[414] A. A. Mikhalev and A. A. Zolotykh, Rank of an element of the free Lie (p-)superalgebra. *Doklady Akad. Nauk* **334** (1994), no. 6, 690–693.

[415] A. A. Mikhalev and A. A. Zolotykh, Applications of Fox differential calculus to free Lie superalgebras. *Proc. of the 3rd Internat. Conf. on Non Associative Algebra, Oviedo, 1993*, in press.

[416] A. A. Mikhalev and A. A. Zolotykh, Ranks of elements of free Lie superalgebras. *Summary of Report at Sci. Res. Seminar of Chair of Higher Algebra of Moscow State Univ. Held May 17, 1993. Vestnik Moskov. Univ. Ser. I Mat. Mekh.*, in press.

[417] A. A. Mikhalev and A. A. Zolotykh, Automorphisms and primitive elements of free Lie algebras. *Summary of Report at Sci. Res. Seminar of Chair of Higher Algebra of Moscow State Univ. Held Nov. 15, 1993. Vestnik Moskov. Univ. Ser. I Mat. Mekh.*, in press.

[418] A. A. Mikhalev and A. A. Zolotykh, Rank and primitivity of elements of free color Lie (p-)superalgebras. *Internat. J. Algebra Comput.* **4** (1994), 617–656.

[419] A. A. Mikhalev and A. A. Zolotykh, Automorphisms and primitive elements of free Lie superalgebras. *Comm. Algebra* **22** (1994), 5889–5901.

[420] A. A. Mikhalev and A. A. Zolotykh, Endomorphisms of free Lie superalgebras and their Jacobian matrices. *Proc. of the 2nd Math. Conf. of Moscow State Social Univ.*, 1994, 58–61.

[421] A. A. Mikhalev and A. A. Zolotykh, Endomorphisms of free associative algebras over commutative rings and their Jacobian matrices. *Fundamental'naya i Prikladnaya Matematika* **1** (1995), 177–190.

[422] A. A. Mikhalev and A. A. Zolotykh, Bases of free super shuffle algebras. *Uspekhi Mat. Nauk* **50** (1995), in press.

[423] A. A. Mikhalev and A. A. Zolotykh, An inverse function theorem for free Lie algebras over commutative rings. *Algebra Colloquim*, it press.

[424] A. A. Mikhalev and A. A. Zolotykh, Test elements for monomorphisms of free Lie algebras and Lie superalgebras. *Comm. Algebra*, in press.

[425] A. A. Mikhalev and A. A. Zolotykh, Ranks of subalgebras of free Lie superalgebras. *Vestnik Moskov. Univ. Ser. I Mat. Mekh.*, in press.

[426] A. A. Mikhalev and A. A. Zolotykh, An inverse function theorem for free associative algebras over commutative rings. *Int. Conf. Semigroups with Applications, Short Communications.* S.-Peterburg, 1995, in press.

[427] J. W. Milnor and J. C. Moore, On the structure of Hopf algebras. *Ann. of Math. (2)* **81** (1965), 211–264.

[428] R. Mines, F. Richman, and W. Ruitenberg, *A Course in Constructive Algebra*. Springer-Verlag, New York, 1988.

[429] S. P. Mishchenko, To Engel problem. *Mat. Sb.* **124** (1984), no. 1, 56–67.

[430] B. Mitra, Finite-dimensional indecomposable representations of the generalized Lie algebras. *J. Phys. A.* **20** (1987), 4065–4073.

[431] A. I. Molev and L. M. Tsalenko, Representation of the symmetric group in the free Lie (super)algebra and in the space of harmonic polynomials. *Funktsional. Anal. i Prilozhen.* **20** (1986), no. 2, 76–77.

[432] S. Montgomery, *Fixed Rings of Finite Automorphism Groups of Associative Rings*. Lecture Notes in Math. **818** (1980).

[433] S. Montgomery, *Hopf Algebras and Their Action on Rings*. Amer. Math. Soc., 1992.

[434] A. Montorsi and M. Rasetti, Symbolic manipulation techniques for ($\mathbb{Z}_2$-graded) Lie, Hopf and Hecke algebras. *Rend. Sem. Fac. Sci. Univ. Cagliari* **61** (1991), no. 1, 107–110.

[435] T. Mora, *Gröbner Basis for Non-Commutative Polynomial Rings*. Lecture Notes in Comput. Sci. **229** (1986), 353–362.

[436] T. Mora, *Seven Varitions on Standard Bases*. Report Dip. Mat. Univ. Genoa, 1988.

[437] N. Morales and L. F. Urrutia, The Cayley-Hamilton theorem for supermatrices. *J. Phys. A* **27** (1994), 1981–1997.

[438] M. V. Mosolova, Functions on non commuting operators that generate a graded Lie algebra. *Mat. Zametki* **29** (1981), 35–44.

[439] W. Nahm, V. Rittenberg, and M. Scheunert, Generalized superalgebras. *Phys. Lett. B* **61** (1976).

[440] C. Năstăsescu and F. van Oystaeyen, *Graded Ring Theory*. North-Holland, Amsterdam, 1982.

[441] Y. Neeman, The application of graded Lie algebras to invariance considerations in particle physics. *Lecture Notes in Math.* **570** (1977), 109–144.

[442] H. Neumann, On the intersection of finitely generated free groups, *Publ. Math. Debrecen* **4** (1955–56), 186–189.

[443] H. Neumann, On the intersection of finitely generated free groups: Addendum. *Publ. Math. Debrecen* **5** (1957–58), p. 128.

[444] W. D. Neumann, On intersections of finitely generated subgroups of free groups. *Lecture Notes in Math.* **1456** (1989), 161–170.

[445] M. H. A. Newman, On theories with a combinatorial definition of "equivalence". *Ann. of Math. (2)* **43/2** (1942), 223–243.

[446] P. Nickolas, Intersections of finitely generated free groups. *Bull. Austral. Math. Soc.* **31** (1985), 339–348.

[447] J. Nielsen, Om Regning med ikke-kommutative Faktoren og dens Anvendelse i Gruppetheorien. *Mat. Tidskr. B* 1921, 77–94.

[448] J. Nielsen, A basis for subgroups of free groups. *Math. Scand.* **3** (1955), 31–43.

[449] A. Nijenhuis and R. W. Richardson, Cohomology and deformations of algebraic structures. *Bull. Amer. Math. Soc. (N. S.)* **70** (1964), 406–411.

[450] A. Nijenhuis and R. W. Richardson, Cohomology and deformations in graded Lie algebras. *Bull. Amer. Math. Soc. (N. S.)* **72** (1966), no. 1, part 1, 1–29.

[451] A. Nijenhuis and R. W. Richardson, Deformation of Lie algebra structures. *J. Math. Mech.* **17** (1967), no. 1, 89–105.

[452] K. Nishiyama, Oscillator representations for orthosympletic algebras. *J. Algebra* **129** (1990), 231–262.

[453] Y. Nouaze, *Un Cas Particulier de Théorème de Poincaré-Birkhoff-Witt*. Thesis of D.Sci., Univ. de Montpellier, 1973.

[454] Y. Nouaze and Ph. Revoy, Un cas particulier de théorème de Poincaré-Birkhoff-Witt, *C. R. Acad. Sci. Paris Sér. I Math.* **273** (1971), 329–331.

[455] L. M. Nurieva, Finitely generated subalgebras of a free product of Lie algebras. *Izv. Vyssh. Uchebn. Zaved. Mat.* **28** (1984), no. 4, 32–37.

[456] A. L. Onishchik, Lie superalgebras of vector fields. *Amer. Math. Soc. Transl. Ser. 2* **148** (1991), 87–97.

[457] A. I. Papistas, Automorphisms of free polynilpotent Lie algebras. *Comm. Algebra* **21** (1993), 4391–4395.

[458] C. N. Parker and B. D. Winter, On $\mathbb{Z}_n$ graded Baker-Campbell-Hausdorff formulas and coset space parametrizations. *J. Math. Phys.* **24** (1983), 2537–2541.

[459] A. Patsourakos, Sur la représentation adjointe d'une algèbre de Lie libre. *Comm. Algebra* **15** (1987), 2199–2207.

[460] A. Patsourakos, Sur la représentation adjointe d'une algèbre de Lie libre. II. *Ann. Inst. Fourier (Grenoble)* **44** (1994), no. 2, 387–400.

[461] V. M. Petrogradsky, Identities in the enveloping algebras for modular Lie superalgebras. *J. Algebra* **145** (1992), 1–21.

[462] V. M. Petrogradsky, On some types of growth in Lie algebras. *Uspekhi Mat. Nauk* **48** (1993), no. 5, 181–182.

[463] V. M. Petrogradsky, Intermediate growth in Lie algebras and their envelopantes. *J. Algebra*, in press.

[464] G. Pinczon, The enveloping algebra of the Lie superalgebra osp(1, 2). *J. Algebra* **132** (1990), 219–242.

[465] X. G. Poetsch and W. von Waldenfels, *Maschinelle Berechnung der Koeffizienten der Formalen Potenzreihe* $\log e^x e^y$. Bericht der Kernforschungsanlage Jülich HÜL-206-PP.

[466] V. L. Popov, Automorphism groups of polynomial algebras. *Voprosy algebry* **4** (1989), 4–16.

[467] D. Quillen, Rational homotopy theory. *Ann. of Math. (2)* **90** (1969), 205–295.

[468] D. E. Radford, A natural ring basis for the shuffle algebra and an application to group schemes. *J. Algebra* **58** (1979), 432–454.

[469] D. W. Rand, Pascal programs for identification of Lie algebras. 3. Levi decomposition and canonical basis. *Comput. Phys. Comm.* **46** (1987), 311–322.

[470] Yu. P. Razmyslov, Trace identities of full matrix algebras over a field of zero characteristic. *Izv. Akad. Nauk SSSR Ser. Mat.* **38** (1974), no. 4, 723–756.

[471] Yu. P. Razmyslov, Trace identities and central polynomials in matrix superalgebras $M_{n,k}$. *Mat. Sb.* **128** (1985), no. 2, 194–215.

[472] Yu. P. Razmyslov, Polynomial c-dual sets and central polynomials in matrix superalgebras $M_{n,k}$. *Vestnik Moskov. Univ. Ser. I Mat. Mekh.* 1986, no. 1, 49–52.

[473] Yu. P. Razmyslov, *Identities of Algebras and Their Representations*. Amer. Math. Soc., 1994.

[474] Yu. P. Razmyslov, On the duality of varieties of representations for triple Lie and super Lie systems. I. *Vestnik Moskov. Univ. Ser. I Mat. Mekh.* 1994, no. 6, 32–39; II. *Vestnik Moskov. Univ. Ser. I Mat. Mekh.*, in press.

[475] R. Ree, Lie elements and an algebra associated with shuffles. *Ann. of Math. (2)* **68** (1958), 210–220.

[476] R. Ree, Generalized Lie elements. *Canad. J. Math.* **12** (1960), 493–502.

[477] Ch. Reutenauer, The shuffle algebra on the factor of a word is free. *J. Combin. Theory Ser. A* **38** (1985), 48–57.

[478] Ch. Reutenauer, Theorem of Poincaré-Birkhoff-Witt, logarithm and symmetric group representations of degrees equal to Stirling numbers. *Lecture Notes in Math.* **1234** (1986), 267–293.

[479] Ch. Reutenauer, Applications of a noncommutative Jacobian matrix. *J. Pure Appl. Algebra* **77** (1992), 169–181.

[480] Ch. Reutenauer, *Sur des Fonctions Symétriques Liées aux Vectours de Witt et à l'Algèbre de Lie Libre*. Rapport de Recherche no. 177, Univ. du Québec à Montréal, 1992.

[481] Ch. Reutenauer, *Free Lie Algebras*. Clarendon Press, Oxford, 1993.

[482] G. Rinehart, Differential forms on general commutative algebras. *Trans. Amer. Math. Soc.* **108** (1963), 195–222.

[483] V. Rittenberg and D. Wyler, Generalized superalgebras. *J. Nucl. Phys. B* **139** (1978), 189–202.

[484] V. Rittenberg and D. Wyler, Sequences of $\mathbb{Z}_2 \oplus \mathbb{Z}_2$-graded Lie algebras and superalgebras. *J. Math. Phys.* **19** (1978), 2193–2200.

[485] A. Rosenmann and S. Rosset, Ideals of finite codimension in free algebras and the FC-localization. *Pacific J. Math.* **162** (1994), no. 2, 351–371.

[486] L. E. Ross, Representations of graded Lie algebras. *Trans. Amer. Math. Soc.* **120** (1965), no. 1, 17–23.

[487] H. Scheerer and K. Schuch, A remark on a formula of Zassenhaus. *Geom. Dedicata* **39** (1991), 229–231.

[488] M. Scheunert, Generalized Lie algebras. *J. Math. Phys.* **20** (1979), 712–720.

[489] M. Scheunert, *The Theory of Lie Superalgebras. An Introduction.* Lecture Notes in Math. **716** (1979).

[490] M. Scheunert, Graded tensor calculus. *J. Math. Phys.* **24** (1983), 2658–2670.

[491] M. Scheunert, Casimir elements of ε-Lie algebras. *J. Math. Phys.* **24** (1983), 2671–2680.

[492] M. Scheunert, Serre-type relations for special linear Lie superalgebras. *Lett. Math. Phys.* **24** (1992), no. 3, 173–181.

[493] A. H. Schofield, *Representations of Rings over Skew Fields.* London Math. Soc. Lecture Note Ser. **92** (1985).

[494] O. Schreier, Die Untergruppen der freien Gruppen. *Abh. Math. Sem. Univ. Hamburg* **5** (1927), 161–183.

[495] M.-P. Schützenberger, Sur une propriété combinatoire des algèbres de Lie libres pouvant être utilisée dans un problème de Mathématiques appliquées. *Séminaire Dubreil-Pisot* (1958), Institut Henri Poincaré, Paris.

[496] M.-P. Schützenberger, On a factorization of free monoïds. *Proc. Amer. Math. Soc.* **16** (1965), no. 1, 21–24.

[497] M.-P. Schützenberger, *Sur les Bases de Hall.* Notes Manuscrites, 1971.

[498] D. Segal, Free left-symmetric algebras and an analogue of the Poincaré-Birkhoff-Witt theorem. *J. Algebra* **164** (1994), 750–772.

[499] V. Serganova, Simple Volichenko algebras. *Comtemp. Math.* **131.2** (1992), 155–160.

[500] J.-P. Serre, *Lie Algebras and Lie Groups. 2nd Ed.* Lecture Notes in Math. **1500** (1992).

[501] B. Servantius, A short proof of a theorem of Burns. *Math. Z.* **184** (1983), 133–137.

[502] D. Shannon and M. Sweedler, Using Gröbner bases to determine algebra membership, split surjective algebra homomorphisms, determine birational equivalense. *J. Symbolic Comput.* **6** (1988), 267–273.

[503] V. G. Shantarenko, On the equation $x_4x_3 + x_2x_1 = 0$ in a free Lie algebra. *Internat. Conf. on Algebra. Abstracts of Reports.* Barnaul, 1991, p. 176.

[504] V. G. Shantarenko, On equations in free Lie algebra. *Internat. Conf. on Algebra. Abstracts of Reports.* Krasnoyarsk, 1993, p. 435.

[505] I. P. Shestakov, Prime Malcev superalgebras. *Mat. Sb.* **182** (1991), no. 9, 1357–1366.

[506] I. P. Shestakov, Superalgebras and counterexamples. *Sibirsk. Mat. Zh.* **32** (1991), no. 6, 187–196.

[507] I. P. Shestakov, Quantization of Poison superalgebras and speciality of Jordan superalgebras of Poison type. *Dokl. Akad. Nauk* **334** (1994), no. 1, 29–32.

[508] I. P. Shestakov, A. I. Shirshov, A. M. Slinko, and K. A. Zhevlakov, *Rings That Are Nearly Associative.* Academic Press, New York, 1982.

[509] I. P. Shestakov and E. I. Zelmanov, Prime alternative superalgebras and nilpotence of the radical of a free alternative algebra. *Izv. Akad. Nauk SSSR Ser. Mat.* **54** (1990), no. 4.

[510] A. I. Shirshov, On a presentation of Lie rings in associative rings. *Uspekhi Mat. Nauk* **8** (1953), no. 5, 173–175.

[511] A. I. Shirshov, Subalgebras of free Lie algebras. *Mat. Sb.* **33** (1953), 441–452.

[512] A. I. Shirshov, On free Lie rings. *Mat. Sb.* **45** (1958), 113–122.

[513] A. I. Shirshov, About one conjecture in the theory of Lie algebras. *Sibirsk. Mat. Zh.* **3** (1962), 297–301.

[514] A. I. Shirshov, Algorithmic questions for solvable Lie algebras. *Uspekhi Mat. Nauk* **17** (1962), no. 6, p. 228.

[515] A. I. Shirshov, On bases of free Lie algebras. *Algebra i Logika* **1** (1962), 14–19.

[516] A. I. Shirshov, Some algorithmic problems about Lie algebras. *Sibirsk. Mat. Zh.* **3** (1962), 292–296.

[517] A. I. Shirshov, *Collected Works. Rings and Algebras*. Nauka, Moscow, 1984.

[518] A. L. Shmelkin, Free polynilpotent groups. *Amer. Math. Soc. Transl. Ser. 2* **55** (1966), 270–304.

[519] A. L. Shmelkin, Wreath products of Lie algebras and their applications in the group theory. *Trudy Mosk. Mat. Obshch.* **29** (1973), 247–260.

[520] A. L. Shmelkin and A. D. Tavadze, On free polynilpotent groups. *Issledovaniya po Algebre*, Tbilisi State Univ., 1984, 160–170.

[521] L. M. Shneerson, Conditions under which finitely presented algebras are free. *Izv. Vyssh. Uchebn. Zaved. Mat.* 1979, no. 7, 66–68.

[522] V. Shpilrain, An inverse function theorem for free associative algebras of rank two. *J. Pure Appl. Algebra* **83** (1992), 23–26.

[523] V. Shpilrain, On generators of L/R^2 Lie algebras. *Proc. Amer. Math. Soc.* **119** (1993), 1039–1043.

[524] V. Shpilrain, Recognizing automorphisms of the free groups. *Arch. Math. (Basel)* **62** (1994), 385–392.

[525] V. Shpilrain, On the rank of an element of a free Lie algebra. *Proc. Amer. Math. Soc.*, in press.

[526] V. Shpilrain, Test elements for endomorphisms of free groups and algebras. *Israel J. Math.*, in press.

[527] A. S. Shtern, Free Lie superalgebras. *Sibirsk. Mat. Zh.* **27** (1986), 170–174.

[528] P. E. Singer, Serre relations. Lie algebras and related topics. *Contemp. Math.* **110** (1990), 231–239.

[529] G. Sjödin, Hopf algebras and derivations. *J. Algebra* **64** (1980), 288–329.

[530] O. D. Smirnov, Solvability of alternative $\mathbb{Z}_2$-graded algebras and alternative superalgebras. *Sibirsk. Mat. Zh.* **32** (1991), no. 6, 158–163.

[531] S. P. Smith, Quantum groups. An introduction and survey for ring theoretists. *Noncommutative rings*, Springer-Verlag, New York, 1992, 131–178.

[532] A. Solomon, Algebra and superalgebra. *Irish. Math. Soc. Bull.* **22** (1989), 21–32.

[533] L. Solomon, A decomposition of the group algebra of the finite Coxeter group. *J. Algebra* **9** (1968), 220–239.

[534] L. Solomon, On the Poincaré-Birkhoff-Witt theorem. *J. Combin. Theory Ser. A* **4** (1968), 363–375.

[535] L. Solomon, A. Mackey formula in the group ring of a Coxeter group. *J. Algebra* **41** (1976), 255–268.

[536] W. Specht, Die linearen Beziehungen zwischen höheren Kommutatoren. *Math. Z.* **51** (1948), 367–376.

[537] F. Spitzer, A combinatorial lemma and its applications to probability theory. *Trans. Amer. Math. Soc.* **82** (1956), 323–339.

[538] J. R. Stallings, On torsion-free groups with infinitely many ends. *Ann. of Math. (2)* **88** (1968), no. 2, 312–334.

[539] R. P. Stanley, Unimodality and Lie superalgebras. *Stud. Appl. Math.* **72** (1985), no. 3, 263–281.

[540] R. P. Stanley, Binomial posets, Möbius inversion, and permutation. *J. Combin. Theory Ser. A* **20** (1976), 336–356.

[541] L. Stannarius, *Algorithmen zur Berechnung der Baker-Campbell-Hausdorff-Reihe*. Diploma Thesis, Univ. Leipzig, 1986.

[542] V. V. Stovba, On connections between varieties of Lie algebras and superalgebras and the racionality of Hilbert series of some varieties of superalgebras. *19th All-Union Algebraic Conf. Abstracts of Reports*. Lvov, 1987, part 1, p. 266.

[543] Ch. R. Stover, The equivalence of certain categories of twisted Lie and Hopf algebras over a commutative ring. *J. Pure Appl. Algebra* **86** (1993), 289–326.

[544] A. A. Sukhanov and A. A. Zolotykh, Algorithms of calculations of integer-valued characteristics of classical Lie algebras and Lie groups. *Calculating Invariants in the Theory of Algebraic Systems*, Novosibirsk, 1987, 67–79.

[545] S. Sundaram, Decompositions of some S_n-modules arising in the free Lie algebra. *J. Algebra*.

[546] *Supergravity and Superstrings: a Geometric Perspective (in 3 volumes)*. Ed. L. Castellani, R. D'Auria, and P. Fré. World Scientific, 1991.

[547] R. G. Swan, Groups of cohomological dimension one. *J. Algebra* **12** (1969), 585–610.

[548] M. E. Sweedler, *Hopf Algebras*. Benjamin, New York, 1969.

[549] *Symbolic and Algebraic Computation by Computers*. Ed. N. Inada and T. Soma. World Scientific, 1985.

[550] H. Sze-Tsen, *Homotopy Theory*. Academic Press, New York–London, 1959.

[551] L. A. Takhtajan, Introduction to quantum groups. *Lecture Notes in Phys.* **370** (1990), 3–28.

[552] V. V. Talapov, On solvable Lie algebras with one defining relation. *Sibirsk. Mat. Zh.* **22** (1981), no. 4, 176–181.

[553] V. V. Talapov, On algebraically closed metabelian Lie algebras. *Algebra i Logika* **21** (1982), no. 3, 357–367.

[554] V. V. Talapov, On polynilpotent Lie algebras given by one defining relation. *Sibirsk. Mat. Zh.* **23** (1982), no. 5, 192–204.

[555] R. M. Thrall, On the symmetrized Kronecker powers and the structure of the free Lie ring. *Amer. J. Math.* **64** (1942), 371–388.

[556] H. Tigner, Extensions of Lie-graded algebras. *J. Math. Phys.* **18** (1977), 1987–1991.

[557] S. N. Tronin, On the structure of free Lie superalgebras. *4th All-Union School "Lie Algebras and Their Applications". Abstracts of Reports*. Kazan, 1990, p. 45.

[558] E. C. Turner, *Test Words for Automorphisms of Free Groups*. Preprint, 1993.

[559] V. A. Ufnarovsky, Theorem on independence and its corollaries. *Mat. Sb.* **128** (1985), no. 1, 124–132.

[560] V. A. Ufnarovsky, On regular Shirshov's words. *Internat. Conf. on Algebra. Abstracts in the Theory of Rings, Algebras and Modules*. Novosibirsk, 1989, p. 140.

[561] V. A. Ufnarovsky, Combinatorial and asimptotic methods in algebra. *Sovrem. Probl. Mat. Fund. Naprav.* **57** (1990), 5–177.

[562] V. A. Ufnarovsky, Calculations of growth and Hilbert series by computer. *Comput. Algebra*, Marcel Dekker, 1993.

[563] U. U. Umirbaev, The equality problem for center-metabelian Lie algebras. *Algebra i Logika* **23** (1984), no. 3, 305–318.

[564] U. U. Umirbaev, On Jacobian matrices of Lie algebras. *6th All-Union Conf. on the Varieties of Algebraic Systems. Abstracts of Reports*. Magnitogorsk, 1990, 32–33.

[565] U. U. Umirbaev, On the approximation of free Lie algebras with respect to entry. *Tartu Ül. Toimetised* **878** (1990), 147–152.

[566] U. U. Umirbaev, Partial derivatives and endomorphisms of some relatively free Lie algebras. *Sibirsk. Mat. Zh.* **34** (1993), no. 6, 179–188.

[567] U. U. Umirbaev, The entry problem for Lie algebras. *Algebra i Logika* **32** (1993), no. 3, 326–340.

[568] U. U. Umirbaev, A note on primitive elements of free groups. *Proc. of the 5th Siberian School "Algebra and Analysis"*, in press.

[569] U. U. Umirbaev, *On Jacobian Matrices of Lie Algebras*. Preprint.

[570] U. U. Umirbaev, Primitive elements of free groups. *Uspekhi Mat. Nauk* **49** (1994), no. 2, 175–176.

[571] U. U. Umirbaev, Some algorithmic questions of associative algebras. *Algebra i Logika* **32** (1993), no. 4, 450–470.

[572] U. U. Umirbaev, Universal derivations and subalgebras of free algebras. *Proc. of the Internat. Conf. on Algebra, Krasnoyarsk, 1993*, in press.

[573] U. U. Umirbaev, On ranks of elements of free groups. *Uspekhi Mat. Nauk* **50** (1995), in press.

[574] Y. Ünlü, Graded subalgebras associated with a subalgebra of a graded free Lie algebra. *J. Pure Appl. Algebra* **37** (1985), 103–110.

[575] N. W. Van den Hijligenberg, *Computations and Applications of Lie Superalgebra Cohomology*. Ph.D. Thesis, Univ. of Twente, The Netherlands, 1993.

[576] J. Van der Jeugt, Unimodal polynomials arising from Lie superalgebras. *Stud. Appl. Math.* **77** (1987), 47–59.

[577] B. L. Van der Waerden, *Modern Algebra*. Frederick Ungar, New York, 1953.

[578] B. L. Van der Waerden, *Algebra, I, II*. Springer-Verlag, Berlin, 1971, 1967.

[579] F. Van Oystaeyen and L. Willaert, *Quantum Sections of Color Lie Superalgebras*. Preprint.

[580] M. A. Vasiliev, De Sitter supergravity with positive cosmological constant and generalized Lie superalgebras. *Classical and Quantum Gravity* **2** (1985), 645–652.

[581] M. A. Vasiliev, Some properties of generalized $(\mathbb{Z}_2)^n$-graded Lie superalgebras and generalized de Sitter supersymmetry. *Quantum Field Theory and Quantum Statistics*, Hilger, Bristol, 1987, vol. 2, 273–298.

[582] M. Vaughan-Lee, *The Restricted Burnside Problem.* Oxford Univ. Press, 1990.

[583] M. Vaughan-Lee, An algorithm for computing graded algebras. *J. Symbolic Comput.* **16** (1993), 345–354.

[584] Yu. M. Vazhenin, On elementary theory of Lie rings with one defining relation. *6th All-Union Sympos. on the Theory of Rings, Algebras and Modules. Abstracts of Reports.* Lvov, 1990, p. 31.

[585] A. M. Vershik, On algebras with quadratic relations. *Operator Theory and Infinite Dimensional Analysis*, Kiev, 1984, 32–57.

[586] G. Viennot, Factorisations dichotomiques des monoïdes libres et algèbres de Lie libres. *C. R. Acad. Sci. Paris Sér. I Math.* **276** (1973), 511–514.

[587] G. Viennot, Une généralisation des ensembles de Hall. *C. R. Acad. Sci. Paris Sér. I Math.* **276** (1973), 599–602.

[588] G. Viennot, Factorisations régulières des monoïdes libres et algèbres de Lie libres. *C. R. Acad. Sci. Paris Sér. I Math.* **277** (1973), 493–496.

[589] G. Viennot, Quelques bases et families basiques des algèbres de Lie libres commodes pour les calculs sur ordinateurs. *Bull. Soc. Math. France* **49–50** (1977), 201–209.

[590] G. Viennot, *Algèbres de Lie Libres et Monoïdes Libres.* Lecture Notes in Math. **691** (1978).

[591] I. B. Volichenko, *Nonhomogeneous Subalgebras of Commutative Superalgebras.* Preprint no. 23, Stockholm Univ., 1987.

[592] I. B. Volichenko, Nonhomogeneous subalgebras of Lie superalgebras. *19th All-Union Algebraic Conf. Abstracts of Reports.* Lvov, 1987, part 2, p. 55.

[593] A. Wais, On special varieties of Lie algebras, *Algebra i Logika* **28** (1988), no. 1, 29–40.

[594] W. von Waldenfels, Zur Charakterisierung Liescher Elemente in freien Algebren. *Arch. Math. (Basel)* **17** (1966), 44–48.

[595] M. A. Ward, Basic commutators. *Trans. Roy. Philos. Soc. London A* **264** (1969), 343–412.

[596] A. Weil, Sur certains groupes d'opérateurs unitaires. *Acta Math.* **111** (1964), 143–211.

[597] V. Weispfenning, Gröbner bases for polynomial ideals over commutative regular rings. *Lecture Notes in Comput. Sci.* **378** (1989), 336–347.

[598] P. West, *Introduction to Supersymmetry and Supergravity.* World Scientific, 1990.

[599] F. Wever, Operatoren in Lieshen Ringen. *J. Reine Angew. Math.* **189** (1947), 44–55.

[600] F. Wever, Über Invarianten in Lie'schen Ringen. *Math. Ann.* **120** (1949), 563–580.

[601] G. Whitehead, *Elements of Homotopy Theory.* Springer-Verlag, 1972.

[602] D. Wigner, An identity in the free Lie algebra. *Proc. Amer. Math. Soc.* **106** (1989), 639–640.

[603] E. Witt, Treue Darstellung Lieschen Ringe. *J. Reine Angew. Math.* **177** (1937), 152–160.

[604] E. Witt, Treue Darstellungen beliebier Lieschen Ringe. *Collect. Math.* **6** (1953), 107–114.

[605] E. Witt, Über freie Liesche Ringe und ihre Unterringe. *Math. Z.* **58** (1953), 113–114.

[606] E. Witt, Die Unterringe der freien Lieschen Ringe. *Math. Z.* **64** (1956), 195–216.

[607] X. Xu, Interwining operators for twisted modules of a colored vertex operator superalgebras. *J. Algebra*, in press.

[608] X. Xu, Twisted modules of colored lattice vertex operator superalgebras. *Quart. J. of Math.*, in press.

[609] A. V. Yagzhev, On endomorphisms of free algebras. *Sibirsk. Mat. Zh.* **21** (1980), no. 1, 181–192.

[610] A. V. Yagzhev, On the algorithmic problem of recognizing automorphisms among endomorphisms of free associative algebras of finite rank. *Sibirsk. Mat. Zh.* **21** (1980), no. 1, 193–199.

[611] H. Yamane, A Poincaré-Birkhoff-Witt theorem for quantized universal enveloping algebras of type A_n. *Publ. Res. Inst. Math. Sci.* **25** (1989), 503–520.

[612] I. A. Yunus, On Fox problem for Lie algebras. *Uspekhi Mat. Nauk* **39** (1984), no. 3, 251–252.

[613] M. V. Zaicev, Locally representable varieties of Lie superalgebras. *Mat. Zametki* **52** (1992), no. 3, 33–41.

[614] M. V. Zaicev, On residual separability of relatively free Lie algebras. *Izv. Vyssh. Uchebn. Zaved. Mat.* 1992, no. 10, 16–21.

[615] M. V. Zaicev, Basic rank of varieties of Lie algebras. *Mat. Sb.* **184** (1993), no. 9, 21–40.

[616] M. V. Zaicev, Special Lie algebras. *Uspekhi Mat. Nauk* **48** (1993), no. 6, 103–140.

[617] M. V. Zaicev, The standard identity in special varieties of Lie algebras. *Vestnik Moskov. Univ. Ser. I Mat. Mekh.* 1993, no. 1, 56–59.

[618] M. V. Zaicev, The standard identity and Capelly identities in Lie algebras. *Uspekhi Mat. Nauk* **49** (1994), no. 2, 153–154.

[619] M. V. Zaicev, Finiteness conditions in special Lie algebras. *Sovrem. Mat. i Prilozh.*, in press.

[620] H. Zassenhaus, Über Liesche Ringe mit Primzahlcharakteristik. *Abh. Math. Sem. Univ. Hamburg* **13** (1939), 1–100.

[621] I. A. Zavorotchenko, On homogeneous structure of free Lie superalgebra. *Vestnik Moskov. Univ. Ser. I Mat. Mekh.* 1991, no. 3, 80–82.

[622] I. A. Zavorotchenko, *On Homogeneous Structure of Free Lie Superalgebra.* Diploma Thesis, Moscow State Univ., Moscow, 1991.

[623] E. I. Zelmanov, On the nilpotency of nil algebras. *Lecture Notes in Math.* **1352** (1988), 227–240.

[624] E. I. Zelmanov, On restricted Burnside problem. *Sibirsk. Mat. Zh.* **30** (1989), no 6, 68–73.

[625] E. I. Zelmanov, Superalgebras and identities. *Amer. Math. Soc. Transl. Ser. 2* **148** (1991), 39–46.

[626] E. I. Zelmanov, *Nil Rings and Periodic Groups*. Korean Math. Soc., Seoul, 1992.

[627] R. Zerck, On the fixed points of the Lie ring associated with a free presentation of a finite group. *Bull. London Math. Soc.* **21** (1989), 381–386.

[628] A. I. Zhukov, Nonassociative free decompositions of algebras with finite number of generators. *Mat. Sb.* **26** (1950), 471–478.

[629] A. I. Zhukov, Reduced systems of defining relations in nonassociative algebras. *Mat. Sb.* **27** (1950), 267–280.

[630] V. M. Zhuravlev, Multiplicities of irreducible components of the free Lie algebra as module of general linear group. *Vestnik Moskov. Univ. Ser. I Mat. Mekh.* 1993, no. 2, 31–35.

[631] A. A. Zolotykh, *Decompositions of Tensor Products of Representations of Lie Algebra E_8*. Deposited at VINITI, no. 8360-88.

[632] A. A. Zolotykh, On calculations of integer-valued characteristics of representations of simple Lie algebras. *4th Internat. Conf. on Computer Algebra in Phys. Researches. Abstracts of Reports*. Dubna, 1990, p. 75.

[633] A. A. Zolotykh, Trace identities of matrix superalgebras with an involution. *Izv. Akad. Nauk SSSR Ser. Mat.* **54** (1990), no. 4, 694–725.

[634] A. A. Zolotykh, A package for computations in simple Lie algebra representations. *Proc. of the 1991 Internat. Sympos. on Symbolic and Algebraic Comput.* Bonn, ACM Press, 1991, 237–238.

[635] A. A. Zolotykh, Identities of minimal representations of orthogonal Lie algebras. *Abelevy Gruppy i Moduli (Tomsk)* **10** (1991), 31–45.

[636] A. A. Zolotykh, Trace identities of matrix superalgebras with involution. *Contemp. Math.* **131.2** (1992), 413–432.

[637] A. A. Zolotykh, The equevalence of the irreducible representations of finite dimensional Lie superalgebras with the same identities. *Uspekhi Mat. Nauk* **50** (1995), in press.

Notations

Index